Clean Coal Technologies for Power Generation

Clean Coal Technologies for Power Generation

P. Jayarama Reddy
Independent Energy Consultant, Hyderabad, India

CRC Press
Taylor & Francis Group
Boca Raton London New York

CRC Press is an imprint of the
Taylor & Francis Group, an **informa** business

A BALKEMA BOOK

Cover illustration: Avedøre Power Station, Denmark.

CRC Press
Taylor & Francis Group
6000 Broken Sound Parkway NW, Suite 300
Boca Raton, FL 33487-2742

First issued in hardback 2019

© 2014 by Taylor & Francis Group, LLC
CRC Press is an imprint of Taylor & Francis Group, an Informa business

Typeset by MPS Limited, Chennai, India

No claim to original U.S. Government works

ISBN: 978-1-138-00020-9 (hbk)

Library of Congress Cataloging-in-Publication Data

Reddy, P. Jayarama.
    Clean coal technologies for power generation / P. Jayarama Reddy.
      pages cm
    Includes bibliographical references and index.
    ISBN 978-1-138-00020-9 (hardback : alkaline paper) — ISBN 978-0-203-76886-0 (ebook)
1. Clean coal technologies.   2. Power resources.   I. Title.
    TP325.R35   2013
    333.79—dc23
                                                                    2013029662

Visit the Taylor & Francis Web site at
http://www.taylorandfrancis.com

and the CRC Press Web site at
http://www.crcpress.com

Dedicated to SAI

Table of contents

List of acronyms and abbreviations

AEO	Annual Energy Outlook
AGR	Acid Gas Removal
AR	As Received
ASU	Air Separation Unit
bcm	billion cubic metre
BFW	Boiler Feed Water
BFBC	Bubbling Fluidized Bed Combustion
Btu	British thermal unit
Btu/kWh	British thermal unit per kilowatt hour
Btu/lb	British thermal unit per pound
CBM	Coal Bed Methane
CCS	Carbon Capture and Sequestration
CCT	Clean Coal Technology
CCUS	Carbon Capture, Utilization and Storage
CFBC	Circulating Fluidized Bed Combustion
CGE	Cold Gas Efficiency
CLC	Chemical Looping Combustion
CNG	Compressed Natural Gas
COE	Cost of Electricity
COS	Carbonyl Sulfide
CT	Combustion Turbine
EIA	Energy Information Administration (USA)
EOR	Enhanced Oil Recovery
EPA	Environmental Protection Agency (USA)
EPRI	Electric Power Research Institute
FBC	Fluidized Bed Combustion
GCCSI	Global Carbon Capture and Storage Institute
GCV	Gross Calorific Value
GDP	Gross Domestic Product
GHG	Greenhouse Gas
GOI	Government of India
GT	Gas Turbine
GW	Gigawatt
HHV	Higher Heating Value
HRSG	Heat Recovery Steam Generator
IEO	International Energy Outlook

IGCC	Integrated Gasification Combined Cycle
ISO	International Standards Organization
kg/GJ	Kilogram per Gigajoule
kJ/kg	Kilojoules per Kilogram
kW	Kilowatt
kWe	Kilowatts electric
kWh	Kilowatt-hour
kWth	Kilowatts thermal
LCOE	Levelized Cost of Electricity
LHV	Lower Heating Value
LNB	Low NOx Burner
LSIP	Large Scale Integrated Project
MDEA	Methyldiethanolamine
MEA	Monoethanolamine
MPa	Megapascals
MW	Megawatt
MWe	Megawatts electric
MWh	Megawatt-hour
NCC	National Coal Council
NCV	Net Calorific Value
NDRC	National Development Research Council
NETL	National Energy Technology Laboratory
NGCC	Natural Gas Combined Cycle
NOAK	N^{th}-of-a-kind (plant)
O&M	Operation and Maintenance
PC	Pulverized Coal
PFBC	Pressurized Fluid Bed Combustion
PGCC	Partial Gas Combined Cycle
PM	Particulate Matter
ppm	Parts per million
psia	Pounds per square inch absolute
SC	Supercritical
SNG	Synthetic Natural Gas
SCR	Selective Catalytic Reduction
SOFC	Solid Oxide Fuel Cell System
STG	Steam Turbine Generator
Tcm	Trillion Cubic Metres
Tcf	Trillion cubic feet
TOC	Total Overnight Cost
TPC	Total Plant Cost
Tpd	Tonnes per day
TPI	Total Plant Investment
TWh	Terawatt-hour
UCG	Underground Coal Gasification
USC	Ultra Supercritical
WEO	World Energy Outlook
WGS	Water Gas Shift
WRI	World Resources Institite

Foreword

Coal is the most abundant and inexpensive fossil fuel for power generation. At current consumption rates, proven reserves of coal have been projected to last for two centuries. Coal can be burned in an environmentally-benign manner, by utilizing highly-efficient power-plant designs, effluent treatment and capture and sequestration of carbon dioxide. Promising clean coal technologies for power generation have been developed and are described in this book.

The book contains a wealth of information for the student who is seeking an introduction to the subject of clean coal for power generation, for the professional who is interested in data on coal utilization, for the legislator who is looking for the environmental impact of coal-related technologies and, finally, for the industry planner who is looking for past trends and future projections. Prof. Reddy has painstakingly compiled an abundance of data on coal resources and on coal combustion/gasification/liquefaction applications. At the same time he is presenting the fundamentals that one needs to understand the meaning and the importance of this data. All this is done with exemplary organization of subjects and topics and crystal-clear clarity. This book is well-written and captivating to read, as its importance to supplying clean energy to sustain and enhance the future prosperity of mankind is constantly evident.

Yiannis A. Levendis, PhD
College of Engineering Distinguished Professor
ASME and SAE Fellow
Department of Mechanical and Industrial Engineering
Northeastern University, Boston, MA

Acknowledgements

As the famous Spanish novelist Enrique Poncela put it, 'When something can be read without effort, great effort has gone into its writing.' Presenting a complex technologies-based subject in an understandable manner requires a Himalayan effort. The author sincerely hopes that this book presents the most important energy generating technologies of global relevance in a format that is easy to read and comprehend, enabling the reader to further enhance his/her knowledge of the subject. Several people have helped the author achieve this goal.

First and foremost, the author is highly thankful to Dr. Yiannis A. Levendis, College of Engineering Distinguished Professor, Department of Mechanical & Industrial Engineering, Northeastern University, Boston, MA, for his continued support and helpful suggestions from the inception stage to reviewing the manuscript and writing the foreword. The author is deeply indebted to him for giving so much of his pressing time. The author extends his sincere thanks to Professor Janos M. Beer, Professor Emeritus of Chemical and Fuel Engineering, Massachusetts Institute of Technology, Cambridge, MA, and renowned in the field of coal combustion, for very fruitful discussions on several aspects of the book and for his suggestion to include the recent developments. The author is thankful to Professor Edward Rubin (Carnegie Mellon), Howard Herzog (MIT), J. Dooley (Joint Global Change Res. Inst.), Mike Lynch (KGS), Robert Giglio (Foster-Wheeler), and K.M. Joshi (SSAS Inst. of Technology) who have permitted the use of their figures and material; to IEA (Paris), USEIA (Washington DC), USDOE/NETL, USEPA, WRI, ExxonMobil, BP and several other organizations for providing information. The author is personally thankful to Professor C. Suresh Reddy and his research group and to Gandhi Babu of Sri Venkateswara University for their unstinted help in preparing many of the figures and assisting in computational work. The author acknowledges the constant help and cheerful support of friends and all his family members, particularly Sreeni Bhaireddy, Jagan Yeccaluri, and grand-daughter Hitha Yeccaluri for obtaining references and publications, reading and formatting the manuscript, computing and handling many other odd works. Special thanks go to Alistair Bright and José van der Veer of CRC Press/Balkema for their wonderful cooperation in the publication of the book.

Chapter 1

Introduction

The global energy system faces many challenges in the 21st century. The major challenge is providing access to affordable and secure energy supplies. Today there are 1.3 billion people across the globe, mostly across the developing world, without access to electricity and 2.7 billion people who do not have clean cooking facilities. This problem is particularly severe in sub-Saharan Africa and developing Asia, which together account for 95% of people in energy poverty. Without a commitment to achieve universal energy access, it has been estimated that by 2030, there will be an additional 1.5 million premature deaths per year caused by household pollution from burning biomass/wood and dung and through lack of access to clean water, basic sanitation and healthcare. Modern/advanced energy technologies are essential to meet this challenge (World Coal Association 2012). The IEA projects that over 1,030 million people will be living without electricity in 2030. In addition, hundreds of millions more will have extremely limited access to electricity, which means just a few hours or days a week. For yet others, power will remain seriously unaffordable (NCC 2012).

The power sector around the world, particularly in developing countries, will have to face the challenges of population growth, poverty and degradation of environment. To supply energy for 9 to 10 billion people till 2050, it will be necessary to produce a minimum of 1000 kWh of electricity per year per person. That means every two days one power plant of 1000 MW needs to be set up. One of the ways of dealing with these challenges is probably by using alternative and renewable energy sources in distributive energy production. But it is doubtful whether these sources are currently ready to provide power at the magnitude needed. Modern trends are to transform large centralized systems into a network of small, distributed units. This concept, long used for heat energy production, is now extended to electric energy production.

Access to modern energy services is vital to development. Electricity is one of the most effective and environmentally dependable ways of delivering energy. A reliable source of power without exposure to long- or short-term disruptions is imperative to maintain adequate levels of economic performance, a necessity for poverty alleviation, improving public health, providing modern education, information services and other socio-economic benefits. It is not only the poor and developing countries that are concerned with providing affordable energy to their population; developed countries also face issues relating to energy deficiency. Sustainable energy delivery ensures sustainable socio-economic growth.

Historically, coal has been the major source of energy for the industrial advances and socio-economic well-being in the UK, Europe and the United States since the

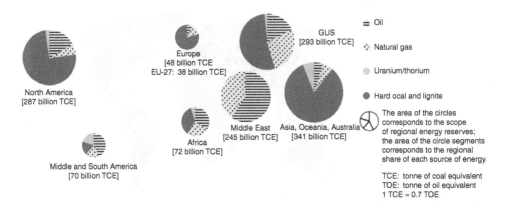

North America
[287 billion TCE]

Europe
[48 billion TCE
EU-27: 38 billion TCE]

GUS
[293 billion TCE]

≈ Oil

⁘ Natural gas

● Uranium/thorium

● Hard ocal and lignite

Middle East
[245 billion TCE]

Asia, Oceania, Australia
[341 billion TCE]

Africa
[72 billion TCE]

Middle and South America
[70 billion TCE]

The area of the circles corresponds to the scope of regional energy reserves; the area of the circle segments corresponds to the regional share of each source of energy.

TCE: tonne of coal equivalent
TOE: tonne of oil equivalent
1 TCE ≈ 0.7 TOE

Figure 1.1   Regional distribution of the worldwide energy reserves for hard coal & lignite, oil, natural gas and uranium/thorium (*Source*: VGB Electricity generation – Figures and Facts, 2012/2013).

mid-18th century. The current developed status of these countries is mostly due to fuel coal. Coal still plays a major role in electricity production in some developed countries, particularly the United States. Worldwide distribution of coal reserves, unlike oil, and its abundance allows coal to occupy the dominant position as energy supplier. Moreover, coal is cheaper compared to other fuels. The regional distribution of world-wide coal, oil, uranium/thorium, and gas reserves is shown in Figure 1.1 (VGB 2012/2013).

Coal is currently the main source of primary energy for the world delivering 29.6% of total energy demand; and 41% of global electricity is generated using coal. To meet the growth in energy demand over the past decade, the coal contribution has exceeded that of any other energy source. The utilization of coal is projected to rise from 149.4 QBtu in 2010 to 209.1 QBtu in 2035, i.e., by over 40.3% (EIA-IEO 2011). Over the first three decades of this century, coal's share of global energy consumption will have increased over 105%. Although the Organization of Economic Cooperation and Development (OECD) countries are not increasing their coal share in the energy mix, the non-OECD Asian countries, especially China and India, will account for 87% of this increase, primarily to meet their mounting electricity needs due to fast growing economic activity and population (EIA-IEO 2011).

Problem with coal

Despite having a high utility value, coal has high potential to pollute the environment right from its production through consumption. The effect on the environment can be grouped as: (1) emissions from fuel-consuming systems; (2) land disturbance and mine subsidence; (3) water, dust and noise pollution; and (4) methane emissions (World Coal Institute-Secure Energy 2005).

Coal production requires mining in either surface (strip) or underground mines. Surface mining creates pits upon coal removal, and to prevent soil erosion and a hideous environment, the land has to be reclaimed by filling the pits and replanting the soil to restore it to its original state. Further, the mining dust is a problem when surface mining takes place near habitats. In the underground mining, acid mine water causes environmental problem. Water that seeps into the mines, sometimes flooding them, and atmospheric oxygen react with pyrite (iron sulfide) present in the coal, producing acid mine water. When pumped out of the mine and into nearby rivers, streams, or lakes, the mine water acidifies them. This acid runoff dissolves heavy metals such as copper, lead, and mercury that are emitted into the ground. Another consequence of underground mining is the release of methane, a powerful greenhouse gas (US EPA 2006). Frondel *et al.* (2007) and Steenblik and Coronyannakis (1995) point out that about 15 tons of methane are emitted for each ton of coal production.

The major source of coal pollution, however, is during its usage, especially when burned. Coal combustion in power plants and factories produces smoke that contains carbon dioxide (CO_2), sulfur dioxide (SO_2), nitrogen oxides (NO_x), particulate matter, mercury, and so on. SO_2 and NO_x create acid rain and smog which are highly harmful to life and the environment. Since the Industrial Revolution, people have witnessed acid rain and smog in many industrial cities. For example, the city of London became notorious for industrial smog during the 19th and 20th centuries, and the worst smog appeared in December 1952, lasting for five days and resulting in about 4000 deaths. Accumulation of CO_2, the major pollutant from coal combustion in high levels in the atmosphere has been confirmed to cause global warming, and the unusual climatic events observed in the last century were attributed to a continuous increase of carbon dioxide levels in the atmosphere (IPCC 2007). Continuous usage of coal as a fuel over a period of two and a half centuries by developed countries has resulted in an increase of greenhouse gases in the atmosphere. The impacts of the climatic change on public health, as well as the physical and biological systems, are evident in terms of sea level rise, melting of glaciers, permafrost thawing, migration of species to higher latitudes, cyclones and floods, increase in heat waves and droughts, disease, and so on. The climatic scientists have projected that the impacts will be more severe resulting in heavy casualties and huge costs if the emission of these gases is not urgently and effectively controlled (IPCC2007; Jayarama Reddy 2010). Coal-fired power plants are one of the major sources of environmental pollution, and the largest source of toxic mercury pollution. Burning one kilogram of coal releases approximately 2.93 kg of carbon dioxide. A 500MW coal-fired power plant releases about 3 million tons of carbon dioxide per year. The electricity sector contributes roughly 60% of total anthropogenic atmospheric emissions of mercury and 35% of sulfur compounds that cause acid rain at the global level (Maruyama and Eckelman 2009). The projection is that carbon emissions from energy use will continue to grow by 26% during 2011–2030 at 1.2% per year. By 2030, 45 billion tonnes of CO_2 will be emitted annually (BP 2012). Post-combustion waste accumulated in large quantities at power plants is another potential polluting source. The dangers of this waste are potential water leakage of lead, mercury and other toxic chemicals from storage sites (NRC 2006).

Coal & energy security

Two aspects are involved in providing a secure energy supply: (a) long-term security or resource availability; and (b) short-term security associated with supply disruptions of the primary fuel or of the electricity generated.

A mix of energy sources, each with different advantages, provides security to an energy system by allowing flexibility in meeting any country's needs. There are many drivers that govern secure energy supply such as diversification of generation capacity, levels of investment required, availability of infrastructure and expertise, interconnection of energy systems, fuel substitution, fuel transformation, prices, and political and social environment (Coal Association of Canada 2010).

Coal reserves are plentiful in dynamic developing economies such as those of China and India; as a result, coal becomes attractive and a natural choice to meet the energy demand for their industrial and welfare activities. the USA has the largest reserves in the world, and they are widely utilized in power generation; they are also planned for use in proposed new power plants. Almost all countries with considerable coal reserves prefer to utilize coal for their energy needs. Only a few countries that use coal for power generation, depend on imports, e.g. as Japan does. Oil is another fossil fuel which has been playing a key role in global energy supply. However, oil reserves, unlike coal, are confined to a few regions on the globe, mostly the Middle East and North Africa (ME-NA) though some reserves exist in other parts of the world (Figure 1.1). As a result, oil prices are hiked (or the production controlled) by the major producing countries for reasons other than technical, creating market uncertainty. In recent years, due to political and social unrest in major oil-producing countries, the oil supplies and costs have become highly irregular. A country like India has to allot huge budgets to import required oil.

Hydroelectricity is an attractive source of energy, but its share of the total energy supply is not likely to grow much further (Howes and Fainberg 1991). Historically it was deployed early in the industrialization of a country because it was very cost-effective and has now reached or exceeded its sustainable potential in many developed countries. In mountainous countries with ample water and low population density, most of the electricity demand is met by hydro. In emerging economies and in poor countries, there is still a large untapped hydro potential left, but there is no reason to believe that these countries could increase the share of the hydro power in their total energy-mix more than the developed nations. By the time they reach the status of a developed country, they too would reach the sustainability limit of hydropower before their internal demand was met (Lackner 2006).

Nuclear energy, though expensive, cannot be ignored as an energy option. It has been one of the secure energy supplies for the last few decades in many countries and will continue to provide stable and large amounts of electricity in future. But the public concerns about the often-mentioned safety and environmental issues need to be addressed and overcome. For example, the recent disaster involving nuclear reactors at Fukushima has raised several questions around the world about its safety. The renewable energy technologies, particularly wind and solar, are mature and environment-friendly, but have practical problems for volume production. Solar fluxes are so large that converting a fraction to provide energy to human needs would hardly be difficult. But the costs of currently available technologies for conversion to electricity are high; either mass production of solar energy panels or the development of novel/cheap photovoltaic technologies that are currently under study – such as organic or dye sensitized solar cells and nano wire-based solar cells – to the commercial level could drastically drive down the cost. Solar electricity at a cost of about 1 cent/kWh could clearly compete with conventional electricity supply. Despite the fact that wind power is already available fairly cheaply it is unclear whether it can be provided at a global scale without environmental consequences and this needs to be investigated. The intermittent nature of wind and solar is often cited as another limitation. Geothermal energy and ocean (tidal) energy are still in their infancy, in spite of their acceptable nature. The contributions of the matured alternative energy sources are currently only a small percentage of the total energy-mix. Hence, these energy sources are still a long way from being practical and competitive, and they simply fall short with respect to the scale on which they can be functional.

As a result, it is essential to develop technologies that keep the fossil energy choice open, especially coal, even if environmental concerns require considerable changes in the way power is produced from fossil fuels. The coal conversion technologies, gasification and coal-to-liquids (CTL) are versatile; CTL provides synthetic fuels which are viable and practical alternatives to petroleum. This was amply proved when Germany and South Africa started producing synthetic fuels from coal decades ago; during the Second World War, Germany ran short of oil and during the apartheid era, South Africa experienced a severe shortage of oil supplies. Therefore, viewed pragmatically, coal is essential for energy security; and coal-based power generation is inevitable at least for the foreseeable future or until other, clean energy, sources are fully developed.

The coal-based power plant technologies, however, needed to be improved significantly to become environment-friendly. The new 'clean coal' technologies (CCTs) constitute process systems that effectively control emissions, SO_2, NO_x, particulate matter, mercury, and so on, as well as resulting in highly efficient combustion with significantly reduced fuel consumption. The widespread deployment of pollution-control equipment to reduce SO_2, NO_x and particulates from industry has brought cleaner air to many countries. Since the 1970s, various policy and regulatory measures worldwide have created a growing commercial market for these emission control technologies, with the result that costs have fallen and performance has improved. With advances in materials research, and developments in the design and production of power plant components, newly established coal-fired power plants, unlike the conventional plants, now operate at higher steam pressures and temperatures (supercritical and ultra-supercritical steam conditions), achieving higher efficiencies and drastically reducing gas emissions and consumption of coal. In addition to efficiency enhancement, another option for limiting future CO_2 emissions from coal combustion was carbon dioxide capture and storage (CCS). CCS is a group of technologies integrated to capture CO_2 from major point sources and transport to a storage site where the CO_2 is injected down wells and then permanently trapped in porous geological formations deep below the surface. Extensive research and development activity has demonstrated that CCS is the most promising and critical technology to control CO_2 emissions and best suited to coal-fired power plants. Applying CCS to highly efficient IGCC technology has great potential for controlling CO_2 emission streams at the pre-combustion stage. However, these clean coal technologies are not as widely deployed around the world as they should be. Much of the challenge is in commercializing these new technologies so that use of coal remains *economically* competitive despite the cost of achieving low and eventually near 'zero emissions'. In essence, coal will continue to be the backbone of global electricity generation for the foreseeable future. The base load power generation through coal is the most cost-effective way to provide affordable, safe and reliable electricity at the scale needed to achieve genuine access to modern electricity services worldwide. Professor Janos Beer, Professor Emeritus of Chemical and Fuel Engineering, Massachussetts Institute of Technology, has observed that clean coal is a reality, and it will provide clean power and answer environmental concerns (March 2013).

There is another important benefit associated with the application of advanced coal technologies. CO_2 is a valuable byproduct (a commodity!) of fossil fuel consumption; for the newly emerged 'carbon capture, utilization, and storage (CCUS)' initiative, coal will serve as the primary source to obtain adequate supplies of carbon dioxide. As the European Center for Energy and Resource Security (2011) recently noted, 'CO_2 should not be looked at as a waste product … it can have economic value'. The wealth creating opportunities will soon supplant the view that the geological sequestration of CO_2 is no? more than a waste disposal business. CO_2 utilization for enhancing oil recovery (EOR) from depleted oil wells is an example of the CCUS approach. The primary theme of the CCUS initiative is to develop a process, motivated by business economics. By putting the captured CO_2 to commercial use, CCUS provides an additional business and market case for organizations/industry to pursue the environmental benefits of CCS (USDOE 2012). The 2012 National Clean Council report 'Harnessing Coal's carbon content to Advance the Economy, Environment and Energy security' prepared with reference to the USA, states that CCUS is a key clean energy technology approach which is an essential part of any strategy to pursue a sustainable low carbon future.

Many leading research institutions (e.g., MIT, Carnegie-Mellon) and coal institutes, associations and industry in North America, the UK, Europe and Asia, the International Energy Agency (IEA), US DOE–EIA, DOE/NETL, the World Resources Institute, specially created Global Carbon Capture Storage Institute (GCCSI), prominent voluntary organizations and many other institutions, are dedicated to the development of different aspects of clean-coal technologies – the technology development, R&D, demonstration, deployment, performance monitoring, economics, viability and reliability, policy, legal and regulatory measures, public response, international collaborations etc., to ensure that CCTs provide the answer for both the growing global energy demand as well as climate policy goals.

This book presents an overview of how these new clean coal technologies provide highly efficient, clean, electric power and global energy security with reduced emissions. The book covers global primary energy consumption along with coal's share, the basic aspects of coal and its availability, pollution from coal combustion and control methods, conventional and advanced coal-based power generating technologies and their status, gasification of coal, IGCC and production of synthetic liquid fuels, underground in-situ gasification, CCS and its prospects, problems and global status, the operational aspects, on-going developments and outlook, and status of CCTs in developing countries with coal as major share in their energy-mix, particularly China and India. The novel approach of exploring the economic advantages of the captured CO_2 and developing commercially viable CCUS technologies while reducing GHG emissions is briefly explained. The engineering details and theories of the technologies (which are outside the objectives of the book) are not covered, though the basic principles, concepts and developments are amply explained. These technologies are of paramount importance for China and India, the first and third highest carbon dioxide emitters, respectively, in the world, in order to cut down carbon pollution.

China has recently started indigenous R&D efforts, as well as some manufacturing and the deployment of clean coal technologies with international collaborations; India has initiated a clean coal program with research into and the implementation of ultra mega projects deploying and planning more large and efficient coal-fired supercritical units (Clemente 2012).

It is expected that this book may cater well to the needs of first level students of engineering and physical science taking a course on coal combustion/power generation, and at the same time, serve as a comprehensive source of technical material on CCTs for those involved in consultancy, management, policy/decision making and so on, related to the power sector/coal-based power generation, especially in developing countries.

REFERENCES

Almendra, F., West, L., Zheng, L., & Forbes, S. (2011): CCS Demonstration in Developing Countries: Priorities for a Financing Mechanism for Carbon dioxide Capture and Storage, WRI Working Paper; World Resources Institute, Washington DC; available at www.wri.org/publication/ccs-demonstration-in-developing-countries

Beer, Janos (2013): Personal discussion at MIT on March 20, 2013.

BP (2008): BP Statistical review of World Energy, 2008, London, UK. BP (2013): BP Energy Outlook 2030, released in January 2013, London, UK.

Burnard, K. & Bhattacharys, S., IEA (2011): Power generation from Coal – Ongoing Developments and Outlook, Information Paper, International Energy Agency, Paris, Oct. 2011, © OECD/IEA.

Clemente, J. (2012): China leads the Global Race to Cleaner Coal, POWER Magazine, December 1, 2012; available at www.powermag.com/coal/china-leads-the-global-race-to-cleaner-coal_5192.html

Coal Association of Canada (2011): The Coal Association of Canada National Conference Highlights, news report at www. reuters.com/article/2011/10/20/idUS277339+20-oct-2011+HUG20111020

European Center for Energy and Resource Security (Umbach) (2011): The Future of Coal, Clean Coal Technologies and CCS in the EU and Central East European Countries: Strategic Challenges and Perspectives, December 2011; available at http://www.eucers.eu/wpcontent/uploads/Eucers-Strategy-Paper_No2_Future_of_Coal.pdf

Frondel, M., Ritter, N., Schmidt, C.M., & Vance, C. (2007): Economic impacts from the promotion of Renewable energy technologies – The German experience, RUHR economic papers #156, Bauer, T. K, et al. (eds).

Gruenspecht, H. (2011): IEO 2011 Reference case, USEIA, September 19, 2011.

Howes, R., & Fainberg, A. (eds) (1991): The Energy Sourcebook, New York: API.

IPCC (2005): Special Report on Carbon Dioxide Capture and Storage: Metz, B., Davidson, O., de Coninck, H.C., Loos, M., & Meyer, L.A. (eds), Cambridge University Press, Cambridge, UK.

IPCC (2007): Climate Change 2007 – Synthesis Report, Contribution of Working groups I, II, III to the 4th Assessment Report of IPCC, IPCC, Geneva, Switzerland.

Jayarama Reddy, P. (2010): Pollution and Global Warming, BS Publications, Hyderabad, India, www.bspublications.net

Lackner, K.S. (2006): The Conundrum of Sustainable Energy: Clean Coal as one possible answer, Asian Economic Papers 4:3, The Earth Institute at Columbia University and MIT, 2006.

Maruyama, N. & Eckelman, M.J. (2009): Long-term trends of electric efficiencies in electricity generation in developing countries, Energy Policy, 37(5), 1678–1686.

MIT Report (2007): *The Future of Coal – Options for a Carbon constrained World*, Massachusetts Institute of Technology, Cambridge, MA, August 2007.

McCormick, M. (2012): A Greenhouse Gas Accounting Framework for CCS Projects, Center for Climate & Energy Solutions, C2ES, February 2012, at www.c2es.org/docUploads/CCS-framework.pdf

National Coal Council (2012): Harnessing Coal's carbon content to Advance Economy, Environment and Energy security, June 22, 2012, Study Chair: Richard Bajura, National Coal Council, Washington, DC.

NRC (2006): NRC Committee on Mine Placement of Coal Combustion Wastes, Managing Coal Combustion Residues in Mines, Washington: The National Academies Press, 2006.

OECD/IEA (2009): Carbon Capture & Storage, OECD/IEA, Paris, France.

Society for Mining, Metallurgy & Exploration (2012): Coal's importance in the US and Global energy Supply, September 2012.

Steenblik and Coronyannakis (1995): Reform of Coal Policies in Western and Central Europe: Implications for the Environment, *Energy Policy*, 23(6), 537–553.

USDOE (2012): *Carbon Capture, Utilization and Storage*: Achieving the President's All-of-the-Above Energy Strategy, May 1, 2012; available at http://energy.gov/articles/adding-utilization-carbon-captureand-storage.

USEIA-IEO (2011): Energy Information Administration – International Energy Outlook, US Department of Energy, Washington DC, September 2011.

USEPA (2000): Environmental Protection Agency, US, Washinton DC.

US EPA (2006): US GHG Inventory Reort/ Climate change; at www.epa.gov/climatechange/ghgemissions/usinventoryreport.html

VGB (2012/2013): VGB, Electricity Generation, Figures & Facts, 2012/2013, available at http://www.vgb.org/en/data_powergeneration.html

World Coal Association (2012): Coal – Energy for Sustainable Development 2012, available at http://www.worldcoal.org/blog/coal-%E2%80%93-energy-for-sustainable-development/

World Coal Institute (2005): Coal – Secure Energy 2005; at www.worldcoal.org/.../coal_energy_security_report (03_06_2009).

WEO (2009): World energy Outlook 2009, OECD/IEA, Paris, France.

Chapter 2

Global energy consumption

Energy is a primary driver of economic development and a provider of quality of life. It sustains the living standards of developed countries and, in the developing world, it supports people to overcome poverty and reach a comfortable standard of living. Access to electricity reduces infant mortality, facilitates education, health care and sanitation, and improves productivity. It provides a window to the wider world.

2.1 ENERGY CONSUMPTION BY REGION

The global primary energy consumption has been rising in the last couple of decades. It has grown by 2.5% in 2011, roughly in line with the 10-year average. Primary energy consumption, by region as well as between OECD and non-OECD countries, is shown in Tables 2.1 and 2.2 (BP Report 2012). In 2011, the primary energy consumption was 12,274.6 million tonnes oil-equivalent. Global consumption increased nearly 16% during 1990–2001 but in the next decade, 2001–2011, two-fold, around 30%. If considered by region, the energy consumption is almost unchanged in North America, Europe and Eurasia during this period, whereas other regions have registered an increase in their consumption, the highest growth being in the Middle East (166%) followed by Asia Pacific (156%), South and Central America (89%), and Africa (70%). This growth has been due to several factors such as increased industrial activity, food production, transport and other services, necessitated by growth in population and

Table 2.1 World Energy Consumption by Region (million tonnes oil equivalent).

Region	1981*	1990*	1995*	2001**	2005**	2010**	2011**	2011 share of total
North America	2056.6	2326.5	2517.5	2698.8	2839.2	2763.9	2773.3	22.6%
South & Central America	249.1	323.9	394.7	468.4	521.8	619.0	642.5	5.2%
Europe & Eurasia	2798.0	3191	2779	2852.1	2969.0	2938.7	2923.4	23.9%
Middle East	145.1	264.3	344.1	445.1	562.5	716.5	747.5	6.1%
Africa	158.5	218.5	242.4	280.1	327.0	382.2	384.5	3.1%
Asia Pacific	1170.1	1784.9	2301.0	2689.5	3535.0	4557.6	4803.3	39.1%
World Total	6577.5	8108.7	8578	9434.0	10754.5	11977.8	12274.6	100%

(*Source*: *BP Statistical Review of World Energy 2011 & **BP Statistical Review of World Energy 2012)

Table 2.2 Primary Energy Consumption in OECD, Non-OECD & EU Regions.

Region	2001	2005	2010	2011	2011 share of the total
OECD	5407.4	5668.9	5572.4	5527.7	45%
Non-OECD	4026.6	5085.5	6405.3	6746.9	55%
European Union	1756.4	1808.7	1744.8	1690.7	13.8%

(*Source*: BP Statistical Review of World Energy 2012)

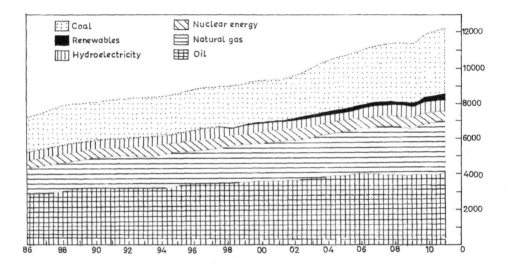

Figure 2.1 World Primary Energy Consumption, fuel-wise (Redrawn from BP Statistical Review of World Energy 2012).

urbanization. Consumption in OECD countries (the richest group of countries) fell by 0.8% in 2011, the third decline in the past four years. Non-OECD (mostly developing countries) consumption grew by 5.3%, in line with the 10-year average. The center of gravity for world energy consumption continues to shift from the OECD countries to emerging economies, especially in Asia. The total global fuel-wise consumption (million tonnes oil-equivalent) since 1986 and up to 2011 is shown in Figure 2.1.

Global consumption growth decelerated in 2011 for all fuels, as did total energy consumption for all regions. Oil remains the world's leading fuel, at 33.1% of global energy consumption, but oil continued to lose market share for the twelfth consecutive year and its current market share is the lowest in our data set, which begins in 1965. Fossil fuels still dominate energy consumption, with a market share of 87%. Renewable energy continues to make gains but today accounts for only 2% of energy consumption globally. Meanwhile, the fossil fuel mix is also changing. Oil, still the leading fuel, has lost market share for 12 consecutive years. Coal was once again the fastest growing fossil fuel, with predictable consequences for carbon emissions. Coal's market share of 30.3% was the highest since 1969.

Table 2.3 Per Capita Energy Consumption in 2006 and 2001 (million Btu).

Region	2006	2001
North America	276.2	277.2
South & Central America	73.2	49.1
Europe	144.5	150.8
Eurasia	160.8	133.4
Middle East	127.2	104.7
Africa	18.1	15.3
Asia Pacific	42.8	32.7

Table 2.3 lists per capita energy consumption for all the regions of the world. The per capita energy consumption ranges from about 276 million Btu for North America to 18 million Btu for Africa. As expected, the per capita consumption is highest for the developed/industrialized countries. The values for 2001 are also given for comparison. Over the five-year period, some regions of the world were fairly steady in per capita energy consumption (e.g., North America, Europe), some exhibited slight increases (e.g., Eurasia, Africa), and others showed significant increases on a percentage basis (e.g., Central and South America, Asia Pacific). Interestingly, it is seen that the regions of the world with the largest increases included the less developed countries that are becoming more industrialized and consuming more energy. On a per capita basis, United Arab Emirates was the largest consumer in the world, with 576 million Btu per person, followed by Kuwait (471 million Btu), Canada (427 million Btu), the United States (334 million Btu) and Australia (277 million Btu).

2.2 PROJECTION INTO THE FUTURE

The US Energy Information Administration (EIA) has released *IEO2011*, which is an assessment of the outlook for international energy markets through 2035. The reference case projection is a business-as-usual trend estimate, given known technology and technological and demographic trends, and the current environmental and energy laws and regulations. Thus, the projections provide policy-neutral baselines that can be used to analyze international energy markets. According to *IEO2011*, world marketed energy consumption might increase by 53% from around 504.7 quadrillion Btu in 2008 to 769.8 quadrillion Btu in 2035. These projections might differ with prospective legislation or policies that might moderate or affect energy markets (Figure 2.2). Please note the slight differences in data between 'BP 2012' and the present 'IEO2011' for the years up to 2010 due to the difference in assumptions and methodologies adopted. Much of the growth in energy consumption occurs in non-OECD countries, where demand is driven by strong long-term economic growth and mounting populations. Energy use in non-OECD nations increases by 85% in the projected period at an annual rate of 2.3%, as compared with an increase of 18% for the OECD economies at an average annual rate of 0.6%.

The energy outlook, however, remains uncertain. First, the recovery from the 2008–2009 global recessions in advanced economies has been slow and uneven, in

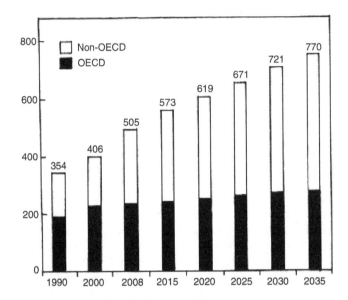

Figure 2.2 World Energy Consumption to 2035 (in quadrillion Btu) (Redrawn from USEIA: IEO 2011).

contrast to high growth in many emerging economies, partly driven by strong capital inflows and high costs of products/services. But, inflation pressures remain a particular concern, and there is a need to rebalance external trade in key developing economies. Second, the rising oil prices – due to increased demand and short supplies – and the political and social unrest in many oil producing countries are disturbing oil production. Further, extensive damage to nuclear reactors as a result of a devastating earthquake and tsunami in Japan, which has severely impacted its economy, have blurred the energy outlook for the coming decades.

The fastest-growing non-OECD economies, China and India, which are least affected by global recession, will emerge as key consumers in the global energy consumption market in the future as their economies expanded by 12.4% and 6.9% respectively in 2009 (Table 2.4). Since 1990, their share of total world energy use has increased significantly, and together they accounted for about 10% of total world energy consumption in 1990 and 21% in 2008. The energy consumption in the USA declined by 5.3% in 2009; and for the first time energy use in China is estimated to have exceeded that of the USA.

Strong economic growth is projected to continue in both China and India through 2035; consequently their combined energy use is projected to more than double and account for 31% of total world energy consumption in 2035. Population growth is yet another factor that influences the increase in energy consumption in these countries. Figure 2.3 shows energy demand growth in the United States, China and India. China's energy demand is projected to grow fast, the growth rate reaching 68% higher than that of the USA in 2035.

According to EIA-IEO 2011, strong growth is also projected for other non-OECD countries. Fast growth in population and availability of abundant indigenous resources

Table 2.4 Projection of World Consumption of Primary Energy & Coal (quadrillion Btu).

Year	Primary energy consumption			Coal consumption		
	OECD	Non-OECD	Total	OECD	Non-OECD	Total
1990	198.6	155.1	353.7			
2000	234.5	171.5	406.0			
2005	243.9	227.2	471.1	46.7	75.6	122.3
2010	238.0	284.1	522.1	43.5	105.9	149.4
2015	250.4	323.1	573.5	42.6	114.7	157.3
2020	260.6	358.9	619.5	43.1	121.4	164.6
2025	269.8	401.7	671.5	44.6	135.1	179.7
2030	278.7	442.8	721.5	45.3	149.4	194.7
2035	288.2	481.6	769.8	46.7	162.5	209.1

(*Source:* USEIA – International Energy Outlook 2011)

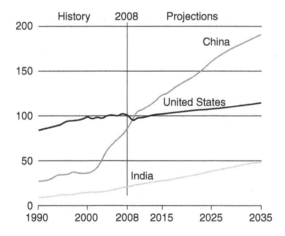

Figure 2.3 Energy consumption in the USA, China and India 1990–2035 (in quadrillion Btu) (*Source:* USEIA-IEO 2011).

increases energy demand in the Middle East by 77%. Energy consumption increases by 72% in Central and South America and by 67% in Africa. The slowest growth is projected for non-OECD Europe and Eurasia, which includes Russia and the other former Soviet blocs, at 16% as its population is declining and substantial gains in energy efficiency are being achieved through the replacement of inefficient capital equipment.

Projection by fuel: The EIA's projected use of the different energy sources over time is shown in Figure 2.4 and Table 2.5 (EIA-IEO 2011). A general increase in the utilization of all energy sources is anticipated. The major role of fossil fuels in meeting global energy needs is also clear. The liquid fuels, mostly petroleum based, remain the largest source of energy. Given the prospect that world oil prices are unlikely to fall in the period of projection, they will be the world's slowest growing energy source

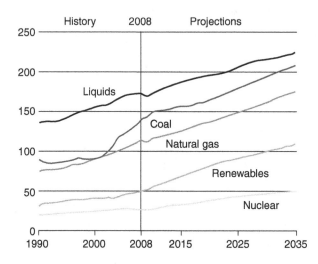

Figure 2.4 Projected world energy consumption by fuel type in quadrillion Btu (*Source:* USEIA/IEO 2011).

Table 2.5 Projection through 2035: World Primary Energy Consumption by fuel (Quadrillion Btu).

Year	Liquids	Natural gas	Coal	Nuclear	Others	Total
2005	170.8	105.0	122.3	27.5	45.4	471.1
2010	173.2	116.7	149.4	27.6	55.2	522
2015	187.2	127.3	157.3	33.2	68.5	573.5
2020	195.8	138.0	164.6	38.9	82.2	619.5
2025	207	149.4	179.7	43.7	91.7	671.5
2030	216.6	162.3	194.7	47.4	100.6	721.5
2035	225.2	174.7	209.1	51.2	109.5	769.8

(*Source:* USEIA/IEO 2011)

at an average rate of 1% per year. Further, their share of global marketed energy consumption will fall from 34% in 2008 to 29% in 2035, as many countries may switch away from liquid fuels when feasible. Renewable energy and coal are the fastest growing energy sources, with consumption increasing by 3.0% and 1.7% respectively. The prospect for renewable energy sources is improving due to the projected high prices for oil and natural gas, the growing concerns about the environmental impact of fossil fuel usage and the attractive incentives for their production and application all over the world.

Liquid fuels: World use of petroleum and other liquids (including petroleum-derived fuels and non-petroleum-derived liquid fuels such as ethanol and biodiesel, coal-to-liquids (CTL), and gas-to-liquids (GTL), and petroleum coke, which is a solid; also natural gas liquids, crude oil consumed as a fuel, and liquid hydrogen) grows from 85.7 million barrels/day in 2008 to 97.6 million barrels/day in 2020 and 112.2 million barrels/day in 2035. Most of the growth in liquids use is in the transportation sector accounting for 82%, where, in the absence of significant technological advances, liquids continue to provide much of the energy consumed. The remaining portion of the growth is attributed to the industrial sector. Despite rising fuel prices, the use of liquids for transportation increases by an average of 1.4% per year or 46% overall from 2008 to 2035. To meet this increase in global demand, liquids production (including both conventional and unconventional liquids supplies) increases by a total of 26.6 million barrels/day from 2008 to 2035.

The estimates are based on the best guess that OPEC countries will invest in incremental production capacity to retain their share of around 40%, over the past 15 years, of total world liquids production through 2035. OPEC producers contribute 10.3 million barrels of conventional liquids (crude oil and lease condensate, natural gas plant liquids, and refinery gain) per day to the total increase in world liquids production, whereas non-OPEC countries add conventional supplies of another 7.1 million barrels/day (EIA/IEO 2011). Unconventional resources (including oil sands, extra-heavy oil, bio-fuels, coal-to-liquids, gas-to-liquids, and shale oil) from both OPEC and non-OPEC sources grow on average by 4.6% per year over the projection period. Sustained high oil prices allow unconventional resources to become economically competitive, particularly when geopolitical or other 'above ground' constraints limit access to prospective conventional resources. ('Above-ground' constraints refer to those non-geological factors that might affect supply, such as a country's policies that limit access to resources or conflict or terrorist activity or lack of technological advances and access to technology or price constraints on the economical advance of resources or labor/materials shortages or weather or environmental concerns and other short- and long-term geopolitical factors). World production of unconventional liquid fuels, which are only 3.9 million barrels/day in 2008, increases to 13.1 million barrels/day and accounts for 12% of total world liquids supply in 2035. The largest contributions to unconventional liquids supply which account for almost three-quarters of the increase over the projection period are the production of 4.8 million barrels/day of Canadian oil sands, 2.2 million barrels/day of US bio-fuels, 1.7 million barrels/day of Brazilian bio-fuels, and 1.4 million barrels/day of Venezuelan extra-heavy oil (EIA/IEO 2011).

Natural gas: Natural gas continues to be the fuel of choice for many regions of the world in the power and industrial sectors because many countries that are interested in reducing GHGs opt for it due to its relatively low carbon intensity compared with oil and coal. Additionally, low capital costs and fuel efficiency favor natural gas in the power sector.

Significant changes in natural gas supplies and global markets are caused by the expansion of liquefied natural gas (LNG) production capacity. Further, new drilling techniques and other advances allow production from many shale basins to be economical, though it has its own environmental concerns. The net result is a significant increase in resource availability contributing to lower prices and higher demand for natural gas in the projection period (EIA-IEO 2011). Although the extent of the world's unconventional gas resources (tight gas, shale gas, and coal-bed methane) are not yet fully assessed, the IEO2011 reference case projects a substantial increase, especially from the United States, Canada and China.

Rising estimates of shale gas resources have helped to increase the total natural gas reserves of the United States by almost 50% over the past decade, and shale gas rises to 47% of the United States' natural gas production in 2035 in the 'IEO2011 reference case'. If the production of tight gas and coal-bed methane is added, the US unconventional natural gas production rises from 10.9 tcf in 2008 to 19.8 tcf in 2035. Unconventional natural gas resources are even more important for the future of domestic gas supplies in Canada and China, where they account for 50% and 72% of total domestic production respectively, in 2035 in the reference case (EIA/IEO 2011).

In the coming decades, world natural gas trade in the form of LNG is projected to increase. Most of the increase in LNG supply comes from the Middle East and Australia, where several new liquefaction projects are expected to become operational in the coming years. In addition, several LNG export projects have been proposed for Western Canada; there are also proposals to convert underutilized LNG import facilities in the USA to liquefaction and export facilities for domestically sourced natural gas. In the IEO2011 reference case, world liquefaction capacity increases more than twofold, from about 8 tcf in 2008 to 19 tcf in 2035. In addition, natural gas exports from Africa to Europe and from Eurasia to China are likely to increase. This is discussed in more detail in chapter 4.

Coal: During the projection period, the world's consumption of coal is expected to increase from 122.3 QBtu in 2005 to 209.1 QBtu in 2035; i.e., by over 71%, with developing countries accounting for most of this increase. Coal continues to be a major source of fuel, especially in non-OECD Asia, where both fast-paced economic growth and large domestic reserves support growth in coal usage. Coal global usage is uneven with OECD countries maintaining the same level as 2008. While world coal consumption increases by 1.5% per year on average from 2008 to 2035, in non-OECD Asia, the annual average increase is 2.1% (EIA-IEO 2011).

Although the global recession had a negative impact on coal use in almost every other part of the world, in 2009, coal consumption continued to increase in China. In the absence of policies or legislation that would limit the growth of coal use, China and, to a lesser extent, India and the other non-OECD Asian countries consume coal in place of more expensive fuels. In the projection period, China alone accounts for 76%, and India and the rest of non-OECD Asia account for 19% of the net increase in world coal consumption.

Coal has fuelled the industrial revolution that started in England and spread to Europe, Japan, and the USA. In the present day, coal is vital for secure and cost-effective power generation. Coal can provide usable energy at a cost of between $1 and $2/MMBtu (million metric British thermal units) compared to $6 to $12/MMBtu for oil and natural gas (MIT Report 2007). Coal prices are relatively stable too. Coal has the advantage of burning with a relatively high efficiency of 33% energy converted to electricity compared to other major fuels, thus becoming an economically viable source of energy to power the economic expansion. The environmental pollution concerns in using coal are, of course, very catastrophic. Coal use in the USA was 22.8 quadrillion Btu in 2005, nearly half (49%) of the consumption by OECD countries. Coal demand in the USA is projected to rise to 24.3 quadrillion Btu (665 million tonnes oil-eq.) in 2035, an increase of 18%. This projection is due to substantial coal reserves and the heavy reliance on it for electricity generation at the new and existing plants. Despite the fact that investments in new coal-fired power plants may appear unfavorable due to an implicit risk factor for carbon-intensive technologies, increased generation from coal-fired power plants accounts for 39% of the growth in total country's power generation

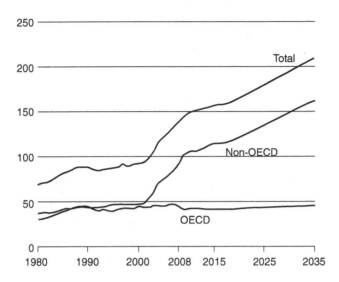

Figure 2.5 World coal consumption by region, 1980–2035 (*Source:* USEIA-IEO 2011).

during the period of projection. The USA, through the implementation of *advanced* coal technologies, may use coal to increase the domestic energy supply even more than projected. Total coal consumption in OECD European countries is projected to decline slightly from 12.8 quadrillion Btu in 2005 to 10.4 quadrillion Btu in 2035 (Figure 2.5). The Czech Republic, Germany, Italy, Poland, the UK, Spain and Turkey are the major consumers of coal in this group. Low-Btu brown coal is an important domestic source of energy in this region, accounting for 47% of their total coal consumption in terms of weight in 2006. In spite of many countries enacting policies to discourage the use of coal due to environmental concerns, the plans to replace or renew existing coal-fired generating capacity are in progress, suggesting that coal will continue to play a vital role in the overall energy mix. Germany is not only one of the world's largest energy consumers but one of the largest coal-consuming countries. However, the country has limited energy resources and the security of supply remains a concern. In Poland, the second largest consumer of coal in OECD Europe, coal plays a dominant role in the country's energy supply, meeting around 63% of the country's primary energy demand, mainly its power. This situation is expected to persist as the country possesses vast reserves of hard coal and lignite, and an increasingly efficient infrastructure for its recovery (Mills 2004).

OECD Asian countries are prominent consumers of coal. In 2005, they utilized 9.1 quadrillion Btu of coal – 19.4% of total OECD coal consumption. OECD Asia's coal demand is anticipated to increase by around 6%, to 9.7 quadrillion Btu in 2035.

Non-OECD coal consumption is projected to rise to 162.5 quadrillion Btu in 2035, an increase of 115% over 2005 consumption (EIA-IEO 2011), primarily due to strong economic development and rising energy demand in China and India (Figure 2.6). This increase in the total world coal consumption further signifies the importance of coal in meeting overall energy demand in the non-OECD countries.

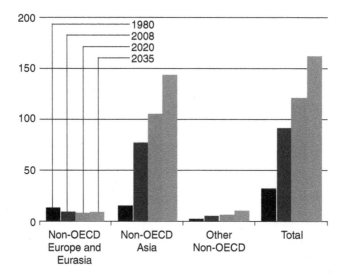

Figure 2.6 Non-OECD coal consumption by region in quadrillion Btu (*Source*: USEIA-IEO 2011).

Their primary energy consumption, less than 8% of the world's consumption in 1980, is projected to increase to 31% in 2035. China has an abundance of coal reserves, and its coal output increased from 1384 million tons in 2000 to 3240 million tons in 2010 making China by far the world's largest coal producer, the next largest being the USA with 985 million tons produced in 2010 (BP 2011).

2.3 WORLD ELECTRICITY GENERATION

Electricity is the world's fastest-growing form of end-use energy consumption in the present IEO assessment, as it has been for the past few decades. Net electricity generation worldwide is expected to riseon average by 2.3% per year from 19.1 trillion kWh in 2008 to 25.5 trillion kWh in 2020 and 35.2 trillion kWh in 2035, while total world energy demand grows by 1.6% per year. The strongest growth in electricity generation is projected for the non-OECD countries at an average annual rate of 3.3%. This growth is because of the increase in domestic demand due to rising standards of living as well as the expansion of commercial services including hospitals (Figure 2.7).

Non-OECD Asia (including China and India) leads with a projected annual average increase of 4.0% from 2008 to 2035 (Figure 2.8). In the OECD countries, where consumption patterns are more mature and population growth is relatively slow or declining, the growth in power generation is projected to be much slower, averaging 1.2% per year from 2008 to 2035 (EIA/IEO 2011).

The net electricity generation by fuel type is shown in Figure 2.9, and the actual values are tabulated in Table 2.6.

In many countries, concerns about energy security and the environmental penalty of greenhouse gas emissions have resulted in government policies that support a projected increase in renewable energy sources. As a result, renewable energy sources are

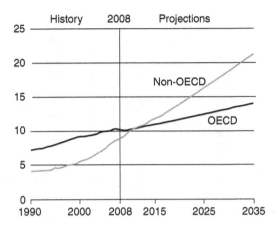

Figure 2.7 OECD and non-OECD net electricity generation, 1990–2035 (trillion kilowatt hours) (*Source:* USEIA-IEO 2012).

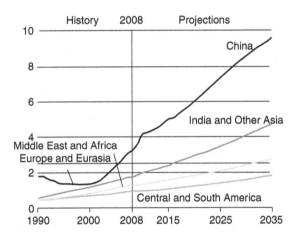

Figure 2.8 Non-OECD net electricity generation by region, 1990–2035 (trillion kilowatt hours) (*Source:* USEIA-IEO 2012).

the fastest growing sources of electricity generation at 3.1% per year from 2008 to 2035.

Natural gas is the second fastest growing power generation source at 2.6% per year. An increase in unconventional gas resources in many parts of the world, particularly in North America, allows pumping adequate supplies into global markets, keeping prices competitive. Power generation from renewable sources, natural gas and to a lesser extent nuclear power, in future largely decreases coal-fired generation, although coal will remain the largest source of world electricity through to 2035.

More than 82% of the increase in renewable generation is in the form of hydro-electric power and wind power. The contribution of wind energy, in particular, has

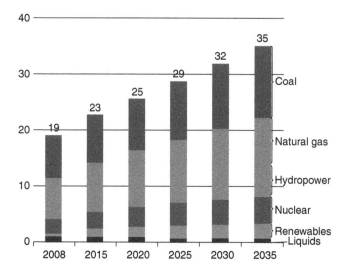

Figure 2.9 World net electricity generation by fuel, 2008–2035 (trillion kWh) (*Source:* USEIA-IEO 2012).

Table 2.6 OECD and non-OECD net Electricity generation by fuel type, 2008–2035 (trillion kilowatt hours).

Region	Fuel	2008	2015	2020	2025	2030	2035	Ave. annual % change (2008–2035)
OECD	Liquids	0.4	0.3	0.3	0.3	0.3	0.3	−0.8
	Natural gas	2.3	2.5	2.7	2.9	3.4	3.8	1.8
	Coal	3.6	3.3	3.4	3.5	3.6	3.8	0.2
	Nuclear	2.2	2.4	2.6	2.7	2.8	2.9	1.0
	Renewables	1.8	2.3	2.7	2.9	3.1	3.2	2.2
	Total	**10.2**	**10.9**	**11.6**	**12.4**	**13.2**	**13.9**	**1.2**
Non-OECD	Liquids	0.7	0.6	0.6	0.6	0.5	0.5	−1.0
	Natural gas	1.8	2.4	3.0	3.5	4.1	4.6	3.4
	Coal	4.1	5.2	5.6	6.7	7.9	9.1	3.0
	Nuclear	0.4	0.7	1.2	1.5	1.7	2.0	6.0
	Renewables	1.9	2.8	3.6	4.0	4.5	5.0	3.7
	Total	**8.9**	**11.8**	**13.9**	**16.3**	**18.8**	**21.2**	**3.3**
World	Liquids	1.0	0.9	0.9	0.9	0.8	0.8	−0.9
	Natural gas	4.2	4.9	5.6	6.5	7.5	8.4	2.6
	Coal	7.7	8.5	8.9	10.2	11.5	12.9	1.9
	Nuclear	2.6	3.2	3.7	4.2	4.5	4.9	2.4
	Renewables	3.7	5.1	6.3	7.0	7.6	8.2	3.1
	Total	**19.1**	**22.7**	**25.5**	**28.7**	**31.9**	**35.2**	**2.3**

(*Source:* USEIA/IEO 2011 Report, Table 11)

grown swiftly over the past decade, from 18 GW of net installed capacity at the end of 2000 to 121 GW at the end of 2008 – a trend that will continue into the future. Of the 4.6 trillion kWh of new renewable generation added over the projection period, 2.5 trillion kWh (55%) is attributed to hydroelectric power and 1.3 trillion kWh (27%) to wind.

The majority of the hydroelectric growth (85%) occurs in the non-OECD countries, while a slight majority of wind generation growth (58%) occurs in the OECD. The total cost to build and operate renewable generators is higher than those for conventional plants. The intermittent nature of wind and solar, which is not operator-controlled, can further reduce the economic competitiveness of these resources. However, an improved battery storage technology and dispersion of wind and solar generating facilities over wide geographic areas could lessen many of the problems associated with intermittency over the projection period.

Electricity generation from nuclear power worldwide increases from 2.6 trillion kWh in 2008 to 4.9 trillion kWh in 2035 as concerns about energy security and climate change support the development of new nuclear generating capacity. Further, there has been a continuous increase in the global average capacity utilization rate from about 65% in 1990 to about 80% today, with some anticipated increases in the future. The future of nuclear power sometimes appears uncertain as a number of issues – plant safety, radioactive waste disposal, and proliferation of nuclear materials – continues to raise public concerns in several countries which may obstruct or delay plans for new installations. In addition to high capital and maintenance costs, lack of trained manpower as well as limited global manufacturing capacity for certain components could keep national nuclear programs from advancing quickly. In the wake of the disaster at Japan's Fukushima nuclear power plant, Germany, Switzerland and Italy have announced plans to phase out or cancel all their existing and future nuclear reactors. Such plans and new policies that other countries may adopt in response to the Fukushima disaster which are not reflected in the given projections may affect the projection for nuclear power and some reduction should be expected.

In the present case, 75% of the world expansion in installed nuclear power capacity occurs in non-OECD countries (Figure 2.10). China, Russia, and India account for the largest increment in world net installed nuclear power from 2008 to 2035; China adds 106 GW of nuclear capacity, Russia 28 GW, and India 24 GW.

Coal provides the largest share of world electricity generation, although its share declines over the projection period, 40% in 2008 to 37% in 2035. Coal-fired thermal power plants are the principal means of producing electricity all over the world, particularly in the United States, China, Russia, India, Germany, South Africa, Australia, and Japan; and around 100 GW out of 185 GW of the total projected addition will be coal-fired generation, mostly due to low and fairly stable costs of coal.

China's 80% of electricity generation fueled by coal is projected to increase at a 3% annual rate, from 2008 to 2035, against 0.2% in the USA during this period (EIA-IEO 2011). At the end of 2008, China had an estimated 557 GW of operating coal-fired capacity, and another 485 GW of coal-fired capacity are projected through 2035 to meet the increasing demand for electric power. This new capacity addition represents, on average, 18 GW of annual installations which is a considerably slower rate of addition compared to that during the five-year period ending in 2008. However, coal's share of total electricity generation in China declines from 80% in 2008 to 66%

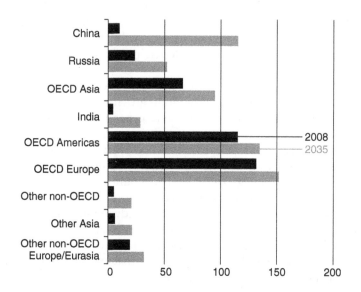

Figure 2.10 World nuclear generating capacity, 2008 and 2035 (gigawatts) (*Source*: USEIA-IEO 2011).

in 2035, as generation grows more rapidly from nuclear, natural gas and renewables than generation from coal. In the industrial sector, especially in steel and pig iron production, coal remains the primary source of energy because of limited oil and natural gas reserves.

In India, 54% of the projected growth in coal consumption is mostly in the electric power and industrial sectors. In 2008, India's coal-fired power plants consumed 6.7 quadrillion Btu of coal, representing 62% of the country's total coal demand. Coal-based electricity generation in India grows annually by 2.0% on average to 11.4 quadrillion Btu in 2035, requiring an additional 72 GW of coal-fired capacity (net of retirements), i.e., India's coal-fired generating capacity increases from 99 GW in 2008 to 171 GW in 2035. However, with more rapid generation growth from natural gas, nuclear power, and renewable energy sources than in China, the coal share of India's total generation declines from 68% in 2008 to 51% in 2035 (EIA-IEO 2011).

The projected growth in coal consumption in the other non-OECD Asian countries, from 6.3 quadrillion Btu in 2008 to 11.0 quadrillion Btu in 2035 at 2.1% per year, will be in the power generation and industrial sectors. In Vietnam, Indonesia, Malaysia and Taiwan a significant growth in coal-fired electric power generation is anticipated.

Coal consumption in non-OECD Europe and Eurasia is projected to decline slightly from 8.9 quadrillion Btu in 2008 (EIA-IEO 2011). Russia, which consumed 4.5 quadrillion Btu of coal in 2008, nearly half of that of the region, leads the region's coal consumption, and Russia's coal-fired power plants have provided 18% of its electricity; in 2035 this share would be slightly lower, at 16%. In the projection period, Russia's coal consumption increases to 4.9 quadrillion Btu in 2035. In the other countries of this region, coal usage declines from 4.5 quadrillion Btu in 2008 to 3.7 quadrillion Btu in 2035. For the region as a whole, coal-fired electricity generation remains near its current level resulting in the decline of the coal share of total generation from 34% in 2008 to 24% in 2035. During this period, nuclear power meets about 38% and natural gas meets around 39% of the additional electricity demand for non-OECD Europe and Eurasia (excluding Russia). There are plans for the reconstruction of the existing coal-fired power plants in Albania, Bosnia and Herzegovina, Bulgaria, Montenegro, Romania, Serbia, and Ukraine.

South Africa, which currently accounts for 93% of coal consumption on the continent, is expected to increase its consumption in the projection period. The total consumption across the continent is projected to increase by 2.5 quadrillion Btu from 2008 to 2035. As the demand for electricity has increased in recent years, South Africa has decided to restart three large coal-fired power plants with 3.8 giga watts capacity closed for more than a decade (EIA-IEO 2011). A further addition of 9.6 GW of coal-fired power generation is planned. Coal reserves in Botswana and Mozambique are presently developed for supplying coal to domestic coal-fired plants to meet the power shortage in southern Africa and to export to international markets. An increase in coal consumption in the industry sector, of 0.6 quadrillion Btu (around 26% of the total increase for Africa) from 2008 to 2035, is due to the demand for coal for the production of steam and process heat, production of coke for the steel industry, and production of synthetic liquid fuels. Approximately 25% of the coal consumed in South Africa is for the production of liquid fuels.

Central and South America consumed 0.8 quadrillion Btu of coal in 2008. Brazil having one of the world's largest steel industry accounts for 61% of the region's utilization. The rest was mostly consumed by Argentina, Colombia, Peru and Puerto Rico. Coal consumption in this region is projected to increase by 1.5 quadrillion Btu from 2008 to 2035. Most of Brazil's increase is primarily coke manufacture and the rest for three new power plants to generate 1.4 GW of electricity.

Projected coal consumption in the Middle East in 2008 was 0.4 quadrillion Btu, and remains relatively low to 2035. Israel accounts for about 85% of the total, and Iran accounts for the rest.

BP Energy Outlook 2030 released in January 2013 projects a rise of roughly 4.5 btoe from 2011 to 2030, about 36%, and the increase will be dominated by fossil fuels. Coal consumption rises by 26% (3.7 to 4.7 btoe), total liquids by 17% (4.1 to 4.8 btoe), and gas by 46% (2.9 to 4.3 btoe). In due course, large shifts are expected in the fuel mix for power generation, driven by relative prices, policy, and technology developments (Figure 2.11). In the 1970s and 1980s, high priced oil was replaced by nuclear and to a lesser extent by coal. In the 1990s and 2000s, gas increased its share as CCGT technology was deployed, and coal's share also rose, reflecting the growing influence of Asia's coal-intensive power sector in global power generation. From 2011 to 2030 coal loses share and natural gas gains share only marginally, as renewables start to penetrate the market at scale. The impact on the growth of fuels for power, in volume terms, is particularly striking in the final decade of the outlook. After 2020, very little growth in coal usage in power generation is seen, in contrast to the previous two decades. This is the result of the slowdown in total power growth, and increased role of both renewables and nuclear energy. Gas growth is also reduced, but very much less compared to coal.

However, the aggregate projection for world energy demand and supply is little changed since BP's last Outlook, up about 0.5% by 2030. The oil and natural

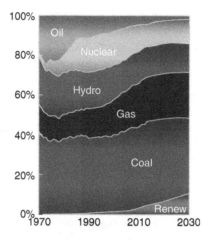

Figure 2.11 Share of power output (*Source*: BP Energy Outlook 2030, January 2013).

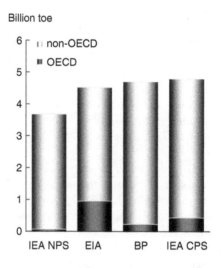

Figure 2.12 Energy Consumption Growth 2010–2030: Comparison between different Outlooks (*Source*: BP Energy Outlook 2030, January 2013).

gas supply outlook in North America has been revised higher, around 14%, due to evolving expectations for shale gas/tight oil plays; and demand for power generation has been revised higher due mainly to increased demand for electricity in non-OECD Asia Pacific, where fossil fuel useage is impacted by a re-evaluation of that region's potential for economic development.

Comparison among different Outlooks: BP Energy Outlook 2030 is compared with those of USEIA and IEA (shown in Figure 2.12). BP EO 2030 is based on a 'most likely' assessment of future policy trends. In that respect it differs from the energy projections of the IEA and the EIA, which are based on specific policy scenarios.

IEA's 'New Policies Scenario' (NPS) assumes that governments will translate the announced fossil fuel mitigation commitments to actual concrete actions, and the Current Policies Scenario (CPS) assumes no change in policy settings, i.e., business-as-usual. Although BP 2030 assumes scenario similar to NPS, the outcomes are closest to CPS scenario. More growth in non-OECD energy demand is seen in BP's than in the IEA's NPS; and BP's assessment also shows more growth for fossil fuels, especially for coal. This probably reflects differing views on the outlook for rapidly industrializing economies, in particular on the speed with which they can shift to a less energy-intensive growth path.

Regards to carbon emissions, they continue to grow from energy use by 26% from 2011 to 2030 at 1.2% per year. This means, emissions remain well above the required path to stabilize the concentration of GHG at the recommended level of 450 ppm. By 2030, 45 billion tonnes of annual CO_2 emissions will be reached.

This prediction of carbon dioxide emissions is quite similar to the IEA's CPS. The implication of this development, according to IEA 2011, is 'in the absence of efforts to stabilize atmospheric concentration of GHG's, and the average global temperature rise is projected to be atleast 6°C in the long term'. The IEA NPS projects CO_2 levelling off above 35 btonnes, far below BP projection. Its implication is projecting a temperature rise of 4°C that demands significant changes in policy and technologies. Substantial additional cuts in emissions after 2050 are essential to cap it at 4°C. These projections see a world on a path for a rise of 4°C to 6°C. According to World Bank Report, 'Turndown the Heat' released in November 2012, a temperature increase of 4°C will be devastating to the Earth.

REFERENCES

BP (2008): BP Statistical Review of World Energy, June 2008, London, UK.
BP (2011): BP Statistical Review of World Energy, June 2011, London, UK.
BP (2012): BP Statistical Review of World Energy, June 2012, London, UK.
Grimston, M.C. (1999): Coal as an Energy Source, IEA Coal Research, 1999.
IEA (2011): Energy Technology Policy Division, Power Generation from Coal – Ongoing Developments and Outlook, October 2011, Paris, OECD/IEA.
Macalister, T. (2011): news item, at guardian.co.uk, Dec., 29, 2011.
MIT Study (2007): The Future of Coal – Options for a carbon-constrained World, MIT, Cambrideg, MA, August 2007.
Mills, S.J. (2004): Coal in an Enlarged European Union, IEA Coal Research, 2004.
NEPD (2001): National Energy Policy Development Group, National Energy Policy, U.S. Govt Printing Office, May 2001.
Saraf, S. (2011): Platt.com News feature, June 28, 2011.
USEIA (2002): EIA-International Energy Outlook 2002, USDOE, March 2002.
USEIA-IEO (2008): EIA-International Energy Outlook 2008, US DOE, Sept. 2008, Washington DC.
USEIA-IEO (2009): EIA-International Energy Outlook 2009, US DOE, Sept. 2009, Washington DC.
USEIA (2009): Annual Coal Report 2008, USDOE, Sept. 2009, Washington DC.

USEIA-IEO (2011): EIA-International Energy Outlook 2011, US DOE, Sept. 2011, Washinton DC.

USEIA (2011): *World Shale Gas Resources: an Initial Assessment of 14 Regions outside the United States*, Washington DC, April 2011.

Walker, S. (2000): Major Coalfields of the World, *IEA Coal Research*, 2000.

WEC (2007): World Energy Council 2007, *Survey of Energy Sources*, 2007 ed., London: WEC.

World Bank (2012): 'Turndown the Heat – Why a 4°C Warmer World must be avoided', Report released in November 2012, Washington DC.

Worldwide Look at Reserves and Production (2010): *Oil & Gas Journal*, **106** (47) 2010, pp. 46–49, at www.ogj.com, adjusted with the EIA release of proved reserve estimates as of December 31, 2010.

Chapter 3

Coal: Formation, classification, reserves and production

3.1 COAL FORMATION

Geologists believe that underground coal deposits formed about 290–360 million years ago during Carboniferous geologic era, when much of Earth was covered with muddy ground, swamps and thick forests (Moore 1922, Jeffry 1924). As the plants and trees grew and perished, the plant debris was buried under Earth's wet surface, where insufficient oxygen slowed their decay and led to the formation of peat. The plant debris converting into peat was through a biochemical process. Over millions of years, these layers of peat are altered physically and chemically; this process is called 'coalification'. Covered by newly formed oceans and land masses, peat started to become a tightly packed and compressed sediment under combined effect of heat (from the Earth's interior or near volcanic source) and pressure. The accumulation and sediment deposition continued repeatedly and was followed by geochemical process, i.e., biological and geological actions. This means, peat underwent several changes as a result of bacterial decay, compaction, heat, and time. Depending upon the extent of temperature, time, and pressure exerted, different types of coal (from peat to lignite coal, to sub-bituminous coal, to bituminous coal, and to anthracite coal) were formed and are mined today (e.g., Lieske 1930, Berl *et al.* 1932, Stadnikov 1937, Hendricks 1945, Fuchs 1952). The sequence of formation is represented in Figure 3.1 (Kentucky Geological Survey: How is coal formed?). It is estimated that three to seven feet of compacted plant matter is needed to form one foot of coal.

Some coal is of more recent times, formed 15 to 100 million years ago, and the newest coal has an estimated age of 1 million years (Nersesian 2010). The considerable diversity of various coals is due to the differing climatic and botanical conditions that existed during the main coal-forming periods, along with subsequent geophysical actions. The vegetation of various geological periods differed both biologically and chemically. The depth, temperature, degree of acidity, and the natural movement of water in the original swamp are also contributory factors in the formation of the coal (Schobert 1987, Van Krevelen 1993, Mitchell 1997).

Due to its origin in ancient living matter like oil and gas, it is known as a fossil fuel. The coalification can be described geochemically as consisting of three processes: the microbiological degradation of the cellulose of the initial plant material; the conversion of the lignin of the plants into humic substances; and the reduction of these humic substances into larger coal molecules (Tatsch 1980).

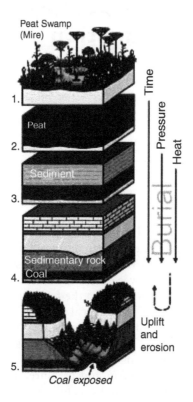

Figure 3.1 Formation of Coal (*Source*: Kentucky Geological Survey, artist: Steve Greb. Reproduced with permission from Mike Lynch, KGS).

3.2   RANK OF COAL

The rank is a measure of the amount of transformation coal has undergone from the original plant material during its formation. The rank of coal is influenced by physical and chemical properties of coal. The consecutive stages in evolution of rank, from the initial peat stage are: lignite (brown coal), sub-bituminous coal, bituminous coal, and anthracite. These stages, along with the main chemical reactions, are shown in Table 3.1. Coal first formed from peat has high moisture content and a relatively low heating value. Initially formed 'lignite' is a soft, brownish-black material that has a low organic maturity and heating value. Among the world's coal reserves, lignite is the largest portion. Over many more millions of years, the continuing effects of temperature and pressure produces further change in the lignite, progressively increasing its organic maturity and transforming it into the range called 'sub-bituminous' coal, a dull black coal. With further chemical and physical changes occurring, these coals became harder and blacker, forming the 'bituminous coals'; and with the progressive increase in the organic maturity, anthracite is finally formed. The first two types are soft coals, primarily thermal coals, used for power generation. The last two types, bituminous coal and anthracite, are 'black coals' or 'hard coals'.

Table 3.1 The stages in coalification process.

Stage	Process	Main Chemical reaction
Decaying Vegetation → Peat	Peatification	Bacterial and fungal life cycles
Peat → lignite	Lignitification	Air oxidation followed by decarboxylation and dehydration
Lignite → bituminous coal	Bituminization	Decarboxylation and hydrogen disproportioning
Bituminous coal → semi-anthracite	Preanthracitization	Condensation to small aromatic ring systems
Semi-anthracite → anthracite	(a) Anthracitization	Condensation of small aromatic ring systems to larger ones; dehydrogenation
	(b) Graphitization	Complete carbonification

(Drawn from Van Krevelen 1993)

Coal occurs in seams or veins in sedimentary rocks; formations vary in thickness, with those in underground mines 0.7–2.4 meters (2.5–8 feet) thick and those in surface mines, as in the western United States, sometimes around 30 meters (100 feet) thick.

3.3 COMPOSITION

Coal is a combustible rock composed of complex mixtures of organic and inorganic compounds. The organic component consists primarily of carbon, hydrogen, and oxygen, with lesser amounts of sulfur and nitrogen. Trace amounts of a range of other elements inherited from the vegetation that existed and perished in the swamps are also present. The inorganic material consisting of a varied range of ash-forming compounds is distributed throughout the coal. This inorganic material was either introduced into the peat swamp during coalification by residues brought by water or wind flows, or was derived from elements in the original vegetation or through the mobility of solutions in cracks and cavities (Mackowsky 1968). Coal mineralogy can influence the ability to remove minerals during coal preparation/cleaning and coal combustion and conversion (i.e., the production of liquid fuels or chemicals).

Some of these elements also combine to form discrete minerals, such as pyrite. The inorganic constituents can widely vary in concentration from several percentage points down to parts per billion of the coal. Although as many as 120 different minerals have been identified in coal, only about 33 of them are commonly found in coal, and of these, only about 8 (quartz, kaolinite, illite, montmorillonite, chlorite, pyrite, calcite, and siderite) are abundant enough to be considered major constituents.

3.4 STRUCTURE OF COAL

The structure of coal is extremely complex and depends on the origin, history, age, and rank of the coal. As late as the 1920s, chemists believed that coal consisted of carbon mixed with hydrogen-containing impurities. The analysis using destructive distillation (heating in the absence of air) and solvent extraction (reacting with different organic solvents) methods has revealed that coal contained significant portion of carbon, and smaller amounts of hydrogen, oxygen, nitrogen, and sulfur; and the ash contained inorganic compounds such as aluminum and silicon oxides. The distillation produced tar, water, and gases. Hydrogen was the chief component of the gases released, although ammonia, carbon monoxide and dioxide gases, benzene and other hydrocarbon vapors are present.

Richard Wheeler at the Imperial College, London, Friedrich Bergius in Mannheim, and Franz Fischer in Mülheim made important contributions that revealed the presence of benzenoid (benzenelike) compounds in coal. It was William Bone at Imperial College that confirmed coal's benzenoid structure in 1925. The benzene tri-, tetra-, and other higher carboxylic acids they obtained as oxidation products indicated a prevalence of aromatic structures with three-, four-, and five-fused benzene rings, and other structures with a single benzene ring. The simplest structures consisted of eight or ten carbon atoms, the fused-ring structures contained 15 or 20 carbon atoms.

The molecular (chemical) and conformational structures of coal are studied to determine its reactivity during combustion, pyrolysis, and liquefaction processes. Several papers have been published on these aspects (e.g., WCI, WEC 2004, Smith & Smoot 1990, Bhatia 1987, Xuguang 2005, Carlson 1991, so on). Carlson (1992) studied the three dimensional structures of coal using computer simulation and further analyzed the structures suggested by Given (1960), Solomon (1981), Shinn (1984) and Wiser (1984).

3.5 CLASSIFICATION AND ANALYSIS

Several classification systems for coal have been proposed and are generally based on (a) ash content; (b) its structure: anthracite, bituminous, sub-bituminous, lignite; (c) heating values, (d) sulfur content, low or high; (e) coke grade; and (f) caking properties.

Henri-Victor Regnault has classified coal by *proximate analysis*, by determining the percentage of moisture content, combustible matter (gases released when coal is heated), fixed carbon (solid fuel left after the volatile matter is removed) and ash (impurities such as iron, silica, alumina, other unburned matter). These factors are determined on 'as-received basis'.

Clarence Seyler's *ultimate analysis* determines constituent chemical elements, carbon, hydrogen, and oxygen, and nitrogen, exclusive of sulfur and dry ash. British scientist Marie Stopes has classified coal based on their microscopic constituents: clarain (ordinary bright coal), vitrain (glossy black coal), durain (dull rough coal), and fusain, also called mineral charcoal (soft powdery coal). Since moisture and mineral matter or ash are not pertinent to the coal substance, analytical data can be expressed on several different bases to reflect the composition of *as-received*, *air-dried*, or *dry coal*, or of *dry ash-free* or *dry mineral matter-free* coal. The most commonly used bases with coal components in the various classification schemes are mentioned by Ward (1984):

(i) *as-received coal*: contains fixed carbon, volatile matter (both organic and mineral), ash, moisture (both inherent and surface);
(ii) *air-dried coal*: fixed carbon, volatile matter, ash, and inherent moisture;
(iii) *dry coal*: fixed carbon, volatile matter, and ash;
(iv) *dry, ash-free coal*: fixed carbon, and volatile matter; and
(v) *dry, mineral matter-free coal*: fixed carbon and volatile organic matter; this is also called pure coal.

A classification system for commercial applications was developed applying guidelines established by American Standards Association (ASA) and American Society for Testing Materials (ASTM) which is used for scientific purpose also. This ASTM system (ASTM D388) established the four ranks mentioned above each of which is subdivided into several groups. High-rank coals are high in fixed carbon (heat value) and volatile matter (expressed on a dmmf basis), and low in hydrogen and oxygen. Low-rank coals are low in carbon but high in hydrogen and oxygen content and are classified in terms of their heating value (expressed on mmmf basis).

Anthracite is the highest ranked coal with little flame and smoke. It has the highest fixed-carbon content, 86–98%, and high energy content. The heating value range is 13500–15600 Btu/pound (equivalent to 14.2–16.5 million joules/pound). It has relatively little moisture or volatiles. It is mainly used to provide home heating, though used in power generation, and for the iron and steel industry. It is a scarce type of coal. *Bituminous coal* is second-ranked soft coal that produces smoke and ash when burned. It has 46–86% fixed-carbon content and a high heating value, 10500–15000 Btu/pound. It is the most abundant economically recoverable coal globally, and the main fuel in steam turbine-powered power generating plants. Some bituminous coals have the right characteristics suitable for conversion to metallurgical coke used in steel industry. *Sub-bituminous* coal is next in rank and has 35–45% fixed-carbon content and more moisture than bituminous coal. It has a heating value of 8300–13000 Btu/pound. Despite its lower heat value, its lower sulfur content than other types makes it attractive for use in power generation because it ensures environmental advantage. *Lignite* is a soft coal and has 25–35% fixed-carbon content, and with lowest heating value, 4800–8300 Btu/pound. The lignite coal is mainly used for electric power generation.

With increasing rank of coal, the moisture decreases, fixed carbon increases, volatile matter decreases, and heating value increases (optimum Btu at low-volatile bituminous). Figure 3.2 provides a more detailed overview of various US-, UN-, and German based classification methods (EURACOAL website).

Apart from producing heat and generating electricity, coal is an important source of raw materials used in manufacturing. Its destructive distillation (carbonization) produces hydrocarbon gases and coal tar, which are synthesized for obtaining drugs, dyes, plastics, solvents, and numerous other organic chemicals. High pressure coal hydrogenation or *liquefaction*, and the indirect liquefaction of coal using Fischer–Tropsch syntheses are also potential sources of clean-burning liquid fuels and lubricants.

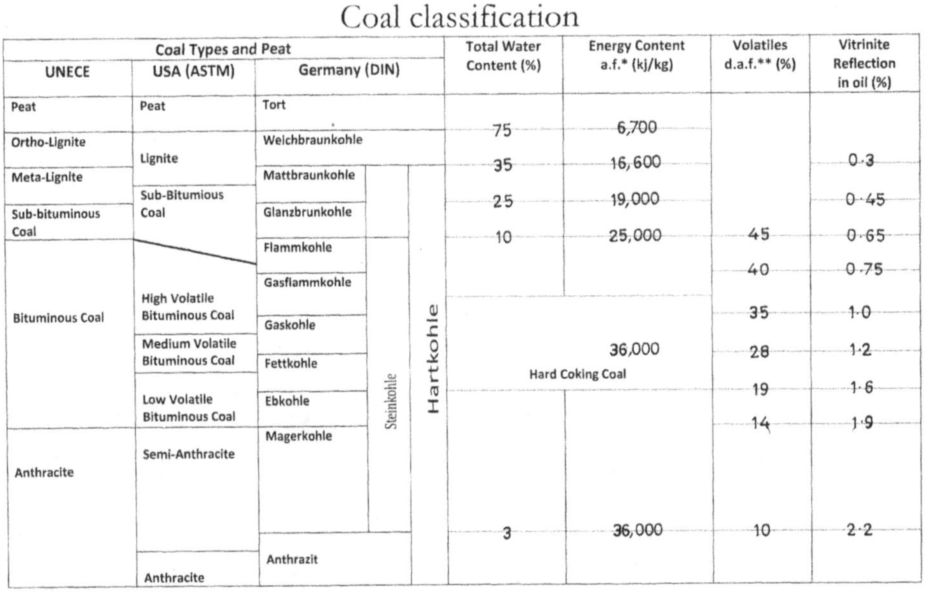

Coal classification

Coal Types and Peat			Total Water Content (%)	Energy Content a.f.* (kj/kg)	Volatiles d.a.f.** (%)	Vitrinite Reflection in oil (%)
UNECE	USA (ASTM)	Germany (DIN)				
Peat	Peat	Tort				
Ortho-Lignite	Lignite	Weichbraunkohle	75	6,700		
Meta-Lignite		Mattbraunkohle	35	16,600		0·3
Sub-bituminous Coal	Sub-Bitumious Coal	Glanzbrunkohle	25	19,000		0·45
		Flammkohle	10	25,000	45	0·65
		Gasflammkohle			40	0·75
Bituminous Coal	High Volatile Bituminous Coal	Gaskohle			35	1·0
	Medium Volatile Bituminous Coal	Fettkohle		36,000 Hard Coking Coal	28	1·2
	Low Volatile Bituminous Coal	Ebkohle			19	1·6
		Magerkohle			14	1·9
Anthracite	Semi-Anthracite					
	Anthracite	Anthrazit	3	36,000	10	2·2

(vertical labels: Steinkohle, Hartkohle)

Figure 3.2 Overview of classification approaches (*Source*: EURACOAL, Redrawn).

3.6 USES

Coal has come to stay as the most preferred energy source despite its established contribution to global warming. The most significant uses, in addition to electricity generation are steel production, cement manufacturing and as a liquid fuel. Since 2000, global coal consumption has grown, and around 6.1 billion tonnes of hard coal and 1 billion tonnes of brown coal were used worldwide in 2010. Different types of coal have different uses. Steam coal – also known as thermal coal – is hard coal, mainly used in power and heat generation. Coking coal – also known as metallurgical coal – is also hard coal, mainly used to produce coke (mostly bituminous coal) and in steel production. Alumina refineries, paper manufacturers, chemical and pharmaceutical industries are other important users of coal. Several chemical products can be produced from the by-products of coal. Refined coal tar is used in the manufacture of chemicals, such as creosote oil, naphthalene, phenol, and benzene. Ammonia gas recovered from coke ovens is used to manufacture ammonia salts, nitric acid and agricultural fertilizers. Thousands of different products have coal or coal by-products as components: soap, aspirins, solvents, dyes, plastics and fibres, such as rayon and nylon. Coal is also an essential ingredient in the production of special products such as 'activated carbon' used in filters for water and air purification and in kidney dialysis machines; carbon fibre, an extremely strong but light weight reinforcement material used in construction, mountain bikes and tennis rackets; and 'silicon metal', used to produce silicones and silanes, which are in turn used to make lubricants, water repellents, resins, cosmetics, hair shampoos and tooth pastes.

3.7   ENERGY DENSITY

The *energy density* of coal, i.e. its *heating value*, is roughly 24 *megajoules* per kilogram. The energy density of coal can also be expressed in *kilowatt-hours*, the units of electricity. One kWh is 3.6 MJ, so the energy density of coal is 6.67 kWh/kg. The typical thermodynamic efficiency of coal power plants is 30%; that is, $6.67 \times 30\% =$ approx. 2.0 kWh/kg can successfully be turned into electricity, and the rest is waste heat. So coal power plants obtain *approximately 2.0 kWh per kilogram of burned coal* (Glen Elert).

As an example: running one 100-watt light bulb for one year requires 876 kWh ($100 W \times 24$ h/day $\times 365$ day/year $= 876000$ Wh $= 876$ kWh). Converting this power usage into physical coal consumption:

876 kWh/2.0 kWh per kg $= 438$ kg of coal

If the coal-fired power plant's efficiency is high, it takes lesser amount of coal to power a 100-watt light bulb. The *transmission and distribution losses* caused by resistance and heating in the *power lines*, which are in the order of 5–10%, depending on distance from the power station and other factors have to be taken into account.

Carbon intensity

Commercial coal generally has a carbon content of at least 70%. Coal with a heating value of 6.67 kWh per kilogram (as mentioned) has a carbon content of roughly 80%, which is:

$(0.8 \text{ kg})/(12 \text{ kg/k mol}) = 2/30 \text{ k mol}$

where 1 mol equals to N_A atoms where N_A is Avogadro number.

Carbon combines with oxygen in the atmosphere during combustion, producing carbon dioxide whose atomic weight is ($12 + 16 \times 2 = 44$ kg/k mol). The CO_2 released to the atmosphere for each kilogram of combusted coal is therefore:

$(2/30) \text{ k mol} \times 44 \text{ kg/k mol} = 88/30 \text{ kg} =$ approx. 2.93 kg.

So, approximately 2.93 kg of carbon dioxide is released by burning 1 kg of coal. This can be used to calculate an *emission factor* for CO_2 from the use of coal power. Since the useful energy output of coal is about 31% of the 6.67 kWh/kg (coal), the burning of 1 kg of coal produces about 2 kWh of electrical energy as seen above. Since 1 kg coal emits 2.93 kg CO_2, *the direct CO_2 emissions from coal power are 1.47 kg/kWh,* or about 0.407 kg/MJ.

Calorific value estimation

The calorific value Q of coal is the heat liberated by its complete *combustion* with oxygen. Q is a complex function of the elemental composition of the coal. Q can be determined *experimentally* using calorimeters. Dulong suggests the following approximate formula for Q when the oxygen content is less than 10%:

$Q = 337C + 1442(H - O/8) + 93S,$

where C is the mass percent of carbon, H is the mass percent of hydrogen, O is the mass percent of oxygen, and S is the mass percent of sulfur in the coal. With these constants, Q is given in kilojoules per gram.

3.8 GLOBAL COAL RESERVES: REGIONAL DISTRIBUTION

Coal is a nonrenewable resource, takes millions of years for formation, and the world's most abundant and geographically broadly distributed fossil fuel. Coal constitutes 70% of the world's proven fossil energy reserves, and very much exceeds those of oil and gas reserves combined. '*Proved reserves of coal*' is generally the term used for those quantities that geological and engineering information indicates, with reasonable certainty, that can be recovered in the future from known deposits under existing economic and operating conditions. All major regions of the world have appreciable quantities of coal with the exception of the Middle East region. Interestingly, the Middle East contains almost two-thirds of the world oil reserves and over 41% of the natural gas reserves. There are two internationally recognized methods for assessing world coal reserves: (1) produced by German Federal Institute for Geosciences and Natural resources (BGR) and is used by IEA as the main source of information about coal reserves, (2) produced by the World Energy Council (WEC) and is used by the BP Statistical Review of World Energy. In Table 3.2 is given the regional distribution of world's proved reserves of coal at the end of 2011 (BP Statistical Review 2012). Quality and geological character of coal deposits are important parameters for coal reserves. Coal is a heterogeneous source of energy, with quality as specified by characteristics such as heat, sulfur, and ash content varying significantly by region and even within individual coal seams. Premium-grade bituminous coals, or coking coals, used to manufacture coke for the steelmaking process are of the top-end quality. Coking coals produced in the USA have an estimated heat content of 26.3 million Btu per ton and relatively low sulfur content of approximately 0.9% by weight. At the other end of the spectrum are reserves of low-Btu lignite. On a Btu basis, lignite reserves show considerable variation. Estimates published by the International Energy Agency for 2008 indicate that the average heat content of lignite in major producing countries varies from a low of 5.9 million Btu per ton in Greece to a high of 13.1 million Btu per ton in Canada (USEIA-IEO 2011).

Table 3.2 Distribution of coal reserves at the end of 2011 by region.

Region	Anthracite and bituminous coal (million tonnes)	Lignite and sub-bituminous coal (million tonnes)	Total (million tonnes)	Share of the world total (%)	R/P
North America	112,835	132,253	245,088	28.5	228
S & C America	6,890	5,618	12,508	1.5	124
Europe & Eurasia	92,990	211,614	304,604	35.4	242
Middle East & Africa	32,721	174	32,895	3.8	126
Asia Pacific	159,326	106,517	265,843	30.9	53

Source: BP Statistical Review of world's Energy, June 2012.

The proven reserves of coal at the end of 1990, 2000 and 2010 are shown in Figure 3.3. While the reserves are almost at the same level up to 2000 there has been decrease of 12.5% in the last decade.

Although coal is available in almost every country worldwide, the recoverable reserves are in around 70 countries. Total proved reserves of coal around the world are estimated at 861 billion tons at the end of 2011 (BP Statistical Review 2012). According to World Energy Council (WEC 2010), at the current consumption levels, this is enough coal to last around 118 years; that is, ratio of reserves(R) to production (P) = 118 (*If the reserves remaining at the end of the year are divided by the production in that year, the result is the length of time that those remaining reserves would last if production were to continue at that rate*). However, this could be extended still further by locating new reserves through the current and improved exploration activities and by advances in mining techniques that will allow previously inaccessible reserves to be reached.

Around 92% of the world's recoverable reserves are located in ten countries, shown in Table 3.3 (BP Statistical Review 2012). the USA leads the world with 27.6% of the world's reserves, with Russia and China far behind in the second and third positions with 18.2% and 13.3% of share respectively. While Saudi Arabia tops in oil reserves (around 20%) in the world, the USA tops with highest coal reserves. It is no wonder the coal producers in America are pushing for more reliance on coal for energy needs.

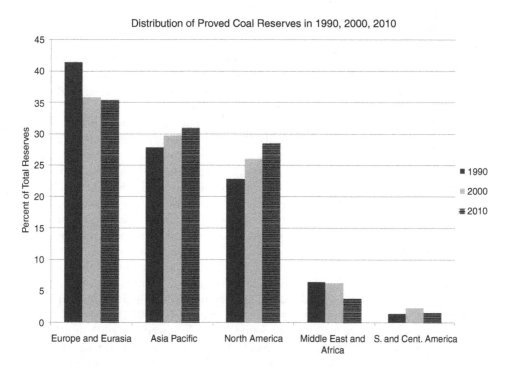

Figure 3.3 Distribution of proved reserves in 1990, 2000, 2010 (Drawn based on data from BP Statistical Review of world energy 2011).

Table 3.3 Top Ten countries with proved reserves of coal at the end of 2011 (in million tonnes).

Country	Anthracite and bituminous coal	Lignite and sub-bituminous coal	Total	Share of the world total (%)	R/P (years)
United States	108,501	128,794	237,295	27.6	241
Russian federation	49,088	107,922	157,010	18.2	495
China	62,200	52,300	114,500	13.3	35
Australia	37,100	39,300	76,400	8.9	180
India	56,100	4,500	60,600	7.0	106
Germany	99	40,600	40,699	4.7	223
Ukraine	15,351	18,522	33,873	3.9	462
Kazakhstan	21,500	12,100	33,600	3.9	300
South Africa	30,156	–	30,156	3.5	119
Poland	4,338	1,371	5,709	0.7	43

(*Source*: BP Statistical review of World Energy, June 2012)

The proportions of 'anthracite and bituminous coal (hard coal)' and 'lignite and sub bituminous coal' differ from country to country. Anthracite and bituminous coal account for 47% of the world's estimated recoverable coal reserves on a tonnage basis, sub-bituminous coal accounts for 30%, and lignite accounts for 23% (USEIA-IEO 2012).

3.9 GLOBAL COAL PRODUCTION

Large scale coal mining has been carried out since the 18th century, though coal mining started over thousand years ago. Coal is mined by two approaches – surface (or open-cut) mining and underground (or deep) mining. There are two main methods of underground mining: 'room-and-pillar' and 'long wall' mining.

With the room-and-pillar approach, coal is mined by cutting a network of 'rooms' into the coal seam and leaving behind 'pillars' of coal. Recovery rates are about 50% to 60%. The long wall approach uses mechanical shearers to cut and remove the coal at the face (100 meters to 250 meters in breadth). Self advancing hydraulically powered supports temporarily hold up the roof while the coal is extracted. The roof area behind the face is then allowed to collapse. Recovery rates are around 75%. Surface mining is economical only when the seam is near the surface. It recovers a higher proportion of the coal than underground methods but the land gets degraded becoming less useful for subsequent purposes.

The global production of coal in 2011 is 7678 million tonnes; out of this total coal produced, around 6637 million tonnes is hard coal and 1041 million tones is brown coal/lignite (WCA 2012). The hard coal, in turn, comprised about 5670 million tonnes of steam coal and 967 million tonnes of coking coal (WCA 2012). The Asia Pacific region accounts for a huge increase of around 120% followed by South and Central America with 78.8% in production. The S & C America and Asia Pacific regions record increases of 13.3% and 7.8% respectively in 2011 over 2010 which are higher than the global increase of 6.1%. Table 3.4 shows the production by region and Table 3.5 shows the major producers in 2001 and 2011. The figures in these Tables are in units

Table 3.4 Coal production in 2001, 2010 and 2011 (million tonnes oil-eq) by region.

Region	2001	2010	2011	Change over 2010
N. America	632.2	592.7	600.0	1.2%
S&C America	36.8	57.2	64.8	13.3%
Europe & Eurasia	439.6	437.3	457.1	4.5%
Middle East	0.7	0.7	0.7	–
Africa	130.2	146.1	146.6	0.3%
Asia Pacific	1220.7	2492.7	2686.3	7.8%
World Total	**2460.2**	**3726.7**	**3955.5**	**6.1%**
OECD	1029.1	1000.0	1004.4	0.4%
Non-OECD	1431.2	2726.6	2951.0	8.2%

(*Source:* BP Statistical review of World energy 2012)

Table 3.5 Coal production by major producers in 2001 and 2011 (in million tonnes oil-eq).

Country	2001	2011
China	809.5	1956.0
United States	590.3	556.8
India	133.6	222.4
Australia	180.2	230.8
Russian Federation	122.6	157.3
Indonesia	56.9	199.8
South Africa	126.1	143.8
Germany	54.1	44.6
Poland	71.7	56.6
Kazakhstan	40.7	58.8
Columbia	28.5	55.8
Ukraine	43.5	45.1

(*Source:* BP Statistical Review of World energy 2012)

of million tonnes oil-eq. In the USA, Germany and Poland, there has been a decreasing trend; but in Asian countries, particularly China, Indonesia and India, a significant increase.

3.10 COAL PRODUCTION PROJECTIONS TO 2035

The EIA Outlook projects the growth in non-OECD countries as 1.8% as against 0.5% in OECD countries. Table 3.6 shows the projected growth in coal production in different regions (EIA-IEO 2011).

China is projected with highest production rise, 67% of the increase in world coal production (45.5 quadrillion Btu out of total increase of 68 quadrillion Btu) from 2008 to 2035. This Outlook further projects that China's domestic production meets much of its own demand for coal. Other substantial increases in coal production from 2008 to 2035 include 6.5 quadrillion Btu in Australia/New Zealand (9% of the increase in

Table 3.6 World Coal production by region, 2008–2035 (million short tons).

Region	2008	2010	2015	2020	2025	2030	2035	Ave. annual growth 2008–2035 (%)
OECD North America	1,260	1,189	1,149	1,206	1,318	1,367	1,443	0.5%
United States	1,172	1,102	1,061	1,114	1,222	1,266	1,331	0.5%
Canada	75	77	77	80	84	87	95	0.9%
Mexico	13	10	11	12	12	13	18	1.0%
OECD Europe	651	613	564	560	537	519	513	−0.9%
OECD Asia	447	479	522	526	573	647	714	1.7%
Japan	0	0	0	0	0	0	0	NA
South Korea	3	3	3	1	3	3	3	−0.3%
Australia/New Zealand	444	476	519	524	570	645	711	1.8%
Total OECD	**2,358**	**2,281**	**2,236**	**2,291**	**2,428**	**2,534**	**2,670**	**0.5%**
Non-OECD Europe and Eurasia	683	621	633	621	611	620	649	−0.2%
Russia	336	328	341	339	339	354	385	0.5%
Other	347	293	292	281	271	266	264	−1.0%
Non-OECD Asia	4,081	4,600	4,965	5,244	5,847	6,457	6,966	2.0%
China	3,086	3,556	3,871	4,103	4,618	5,091	5,431	2.1%
India	568	556	577	620	694	783	885	1.7%
Other non-OECD Asia	426	488	518	521	535	583	650	1.6%
Middle East	2	3	3	2	2	2	3	1.7%
Africa	283	291	346	366	387	417	464	1.8%
Central and South America	97	89	127	158	177	190	206	2.8%
Brazil	7	6	6	6	6	6	6	−0.5%
Other Central and South America	90	83	121	152	171	184	200	3.0%
Total Non-OECD	**5,147**	**5,604**	**6,074**	**6,391**	**7,023**	**7,686**	**8,287**	**1.8%**
Total World	**7,505**	**7,886**	**8,310**	**8,682**	**9,451**	**10,219**	**10,958**	**1.4%**

world coal production), 4.5 quadrillion Btu in India, 4.5 quadrillion Btu in non-OECD Asia (excluding China and India), 3.6 quadrillion Btu in Africa, 2.7 quadrillion Btu in the USA, and 2.5 quadrillion Btu in Central and South America.

Most of the growth in coal production in Australia/New Zealand and Central and South America (excluding Brazil) is based on continuing increases in coal exports, whereas production growth in Africa and non-OECD Asia (excluding China and India) is to meet both rising levels of coal consumption and increasing exports. For the USA, growth in coal production is primarily due to increases in domestic coal consumption.

REFERENCES

Averitt, P. (1976): Coal Resources of the U.S., January 1, 1974, *U.S. Geological Survey Bulletin* No. 1412, 1975 (reprinted 1976), p. 131.
Berkowitz, N. (1979): An Introduction to Coal Technology, Academic Press.
Berl, E., Schmidt, A., & Koch, H. (1932): The origin of coal, *Angew Chem*, 45, 517–519.

Bhatia, S.K. (1987): Modeling the pore structure of coal, *AIChE J*, **33**, 1707–1718.

Brusset, H. (1949): The most recent view of the structures of coal, *Memoires ICF*, **102**, 69–74.

Bustin, R.M., Cameron, A.R., Grieve, D.A., & Kalkreuth, W.D. (1983): Coal Petrology: Its Principles, Methods, and Applications, Geological Assoc. of Canada.

Carlson, G.A. (1991): Molecular modeling studies of bituminous coal structure, Preprints of Papers – *American Chemical Society, Division of Fuel Chemistry*, **36**, 398–404.

Carlson, G.A. (1992): Computer simulation of the molecular structure of bituminous coal, *Energy Fuels*, **6**, 771–778.

Coal – Chemistry Encyclopedia – structure, water, uses, elements, gas' at http://www.chemistry explained.com/Ce-Co/Coal.html#ixzz1ox0eAJIl

Davidson, R.M. (1980): Molecular Structure of Coal, *IEA Coal Research*, London, UK.

Elliott, M.A. (Ed.), (1981): Chemistry of Coal Utilization, Second Suppl. Volume, John Wiley & Sons.

EURACOAL: European Coal Association website, retrieved on March 15, 2012.

Fuchs, W. (1952): Recent investigations on the origin of coal, *Chemiker-Zeitung*, **76**, 61–66.

Ghosh, T.K., & Prelas, M.A. (2009): Energy Resources & Systems – Vol. I: Fundamentals and Non-Renewable Resources, Springer Science + Business Media B.V 2009, pp. 159–279.

Given, P.H. (1960): The distribution of hydrogen in coals and its relation to coal structure, *Fuel*, **39**, 147–153.

Glenn Elert (ed.): The Physics FactBook, An Encyclopedia of scientific essays, started 1995, available at http://hypertextbook.com/facts

Hensel, R.P. (1981): Coal: Classification, Chemistry and Combustion, Coal fired industrial boilers workshop, Raleigh, NC, USA.

Hendricks, T.A. (1945): *The origin of coal & Chemistry of Coal Utilization*, New York: Wiley, vol. 1, 1–24.

Jeffrey, E.C. (1924): Origin and organization of coal, *Mem. Am. Acad. Arts. Sci.*, **15**, 1–52.

Kentucky Geological Survey: How is coal formed, *University of Kentucky*, available at www.uky.edu/kgs/coal/coalform_download.htm, artist: Steve Greb.

Lieske, R. (1930): Origin of coal according to the present position of biological investigation, *Brennstoff-Chemie*, **11**, 101–105.

Mackowsky, M.T. (1968): In: D. Murchson, T.S. Westoll (eds), *Mineral Matter in Coal: in Coal and Coal-Bearing Strata*, Oliver & Boyd, Ltd., pp. 309–321.

Maruyama, N. & Eckelman, M. J. (2009): Long-term trends of electric efficiencies in electricity generation in developing countries, *Energy Policy*, 37, 1678–1686.

Mitchell, G. (1997): Basics of Coal and Coal Characteristics, Iron & Steel Society, Selecting Coals for Quality Coke – Short Course.

MIT Study (2007): *The Future of Coal: Options for a Carbon-constrained World*, Massachusetts Institute of Technology, Cambridge, MA, 2007.

Miller, B.G., & Tillman, D.A. (eds) (2005): *Combustion Engineering Issues for Solid Fuel Systems*, Academic Press, 2005.

Mitchell, G. (1997): Basics of Coal and Coal Characteristics, Iron & Steel Society, Selecting Coals for Quality Coke Short Course, 1997.

Moore, E.S. (1922): *Coal: Its Properties, Analysis, Classification, Geology, Extraction, Uses, and Distribution*, New York: John Wiley & Sons.

Nersesian, R.L. (2010): *Energy for the 21st century*, 2nd edn, Armonk, NY: ME Sharpe, Inc., p. 94.

National Research Council (NRC): Managing Coal Combustion residues in Mines, 2006.

Nomura, M., Iino, M., Sanada, Y., & Kumagai, H. (1994): Advanced studies on coal structure, *Enerugi Shigen*, **15**, 177–184.

Schobert, H.H. (1987): Coal: The Energy Source of the Past and Future, *American Chemical Society*.

Schernikau, L. (2010): *Economics of the International Coal Trade*, SpringerScience + Business Media, DOI 10.1007/978-90-48.

Shinn, J.H. (1984): From coal to single stage and two stage products: a reactive model to coal structure, *Fuel*, **63**, 1187–1196.

Singer, J.G. (ed.) (1981): *Combustion: Fossil Power Systems*, Combustion Engineering, Inc.

Smith, K.L., & Smoot, L.D. (1990): Characteristics of commonly used US coals – towards a set of standard research coals, *Prog Energy & Combust Sci* **16(1)**, 1–53.

Solomon, P.R. (1981): New approaches in coal chemistry, *ACS Symposium Series* No. 169, American Chemical Society, Washington, DC: 61–71.

Stach, E. (1933): The Origin of coal vitrite, *Angew Chem*, **46**, 275–278.

Stadnikov, G.L. (1937): Constitution and the origin of coal, *Brennstoff-Chemie*, **18**, 108–110.

Suárez-Ruiz, I., & Crelling, J.C. (eds) (2008): *Applied Coal Petrology: The Role of Petrology in Coal Utilization*, Elsevier, 2008.

Survey of Energy Resources 2004, World Energy Council, London, UK.

Tatsch, J.H. (1980): *Coal Deposits: Origin, Evolution, and Present Characteristics*, Tatsch Associates.

US EPA: *US Greenhouse gas Emissions and Sinks* 1990–2004, July 2006.

US EPA: EPA to regulate Mercury and other Air toxics emissions from Coal and Oil-fired Power plants, December 2000.

U.S. Geological Survey (2006): U.S. Coal Resource Databases (USCOAL).

Van Krevelen, D.W. (1993): *Coal: Typology, Physics, Chemistry, Constitution*, 3rd edn, New York: Elsevier.

Ward, C. R. (ed.) (1984): *Coal Geology and Coal Technology*, Blackwell, p. 66.

WCA (2012): Coal Facts 2012, World Coal Association, London, UK, at www.worldcoal.org/resources/coalstatistics/coal_facts_2012 (06_08_2012)[1].pdf

Wiser, W.H. (1984): Conversion of bituminous coal to liquids and gases: chemistry and representative processes. NATO ASI Series C, **124**, 325–350.

World Coal Institute, Richmond-upon-Thames, UK.

Xuguang, S. (2005): The investigation of chemical structure of coal macerals via transmitted-light FT-IR microspectroscopy, *Spectrochim Acta: A Mol Biomol Spectrosc*, **62**, 557–564.

Chapter 4

Natural gas: Reserves, growth and costs

4.1 COMPOSITION AND EMISSIONS

Natural gas is a hydrocarbon, formed deep underground, like oil and coal. It is a mixture of several hydrocarbon gases, primarily methane (70%–90%), and heavy hydrocarbons such as ethane, propane, and butane, and carbon dioxide (0–8%), oxygen (0–0.2%), nitrogen (0–5%), hydrogen sulfide (0–5%) and traces of a few rare gases, and these figures can vary widely depending on the gas field. Natural gas is referred as 'wet' in its raw form, and 'dry' when it is almost pure methane having removed the other hydrocarbons; and 'sour' when it contains significant amounts of hydrogen sulfide.

Natural gas is found in the porous spaces in underground rock formations and often associated with oil deposits. After extraction from underground, the natural gas is refined to remove impurities such as water, other gases, sand, and other compounds. Heavier hydrocarbons, ethane, propane and butane are removed and sold separately as natural gas liquids (NGL). Other impurities are also removed, such as hydrogen sulfide (can be refined to produce sulfur which has commercial value). After refining, the clean natural gas is transmitted through a network of pipelines, thousands of miles of which exist in the United States alone. From these pipelines, natural gas is delivered to its point of use after having been compressed to move more efficiently. Compressed natural gas (CNG) is nontoxic, non-corrosive, and non-carcinogenic.

Natural gas is an environment-friendly fuel and by far the cleanest-burning hydrocarbon available, emitting significantly lower CO_2 emissions and fewer pollutants when burned than coal or oil. When used in efficient combined-cycle power plants, natural gas combustion can emit less than half as much CO_2 as coal combustion, per unit of electricity output. The emission levels when oil, coal and natural gas are burned are given in Table 4.1. Natural gas is lighter than air, making it a safe fuel for many applications in residences, industry and commerce. Since methane is a nonreactive hydrocarbon, its emissions do not react with sunlight to create smog.

Measurement of natural gas: Natural gas can be measured in a number of ways. As a gas, it can be measured by the volume it takes up at normal temperatures and pressures, commonly expressed in cubic feet/cubic meters. Production and distribution companies commonly measure natural gas in thousands of cubic feet (Mcf), millions of cubic feet (MMcf), or trillions of cubic feet (Tcf). Natural gas can also be measured as a source of energy, commonly measured and expressed in British thermal units (Btu).

Table 4.1 Fossil fuel emission levels (lbs/billion BTU of energy input).

Pollutant	Natural gas	Oil	Coal
CO_2	117,000	164,000	208,000
CO	40	33	208
NO_x	92	448	457
SO_2	1	1122	2591
Particulates	7	84	2744
Mercury	0	0.007	0.016

(Source: EIA – Natural gas Issues & Trends)

One Btu is the amount of natural gas that will produce enough energy to heat one pound of water by one degree at normal pressure. To give an idea, one cubic foot of natural gas contains about 1,027 Btus. When natural gas is delivered to a residence, it is measured by the gas utility in 'therms' for billing purposes. A therm is equivalent to 100,000 Btu, or just over 97 cubic feet, of natural gas.

4.2 FORMATION AND EXTRACTION OF NATURAL GAS

Natural gas originates from the microscopic plants and animals that lived millions and millions of years ago, like other fossils. When these plants and animals perished, the remains (organic matter) sank below and were covered by layers of sediment and compressed under the Earth's surface at very high pressures and temperatures, found deep underneath the Earth's crust, and converted to gas. Gas generated in the process flows towards the Earth's surface (and into the atmosphere) unless it encounters various types of geological traps in the form of porous rock formations tightly sealed by overlying layers of impermeable rock. The natural gas is trapped in a limited boundary within porous rocks by the impermeable rocks above, thereby creating a seal that prevents it from escaping to the surface and leads to the formation of conventional gas deposits, resembling gas-soaked sponges. The natural gas can be 'associated' (mixed with oil) or 'non-associated' found in reservoirs that do not contain oil. The majority of gas production is non-associated.

Deeper deposits, very far underground, usually contain primarily natural gas and no oil, and in many cases, pure methane, generally referred to as thermogenic methane. This is the mostly accepted hypothesis of gas formation.

There is another way in which methane (and natural gas) may be formed, i.e., through abiogenic processes. Extremely deep under the Earth's crust, there exist hydrogen-rich gases and carbon molecules. As these gases gradually rise towards the surface of the Earth, they may interact with minerals that also exist underground, in the absence of oxygen. This interaction may result in a reaction, forming elements and compounds that are found in the atmosphere (including nitrogen, oxygen, carbon dioxide, argon, and water). These gases are under very high pressure as they move toward the surface of the Earth, they are likely to form methane deposits similar to thermogenic methane.

Natural gas can also be formed through the transformation of organic matter by tiny microorganisms. Methane thus obtained is referred to as biogenic methane, and takes place near the surface of the Earth. Methanogens, tiny methane-producing microorganisms, chemically break down organic matter to produce methane. These microorganisms are commonly found in areas near the surface of the Earth that are void of oxygen. These microorganisms also live in the intestines of most animals, including humans. Formation of methane in this manner usually takes place close to the surface of the Earth, and the methane produced is usually lost in the atmosphere. In certain circumstances, however, this methane can be trapped underground, recoverable as natural gas. An example of biogenic methane is landfill gas. The waste landfills produce a relatively large amount of natural gas from the decomposition of the waste materials that they contain. New technologies are utilized to harvest this gas and used to add to the supply of natural gas. In the present context, thermogenic methane is discussed.

Extraction of natural gas

Gas is found in a variety of subsurface locations, with a gradation of quality. Conventional resources exist in discrete, well-defined subsurface accumulations (reservoirs), with permeability* values greater than a specified lower limit. Such conventional gas is recovered from the formation by an expansion process. Wells drilled into the gas reservoir allow the highly compressed gas to expand through the wells in a controlled manner, to be captured, treated and transported at the surface. This expansion process generally leads to high recovery factors from conventional, good-quality gas reservoirs. If, for example, the average pressure in a gas reservoir is reduced from an initial 300 bar to 60 bar over the lifetime of the field, then approximately 80% of the gas initially in place will be recovered.

By contrast, 'unconventional' resources are found in accumulations where permeability is low. Such accumulations include 'tight' sandstone formations, coal beds (coal bed methane or CBM) and shale formations. Unconventional reservoirs are more continuous and usually require advanced technology such as horizontal wells or artificial stimulation in order to be economically productive; recovery factors are much lower – typically of the order of 15–20% of the gas in place. The various resource types are schematically shown in Figure 4.1.

The term 'resource' will refer to the sum of all gas volumes expected to be recoverable in the future, given specific technological and economic conditions. The resource can be disaggregated into a number of sub-categories; specifically, 'proved reserves', 'reserve growth' (via further development of known fields) and 'undiscovered resources', which represent gas volumes that are expected to be discovered in the future via the exploration process (MIT Report 2011). Different techniques are applied, depending on the type of gas being extracted.

For extracting the gas, the technology commonly used is Hydraulic fracturing, a tested technique in practice in the past 60 years (Figure 4.2). The *fracking process*

---

*Permeability is a measure of the ability of a porous medium, such as that found in a hydrocarbon reservoir, to transmit fluids, such as gas, oil or water, in response to a pressure differential across the medium. In petroleum engineering, permeability is usually measured in units of millidarcies (mD). Unconventional formations, by definition, have permeability less than 0.1 mD – MIT report 2011.

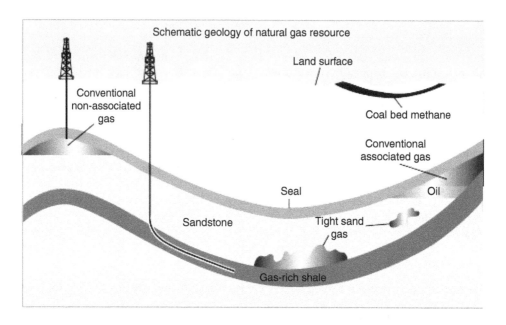

Figure 4.1 Schematic geology of various types of resources (*Source:* US Energy Information Administration).

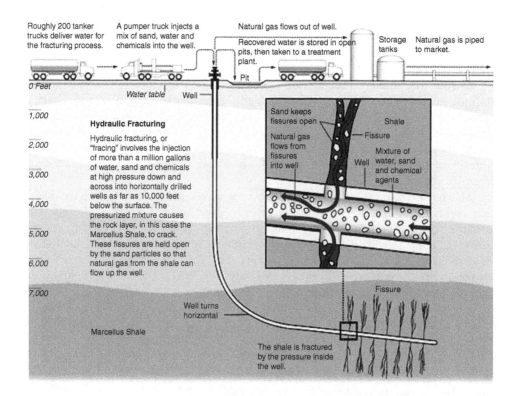

Figure 4.2 Typical Hydraulic Fracturing Operation (*Source:* ProPublica, Graphic by Al Granberg, at http://www.propublica.org/special/hydraulic-fracturing-national).

occurs after a well has been drilled and steel pipe (casing) has been inserted in the well bore. The casing is perforated within the target zones that contain oil or gas, so that when the fracturing fluid is injected into the well it flows through the perforations into the target zones. Eventually, the target formation will not be able to absorb the fluid as quickly as it is being injected. At this point, the pressure created causes the formation to crack or fracture. Once the fractures have been created, injection ceases and the fracturing fluids begin to flow back to the surface. Materials called proppants (usually sand or ceramic beads), which were injected as part of the frac fluid mixture, remain in the target formation to hold open the fractures. Typically, a mixture of water, proppants and chemicals is pumped into the rock or coal formation (see for e.g., Healy 2012). There are, however, other ways to fracture wells. Sometimes fractures are created by injecting gases such as propane or N_2, and sometimes acidizing occurs simultaneously with fracturing.

Acidizing involves pumping acid (usually hydrochloric acid), into the formation to dissolve some of the rock material to clean out pores and enable gas and fluid to flow more readily into the well. Studies have shown that *around 20 to 85% of fracking fluids may remain underground*. Used fracturing fluids that return to the surface are often referred to as flowback, and these wastes are typically stored in open pits or tanks at the well site prior to disposal. Ideally, hydraulic fracture treatment design is aimed at creating long, well-contained fractures for maximum productivity. When used in conjunction with horizontal drilling, hydraulic fracturing enables gas producers to extract shale gas economically. Without these techniques, natural gas does not flow to the well rapidly, and commercial quantities cannot be produced from shale. During the 2000s, the technology evolved to allow single-trip multi-stage hydraulic fracturing systems and zone isolation, which is helping to exploit even difficult reservoirs cost-effectively.

The other technology is horizontal drilling – first started in 1930s to drill wells in Texas, USA – which, with further improvements, became a standard practice from the 1980s. The method enables the well to penetrate significantly more rock in this gas bearing strata, increasing the chances of gas being able to flow into the well. The horizontal drilling method uses vertical drilling from the surface down to a desired level. Then, the direction of drilling turns 90 degrees and bores into a natural gas reservoir horizontally. In late 1990s, the application of horizontal drilling enabled more aggressive development as multiple transverse fractures could be placed along a horizontal lateral well-bore. The technology of horizontal well completions was first adapted for shale gas development to provide increased wellbore exposure to the reservoir area while allowing for a reduced number of surface locations in urban areas (Saurez 2012). This technique facilitates considerable reduction of the production facilities required, thus minimizing the impacts to public and overall environmental footprint.

4.3   GLOBAL NATURAL GAS RESERVES

Analysts mostly tend to refer to proven gas reserves, i.e. volumes that have been discovered and can be produced economically with existing technology at the current gas prices. Worldwide proven gas reserves are estimated at around 208 trillion cubic metres (7346 Tcf) or about 60 times the current annual global gas production (BP 2012a).

Table 4.2 Global 'Proved Reserves' of natural gas (trillion cubic metres).

Region	2001	2010	2011	2011 share of total (%)	R/P
North America	7.7	10.3	10.8	5.2	12.5
South & Central America	7.0	7.5	7.6	3.6	45.2
Europe & Eurasia	56.8	68.0	78.7	37.8	75.9
Middle East	70.9	79.4	80.0	38.4	>100 yrs
Africa	13.1	14.5	14.5	70.0	71.7
Asia Pacific	13.1	16.5	16.8	8.0	35
World Total	**168.5**	**196.1**	**208.4**	**100**	**63.6**
OECD	16.1	18.1	18.7	9.0	16
Non-OECD	152.5	178.0	189.7	91.0	90
EU	3.6	2.3	1.8	0.9	11.8
Former Soviet U	50.9	63.5	74.7	35.8	96.3

(*Source*: Tabulated using the data from BP Statistical Review of World Energy 2012)

Table 4.3 Global natural gas production for 2001, 2010 and 2011 (billion cubic metres*).

Region	2001	2010	2011	Change 2011 over 2010 (%)	2011 share of total (%)
North America	780.3	819.1	864.2	5.5	26.5
South & Central America	104.5	162.8	167.7	3.0	5.1
Europe & Eurasia	945.3	1026.9	1036.4	0.9	31.6
Middle East	233.3	472.3	526.1	11.4	16.0
Africa	131.5	213.6	202.7	−5.1	6.2
Asia Pacific	282.4	483.6	479.1	−0.9	14.6
World Total	**2477.2**	**3178.2**	**3276.2**	**3.1**	**100**

*1 cubic metre = 35.318 cubic feet
(*Source*: Tabulated using data from BP Statistical Review of World Energy 2012)

The estimates of 'proved reserves' given by BP (BP 2012a) are tabulated in Table 4.2, by region. However, recoverable gas resources, i.e. volumes that analysts are confident will be discovered or technology developed to produce them, are much larger, with recoverable conventional resources estimated at around 400 tcm. Recoverable *unconventional* resources are of a similar size. Altogether, this would last around 250 years, based on the current rates of gas consumption.

4.3.1   Global production

Global natural gas production grew by 3.1% from 3178.2 billion cubic metres (112.2 Tcf) in 2010 to 3276.2 bcm (115.7 Tcf) in 2011. The United States recorded the largest volumetric increase, 7.7%, despite lower gas prices, and remained the world's largest producer. The next highest producer is the Russian federation with 607.0 billion cubic metres (21.4 Tcf). Gas output also grew rapidly in Qatar by 25.8%, and Turkmenistan by 40.6%, more than offsetting declines in Libya (−75.6%) and UK (−20.8%). Table 4.3 gives the production volumes of natural gas by region and also world total (BP 2012a).

Table 4.4 Natural gas production in OECD & non-OECD regions (billion cubic metres*).

Region	2001	2010	2011	Change 2011 over 2010 (%)	2011 share of total (%)
OECD	1099.2	1148.2	1168.1	1.7	35.8
Non-OECD	1380.1	2030.0	2108.1	3.8	64.2
European Union	232.8	174.9	155.0	−11.4	4.7
Former Soviet U	655.7	741.9	776.1	4.6	23.6

(*Source:* Tabulated using the data from BP Statistical Review of World Energy 2012)

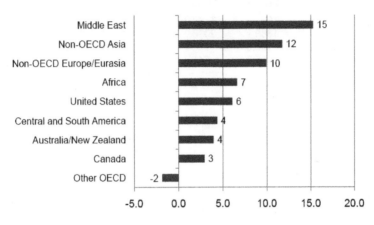

Figure 4.3 World natural gas production increment, 2008–2035 in Tcf (EIA-IEO 2011) (1 cubic foot = 0.0283 cubic metre).

The production of natural gas has been highest in non-OECD countries (64.2%) compared to that of OECD countries (35.8%). Former Soviet Union recorded a maximum growth with 23.6% of the total world production. On the contrary, the European Union recorded the largest decline in gas production (−11.4%), due to a combination of mature fields, maintenance, and weak regional consumption. These are shown in Table 4.4.

4.3.2   Projection through 2035

World natural gas consumption, according to EIA, is projected to increase by 52% from 111 trillion cubic feet in 2008 to 169 trillion cubic feet in 2035. Although the global recession resulted in an estimated decline of 2.0 Tcf in natural gas use in 2009, the demand increased again in 2010 with the consumption exceeding the level witnessed before the decline. In the EIA-IEO 2011 Reference case, the major increase in natural gas production occurs in non-OECD regions between 2008 and 2035, with the largest increments coming from the Middle East (increase of 15.3 Tcf), non-OECD Asia (11.8 Tcf), Africa (7 Tcf), and non-OECD Europe and Eurasia, including Russia and the other former Soviet Republics (9 Tcf) (Figure 4.3). Over the projection period, Iran and Qatar alone increase their natural gas production by 10.7 Tcf, nearly 20% of the total increment in world gas production. A significant share of the increase

Table 4.5 Natural gas consumption in 2001, 2010 and 2011 (in billion cubic metres).

Region	2001	2010	2011	Change 2011 over 2010 (%)	2011 share of total (%)
North America	759.8	836.2	863.8	3.2	26.9
South & Central America	100.7	150.2	154.5	2.9	4.8
Europe & Eurasia	1014.2	1124.6	1101.1	−2.1	34.1
Middle East	233.3	472.3	526.1	11.4	16.0
Africa	131.5	213.6	202.7	−5.1	6.2
Asia Pacific	282.4	483.6	479.1	−0.9	14.6
World Total	**2477.2**	**3178.2**	**3276.2**	**3.1**	**100**
OECD	1097.2	1148.2	1168.1	1.7	35.8
Non-OECD	1380.1	2030.0	2108.1	3.8	64.2
EU	232.8	174.9	155.0	−11.4	4.7
Former Soviet U	655.7	741.9	776.1	4.6	23.6

(*Source*: Tabulated from BP Statistical Review of World Energy 2012); 1 c metre = 35.318 cft

is expected to come from a single offshore field, which is called North Field on the Qatari side and South Pars on the Iranian side. A strong growth outlook for reserves and supplies of natural gas contribute to its strong competitive position among other energy sources (EIA-IEO 2011).

4.3.3   Global consumption of natural gas

World natural gas consumption in 2011 over 2010 grew by 3.1%. Consumption growth was below average in all regions except North America, where low prices drove vigorous growth. Outside North America, the largest volumetric gains in consumption were in China (21.5%), Saudi Arabia (13.2%) and Japan (11.6%). These increases were partly offset by the decline in Europe and Eurasia gas consumption (−2.1%), driven by a weak economy, high gas prices, warm weather and continued growth in renewable power generation (BP 2012a). Table 4.5 provides the global consumption pattern, by region (BP 2012a).

Global Gas demand has increased by around 800 bcm (28.25 Tcf) over the last decade, or 2.7% per year. Gas has a 21% share in the global primary energy mix, behind oil and coal (OECD/IEA 2012).

Projections to 2035

World natural gas consumption is projected to increase by 52% from 111 Tcf in 2008 to 169 Tcf in 2035. Although the global recession resulted in an estimated decline of 2.0 trillion cubic feet in natural gas use in 2009, the demand increased again in 2010 with the consumption exceeding the level witnessed before the decline. The consumption between OECD and non-OECD countries is shown in Table 4.6. There is steady increase in non-OECD consumption reaching about 50% more than the OECD consumption in 2035 (EIA-IEO 2011).

Study by ExxonMobil shows that natural gas will be the fastest-growing major fuel to 2040, with demand rising by more than 60%, shown in Figure 4.4 (ExxonMobil 2012).

Table 4.6 Natural gas consumption projection to 2035 in OECD & non-OECD regions (in trillion cubic feet).

Region	1990	2000	2008	2015	2025	2035
OECD	37	49	55	57	62	68
Non-OECD	37	39	56	66	83	100

(*Source:* EIA-IEO 2012)

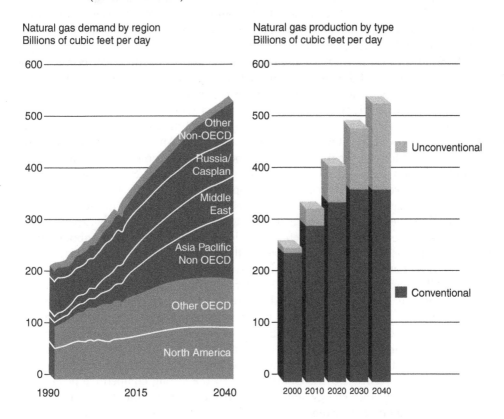

Natural gas demand by region
Billions of cubic feet per day

Natural gas production by type
Billions of cubic feet per day

Figure 4.4 Natural gas demand by region (left), and production by type (right) through 2040 (*Source:* ExxoonMobil, 2012 The Outlook for Energy: A View to 2040).

British Petroleum analysis (BP 2012) shows demand for natural gas is projected to grow globally at an annual rate of 2.1%. Demand is expected to grow in every part of the world, with non-OECD accounting for 80% of global gas demand growth, averaging 2.9% per year growth to 2030. Especially in the non-OECD Asia, demand would be the fastest, tripling over the next 30 years at 4.6% per year, and in the Middle East at 3.7% per year. In China, gas demand grows rapidly at 7.6% per year, to a level of gas use in 2030 at 46 Bcf/d, equal to that of the European Union in 2010. 23% of the global demand increase comes from China. The fraction of gas in China's primary energy consumption expands from 4.0% to 9.5%. The Middle East also will see significant growth (BP 2012, ExxonMobil 2012).

Since natural gas is a versatile fuel that can be used for many purposes, the factors driving these increases in demand vary by region. However, for many regions of the world, natural gas continues to be the fuel of choice in the power and industrial sectors. Global natural gas growth is fastest in power sector at 2.4% per year, and industry at 2.1% per year, consistent with historical patterns. The power sector favors natural gas due to the low capital costs, fuel efficiency and low carbon intensity compared to coal and oil; the natural gas emits up to 60% less CO_2 emissions than coal. Natural gas use in transport is confined to 2% of global gas demand in 2030, despite growing four times from today's level. In the OECD countries, growth is concentrated in the power sector at 1.6% per year. Efficiency gains and low population increase keep gas growth low for the industrial sector at 0.9% per year and other sectors at <0.1% per year. In non-OECD countries, gas use is driven by industrialization, the power sector and the development of domestic resources. Gas consumption expands most strongly in power at 2.9% per year and industry at 2.8% per year (BP 2012).

By 2025 for electricity generation, natural gas will have overtaken coal as the second most popular fuel, after oil. the USA, Russia, Iran and China are the world's largest consumers of gas (OECD/IEA 2013). In North America, natural gas is preferred over coal for electricity generation, especially under policies that impose costs on higher-carbon fuels. Also, advances in unconventional USA's natural gas production are expected to keep domestic supplies ample for the foreseeable future. China's natural gas demand growth will be split between the industrial sector and the residential/commercial sector, where distribution lines are being rapidly expanded and gas is very competitive versus liquefied petroleum gas (LPG). In India, about half of the growth in gas demand through 2040 will come from the industrial sector. And, in the Middle East, gas demand is being driven by the need for electricity generation as well as industrial needs (BP 2012). Growth in gas demand is also due to *fuel substitution*, especially in the OECD, because of regulatory changes and lower relative prices. For instance, about half of all incremental gas demand in the power sector and 75% of incremental demand in the industry sector in OECD countries is substitution for other fuels. But in the non-OECD countries, fuel substitution is much less distinct, where rapidly expanding energy demand is creating opportunity for all fuels to grow (BP 2012). Contributing to the strong competitive position of natural gas among other energy sources is a strong growth outlook for gas reserves and supplies. Significant changes in natural gas supplies and global markets continue with the expansion of liquefied natural gas (LNG) production capacity, while new drilling techniques and other efficiencies have made production from many shale basins economical worldwide as will be seen in the following pages. The net impact has been a significant increase in resource availability, which contributes to lower prices and higher consumption in the *IEO2011* Reference case projection (EIA-IEO 2011).

4.4   UNCONVENTIONAL GAS RESERVES

Unconventional sources of gas are trapped underground by impermeable rocks, such as coal, sandstone and shale. As mentioned earlier, the three types of unconventional resources are shale gas, found in very fine-grained sedimentary rock; coal bed methane, or CBM, also known as coal seam gas (CSG) in Australia, extracted from coal beds; and tight gas, trapped underground in impermeable rock formations.

Shale gas is tightly locked in very small spaces within the reservoir rock requiring advanced technologies to drill and stimulate (fracture) the gas bearing zones. The creation of fractures within the reservoir is critical in allowing the natural gas to flow to the well. Once stimulated, the shale gas reservoirs are produced in the same way as conventional gas wells. The application of these technologies has led to a rapid rise in shale gas production, especially in the United States.

It has been known that global unconventional gas resources may be significant, roughly equal to conventional gas resources (Rogner 1997). Despite their abundance, it was believed that the vast majority of the resource base was too difficult or expensive for commercial extraction. That was why, in all the global estimates of oil and gas reserves, the focus was only on conventional gas resources (NPC 2007). However, two recent developments have shifted the focus onto unconventional gas resources: the rapidly growing demand for energy worldwide pushing the prices of oil and gas upwards, and robust increase in unconventional gas production in the USA driving down the natural gas prices which are expected to continue for some more years (EIA-AEO 2011). There are multiple and substantial uncertainties in assessing the recoverable volumes of shale gas at global level. Even in North America where production is currently taking place, there is significant uncertainty over the size of the resource and considerable variation in the available estimates. For several regions of the world, there are no estimates at all, though some may well contain significant resources. Given the absence of production experiences in most regions of the world, and the number and magnitude of uncertainties, the current resource estimates should be treated with considerable caution.

Current estimates for the global technically recoverable resources suggest: for shale gas just above 200 Tcm (7064 Tcf), for tight gas, 45 Tcm (1589 Tcf), and for CBM, 25 Tcm (883 Tcf). For comparison, the global technically recoverable resources of conventional gas are estimated at 425 Tcm (15010 Tcf) of which around 190 Tcm (6710 Tcf) are currently classified as 'proved' reserves (Pearson *et al.*, EU 2012). It was, nonetheless, possible to obtain 'high, best and low' estimates for shale gas. For e.g., in the USA, the high/best/low estimates are 47/20/13 Tcm and for China, 40/21/16 Tcm. i.e., in the USA, the high and low estimates are 230% and 64% of the 'best' estimate respectively. In the estimates for other regions of the world, the uncertainty may be even greater. Of the several reasons for the variability and uncertainty in the estimates, the main ones are using different methodologies using ambiguous terminology or in assuming high recovery factor or application of 'decline curve analysis' in the case of estimates from extrapolation of production experience (Pearson, EU, 2012). Total natural gas resource potential for gas shale in the USA has been estimated to be between 10 and 25 Tcm (353 to 883 Tcf) of recoverable gas resources (EIA 2011), compared with the actual volume of 7.7 Tcm (978 Tcf) of 'proved' gas reserves in the country (BP 2011).The US domestic gas resources have increased by 40% since 2006, and most of the growth in natural gas production in the US is coming from unconventional sources of natural gas which were believed, until recently, to be non-recoverable resources. New and advanced exploration, well drilling and completion technologies are enabling better access to unconventional gas resources at competitive prices. The US unconventional 'proved' gas resources in 1996 were 48 Tcf as against 'undeveloped

Table 4.7 Estimates of technically recoverable shale gas resources in the 48 shale gas basins that were recently assessed.

Continent	Countries	Technically recoverable (Tcf)
North America	Canada, Mexico	1069
Africa	Morocco, Algeria, Tunisia, Libya, Mauritania, Western Sahara, South Africa	1042
Asia	China, India, Pakistan	1404
Australia		396
Europe	France, Germany, Netherlands, Sweden, Norway, Denmark, UK, Poland, Lithuania, Ukraine & other Eastern Europe countries, Turkey	624
South America	Argentina, Venezuela, Columbia, Bolivia, Brazil, Chile, Uruguay, Paraguay	1225

(*Source:* Howard Gruenspecht, CSIS, USEIA, Sept. 2011)

resource base' of 366 Tcf; and in 2008, these figures increased to 140 Tcf and 917 Tcf respectively (Kuuskraa 2009).

An *EIA-sponsored study* released in April 2011 assessed 48 shale gas basins in 32 countries, containing almost 70 shale gas formations. These assessments cover the most prospective shale gas resources in a select group of countries that demonstrate some level of relatively near-term promise and for basins that have a sufficient amount of geologic data for resource analysis. The study reported initial assessments of 5,760 trillion cubic feet (Tcf) of technically recoverable shale gas resources in these countries, apart from 862 Tcf in the United States, amounting to 6,622 Tcf in total (EIA 2011a, 2011b). These countries and the assessments are listed in the Table 4.7. The total comes to 5760 Tcf, and if the USA's estmate of 862 Tcf is included, the total comes to 6622 Tcf. To put this shale gas resource estimate in some perspective, world proven reserves[1] of natural gas as of January 1, 2010 are about 6,609 Tcf,[2] and world technically recoverable gas resources are roughly 16,000 Tcf,[3] largely excluding shale

---

[1] Reserves refer to gas that is known to exist and is readily producible, which is a subset of the technically recoverable resource base estimate for that source of supply. Those estimates encompass both reserves and that natural gas which is inferred to exist, as well as undiscovered, and can technically be produced using existing technology. For example, EIA's estimate of all forms of technically recoverable natural gas resources in the U.S. for the *Annual Energy Outlook 2011* early release is 2,552 Tcf, of which 827 Tcf consists of unproved shale gas resources and 245 Tcf are proved reserves which consist of all forms of readily producible natural gas including 34 Tcf of shale gas.

[2] 'Total reserves, production climb on mixed results', *Oil and Gas Journal* (Dec. 6, 2010), pp. 46–49.

[3] Includes 6,609 Tcf of world proven gas reserves (*Oil and Gas Journal 2010*); 3,305 Tcf of world mean estimates of inferred gas reserves, excluding the Unites States (USGS, *World Petroleum Assessment 2000*); 4,669 Tcf of world mean estimates of undiscovered natural gas, excluding the United States (USGS, *World Petroleum Assessment 2000*); and U.S. inferred reserves and undiscovered gas resources of 2,307 Tcf in the United States, including 827 Tcf of unproved shale gas (EIA, *AEO 2011*).

gas. Thus, adding the identified shale gas resources to other gas resources increases total world technically recoverable gas resources by over 40% to 22,600 Tcf.

Estimates of shale gas resources in other parts of the world are highly uncertain because many countries are only just beginning to understand how to conduct assessments of the shale gas recoverable resources. Nonetheless, the aggregate estimate is probably quite conservative, since the study excluded several major types of potential shale gas resources such as (a) nations outside the 32 countries studied which include Russia and the Middle East considered to have very large resources of conventional gas, and Central Asia, Southeast Asia, and Central Africa; (b) some shale basins in the countries studied because in many cases, no estimates are possible yet for these basins; and (c) offshore resources. Moreover, coalbed methane, tight gas and other natural gas resources that may exist within these countries were not included in the assessment. Of the 32 countries covered in the EIA-sponsored study, two groups may find shale gas development most attractive. The first group countries that currently depend heavily on natural gas imports but also have significant shale gas resources that include France, Poland, Turkey, Ukraine, South Africa, Morocco, and Chile; and the second group countries that already produce substantial amounts of natural gas and also have large shale resources which includes the USA, Canada, Mexico, China, Australia, Libya, Algeria, Argentina, and Brazil.

4.4.1 Global shale gas production

As of 2010, unconventional gas production reached an estimated 15% of global gas production. Most of it comes from North America, with around 420 billion cubic metres (14.8 Tcf) produced in 2010, half of which is tight gas. Shale gas output has increased by a factor of 11 over the last decade which is just about one-third of total unconventional gas production in 2010.

In the USA, tight gas has been produced for more than four decades and coalbed methane for more than two decades. Production of shale gas in the United States began much more recently, and has rapidly increased from 2005 onwards with advances in extraction technologies. In 2010, shale gas represented more than 20% of total gas production in the USA. Throughout the rest of the world, coal-bed methane (CBM) production is estimated at approximately 10 bcm (350 billion cubic ft) and tight gas at 60 bcm (2.1 Tcf).

Countries that made much progress: Canada produces tight gas, coal bed methane and small amounts of shale gas. Australia has shown good CBM potential, which is already being produced in small quantities. But Australia's future success most likely lies in projects that focus on producing LNG from CBM. Three such projects are committed to start exporting between 2014 and 2016. China, India and Indonesia have produced small amounts of unconventional gas and are aggressively looking at ways to increase their respective volumes.

Despite strong interest throughout Europe, public concern issues such as high population densities and potential unfavorable environmental impact, are hampering progress. As already seen, Argentina, Algeria and Mexico, may also have large shale gas potential.

New drilling techniques and other advances enable economically viable production from many global shale basins. As a result there is a significant increase in resource availability leading to lower prices and higher demand for natural gas in the coming decades (EIA-IEO 2011).

Shale gas as a major natural gas supply in the USA

Natural gas production in the US has increased from 50 billion cubic feet per day in 2005 to 63 billion cubic feet per day in 2011, an increase of 20%, as production has shifted from conventional sandstone basins to shale and tight sandstone formations. In addition, the presence of crude oil and natural gas liquids (such as propane, ethane, and butane), commonly referred to as *wet gas*, is boosting the economic feasibility of shale extraction or the geologic formation where natural gas is being (or can be) produced. Shale development is pushing overall natural gas resources to higher levels. In 2011, the United States had 2,543 Tcf of technically recoverable resources (proved and potential) from all natural gas formations, 827 Tcf of which was from shale formations (California Energy Commission 2012).

Because the production methods are being applied in new ways, and also because shale resources can be located in areas that are not traditional oil and gas producers, the pace of future development will depend on industry, government and local communities working together to build understanding of the potential benefits of unconventional gas production, as well as proven practices used to protect groundwater and air quality, and minimize other environmental impacts. The same will be true in other countries.

While it is less certain that unconventional gas production techniques will be applied as successfully outside the United States, ExxonMobil expects to see unconventional gas become more of a factor in Asia Pacific, south and Central America and Europe in the coming decades.

Rising unconventional production will limit the need for imports in some regions, but Asia Pacific and Europe will need significant imports, both via pipelines and via liquefied natural gas (LNG) tankers, to meet their demand (ExxonMobil 2012).

Reasons for tapping unconventional gas: Two main reasons are mentioned: (i) technological advances over the last two decades, particularly concerning hydraulic fracturing, and (ii) soaring gas prices in the early 2000s. Countries that are importing are keen to explore this, because if they are able to produce significant volumes of unconventional gas, they would have greater energy security and more energy independence, and would rely less on costly energy imports. Alternatively some producers would be able to export more gas.

4.4.2   Projection through 2035

In the Gas Scenario of the IEA World Energy Outlook 2011, the total natural gas production grows from an estimated 3.3 Tcm in 2010 to 5.1 Tcm by 2035, an increase of more than 50%. The average annual growth in gas production is 2% from 2008 to 2020, and then moderates to around 1.6% for the remainder of the Outlook period.

Natural gas production increasingly comes from unconventional sources, and unconventional gas output worldwide is projected to reach 1200 bcm in 2035. As a result, the share of unconventional gas in global gas production is expected to rise from 12% in 2008 to nearly 25% in 2035. Most of the increase in unconventional gas production in the IEA Gas Scenario comes from shale gas and CBM. In particular, the

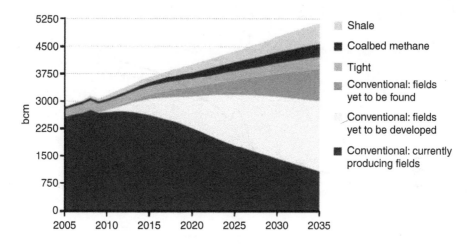

Figure 4.5 Natural gas production by type (*Source:* IEA Energy Outlook 2011).

share of shale gas in global gas production reaches 11% in 2035, while CBM reaches 7% and tight gas 6% (Figure 4.5).

However, in the WEO Golden Rules case (OECD/IEA 2012), released in November 2012, production of unconventional gas, primarily shale gas, more than triples to 1.6 Tcm in 2035. This means, the share of unconventional gas in total gas output rises from 14% in 2012 to 32% in 2035. Most of the increase comes after 2020; and the largest producers of unconventional gas over the projection period are the USA and China, whose large base allows for rapid production starting towards 2020. There are also large increases in Australia, India, Canada, and Indonesia. Unconventional gas production in EU led by Poland is enough after 2020 to offset continued decline in conventional output.

Unconventional gas resources, comprising shale gas, tight gas and coal bed methane are estimated to be as large as conventional resources. IEA analysis suggests that plentiful volumes can be produced at costs similar to those in North America (between 3 $/mmBtu and 7 $/mm Btu).

As already said, new methods have enabled the improvement of estimates of shale gas resources which helped to increase total natural gas reserves in the United States by almost 50% over the past decade; and it is predicted that shale gas will rise to 47% of the total United States natural gas production in 2035 (Stevens, Paul 2012). Adding production of tight gas and coal-bed methane, the US unconventional natural gas production rises from 10.9 Tcf in 2008 to 19.8 Tcf in 2035 (EIA/IEO 2011). Some analysts expect that shale gas will greatly expand worldwide energy supply (Krauss 2009). China is estimated to have the world's largest shale gas reserves (USEIA 2011). The *IEO2011* Reference case projects a substantial increase in those supplies, especially in the USA, Canada and China which are more important for the future of their domestic natural gas supplies.

According to IEA-WEO 2011, by the year 2035, unconventional gas also reaches a significant scale not only in China (CBM and shale), but in Russia (tight gas),

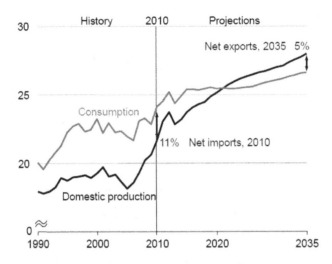

Figure 4.6 Total U.S. natural gas production, consumption, and net imports, 1990–2035 (trillion cubic feet) (*Source*: USEIA – Annual Energy Outlook 2012).

India (shale) and Australia (CBM). The largest gas producer (both conventional & unconventional together) will be Russia with about 900 bcm followed by the USA with a little less than 800 bcm, then China, Iran, Qatar, Canada, Algeria, Australia so on.

A study by the Baker Institute of Public Policy at Rice University concluded that increased shale gas production in the US and Canada could help prevent Russia and the Persian Gulf countries from dictating higher prices for the gas they export to European countries (Rice University 2011).

US Gas Production & Imports/exports: The USA consumed more natural gas than it produced in 2010, importing 2.6 Tcf from other countries. In the *AEO2012* Reference case, domestic natural gas production grows more quickly than consumption, resulting in the United States becoming a net exporter of natural gas by around 2022, and by 2035 the net exports will be about 1.4 Tcf (Figure 4.6). The gas consumption in the USA grows at an annual rate of 0.4% from 2010 to 2035 in the Reference case, or by a total of 2.5 Tcf to 26.6 Tcf in 2035, mostly in power generation, commercial and industrial sectors. The annual consumption in the commercial and industrial sectors grows by less than 0.5% and in electric power generation by 0.8%, while the residential gas consumption declines by a total of 0.3 Tcf during the same period (2010 to 2035). The USA's natural gas production grows by 1.0% per year, to 27.9 Tcf in 2035, more than enough to meet domestic needs for consumption, which allows for exports.

The recent 'EIA-AEO2013 Early Release' report predicts that the natural gas exports from the USA will be more than what was projected in the AEO2012 Reference case. The AEO 2013 Report projects the US natural gas production to increase from 23.0 Tcf in 2011 to 33.1 Tcf in 2040, an increase of 44%. Almost all of this increase in domestic natural gas production is due to projected growth in shale gas

U.S. dry natural gas production
trillion cubic feet

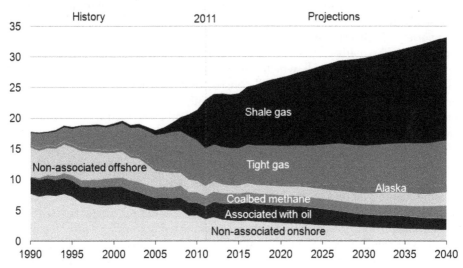

Figure 4.7 Dry natural gas production in the USA by source, 1990–2040 (*Source*: USEIA-AEO 2013 Early Release).

production, which grows from 7.8 Tcf in 2011 to 16.7 Tcf in 2040 (see Figure 4.7) (EIA: Energy Brief, December 2012). Production of higher volumes of shale gas in *AEO2013* enables not only increased total production volumes but an earlier transition to net exports than was projected in the *AEO2012* Reference case. The US exports of LNG from domestic sources rise to approximately 1.6 Tcf in 2027 almost double the 0.8 Tcf projected in *AEO2012*. Industrial production in certain areas grows more rapidly in *AEO2013* due to the strong growth in shale gas production as well as relatively low natural gas prices over an extended period resulting in lower costs of both raw materials and energy, particularly through 2025. For example, industrial production grows by 1.7% per year from 2011 to 2025 in the bulk chemicals industries (which also benefit from increased production of natural gas liquids) and by 2.8% per year in the primary metals industries. The growths are higher compared to what was mentioned in the *AEO2012* Reference case.

In the long term, growing competition from other countries in these industries may limit growth in the output, as other nations develop and install newer, more energy-efficient facilities. The increased level of production also leads to greater natural gas demand for the industry (excluding lease and plant fuel), which grows to more than 8.3 quadrillion Btu in 2035 in *AEO2013*, compared to 7.2 quadrillion Btu in 2035 in *AEO2012*. Most of the increase in industrial energy demand is the result of higher output in the manufacturing sector (EIA-Annual Energy Outlook 2013 Early Release, November 2012).

4.5 ENVIRONMENTAL CONCERNS

The technology of hydraulic fracturing involves injecting fluids at high pressure to break up the rock and hold the new fractures open to allow release of the natural gas. As mentioned already, the fluids are primarily water, plus sand (either natural or synthetic) and a variety of different chemicals. The exact chemical mix differs by formation and by operator. The main concern is the possibility of introducing contaminants into aquifers; the fracturing fluids which contain potentially hazardous chemicals could be released by spills, leaks, faulty well construction or other exposure pathways and contaminate the ground surface, and/or could infiltrate downwards posing a risk to aquifers (Arthur *et al.* 2008). A properly designed and cased well will prevent drilling fluids, hydraulic fracturing fluids, or natural gas from leaking into the permeable aquifer and contaminating ground-water. Fracturing also produces large amounts of wastewater, which may contain dissolved chemicals and other contaminants that could require treatment before disposal or reuse. Because of the quantities of water used, and the complexities inherent in treating some of the wastewater components, treatment and disposal is an important and challenging issue. In addition, seismic activity which was unknown to locations prior to fracking has occurred in some area after the fracking process. According to the United States Geological Survey, this activity has been almost always too small to be a safety concern. Scientists argue that either the fracturing pressures or the disposal injection post-fracking may be the cause. Several studies are underway to understand the reasons. The injection wells typically discharge the wastewater into non-potable salt-water aquifers (EIA: Energy Brief, Dec. 2012).

Shale gas drilling and production taking place near populated areas and/or in areas that previously have not experienced oil and natural gas production have increased public health and environmental concerns.

The usual areas of environmental concerns include surface disturbances, greenhouse gas emissions, water use and disposal, and potential groundwater contamination. Uncertainty about whether and how these concerns are addressed will impact the amount and rate, as well as price of shale gas production going forward (Calif. Energy Commission 2012).

OECD/IEA looked at the concerns related to land use, water use and the potential for contamination of drinking water, and methane and other air emissions and released 'Special *World Energy Outlook (WEO)* Report on unconventional gas' in November 2012, where the environmental impact of unconventional gas production is discussed and came up with 'Golden Rules' that are needed to support a potential 'Golden Age of Gas'. The Report suggests several steps for developing the vast resources of unconventional gases profitably and in an environmentally acceptable manner because these sources would bring a number of benefits in the form of greater energy diversity, more secure supply and reduced energy costs in many countries. For instance, the industry has to apply the highest social and environmental standards at all stages of the process chain, and the governments have to devise appropriate regulatory regimes based on science and high-quality data to maintain or earn public confidence. Full transparency, measuring and monitoring the environmental impacts, and engaging the local community in addressing the concerns such as choosing drilling sites, well design and construction and testing, water handling and disposal, investments for reducing emissions and so on, are underlined in the report.

The US government believes that increased shale gas development will help reduce greenhouse gas emissions (White House 2009). Some studies, however, have felt that the extraction and use of shale gas may result in the release of more greenhouse gases than conventional natural gas (Howarth *et al.* 2011, Shindell *et al.* 2009). Other recent studies point to high decline rates of some shale gas wells as an indication that shale gas production may ultimately be much lower than is currently projected (Hughes 2011, Berman 2011).

4.6 LIQUIFIED NATURAL GAS (LNG)

Liquefied natural gas is natural gas that has been liquefied for transport by ship or truck. Liquefaction requires all heavier hydrocarbons to be removed from the natural gas, which leaves only pure methane. Depending on its exact composition, natural gas becomes liquid at around −162°C at atmos. pressure. Liquifaction reduces gas volume by a factor of 600.

The most important infrastructure needed for LNG production and transportation is an LNG plant consisting of one or more LNG trains, each of which is an independent unit for gas liquefaction. The largest LNG train now in operation is in Qatar. Until recently it was the Train 4 of Atlantic LNG in Trinidad and Tobago with a production capacity of 5.2 million metric ton per annum (mmtpa), followed by the *SEGAS LNG* plant in Egypt with a capacity of 5 mmtpa. The Qatargas II plant has a production capacity of 7.8 mmtpa for each of its two trains. Then LNG is transported in specialized LNG carriers, and finally re-gasified for distribution. Regasification terminals are usually connected to a storage and pipeline distribution network to distribute natural gas to local distributors or independent power plants.

Large Global producers of LNG: The largest producer of LNG is Qatar, whose liquefaction capacity is roughly one-quarter of global LNG liquefaction capacity as of mid-2011. Qatar saw a massive expansion of its capacity, which has increased by 63 bcm (2.2 Tcf) since early 2009 to reach 105 bcm (3.71 Tcf). In 2011, the country supplied 75.5 MT of LNG to the global market; nearly 31% of the total supply. Indonesia, Malaysia, Australia and Algeria are also significant LNG exporters. Russia and Yemen began exporting in 2009 and Peru in 2010. Angola is expected to start exporting in 2012 and Papua New Guinea in 2014 (USIEA website). Australia is set to become the second largest LNG exporter behind Qatar by 2016; six projects are currently committed or under construction, representing 60 bcm (2.1 Tcf) of new capacity.

Global LNG trading: In the coming decades, world natural gas trade in the form of LNG is projected to increase. LNG now accounts for 32.3% of global gas trade. Global natural gas trade has increased by a modest 4% in 2011. With Qatar accounting for 34.8% of virtually all 87.7% of the increase, LNG shipments grew by 10.1% in 2011. Most of the increase in LNG supply comes from the Middle East and Australia, where several new liquefaction projects are expected to become operational in the coming years. In addition, several LNG export projects have been proposed for western Canada. There are proposals to convert underutilized LNG import facilities in the United States to liquefaction and export facilities for domestically sourced natural gas. Among LNG importers, the largest volumetric growth was in Japan and the UK.

Pipeline shipments grew by just 1.3%, with declines in imports by Germany, the UK, the US and Italy offsetting increases in China from Turkmenistan, Ukraine from Russia, and Turkey from Russia and Iran (BP 2012).

In the *IEO2011* Reference case, world liquefaction capacity more than doubles, from about 8 Tcf in 2008 to 19 Tcf in 2035. In addition, natural gas exports from Africa to Europe and from Eurasia to China are likely to increase (EIA website). Global LNG supply is projected to grow 4.5% per year to 2030, more than twice as fast as total global gas production at 2.1% per year and faster than inter-regional pipeline trade at 3.0% per year. LNG contributes 25% of global supply growth 2010–30, compared to 19% for 1990–2010 (BP 2012).

4.7 GAS MARKET ANALYSIS AND PRICES

A special *WEO* report was published in 2011 on market analysis. The report presents a scenario in which global use of gas rises by more than 50% from 2010 levels and accounts for more than one-quarter of global energy demand by 2035. However, it is cautious on the climate benefits of such an expansion, because an increased share of gas in the global energy mix is far from enough on its own to put the world on a carbon emissions path consistent with a global temperature rise of no more than 2°C. There is also a detailed annual report on gas, called the *Medium-Term Gas Market Report*, which analyses the latest trends in global gas markets, and examines investments trends in different parts of the gas value chain, and provides supply and demand forecasts over the upcoming five years. The report also deals with the development of liquefied natural gas (LNG) trade, prices and unconventional gas in detail. The reader may refer to these Reports for details.

Following soaring unconventional gas production in the USA, there has been a sharp drop in imports that resulted in a change in the prospect for the USA in the international gas trade. Further, the recent slump in LNG imports in the USA has had a significant impact on global gas markets. The global economic crisis in 2008 has led to a general drop in global gas demand. With abundant supplies, mostly from Qatar started entering the world market, the USA was in no hurry to buy LNG. As a result, gas spot prices in the USA and in Europe have dropped. Although the global demand has recovered substantially since 2010, due to high growth in domestic gas production, the USA has kept LNG imports very low, while at the same time considering exporting LNG.

Natural gas prices

There are different types of natural gas prices – wholesale prices (such as hub prices, border prices, city gate) and end-user prices which differ depending on the customer served (industrial, or household). Prices typically include the cost of gas supplies, transmission, distribution and storage costs, as well as the retailer's margin and taxes. End-user prices and wholesale prices vary widely across regions.

Some wholesale gas prices are linked to oil prices, through an indexation present in long-term supply contracts in Continental Europe and the OECD Pacific, but this represents only one-fifth of global gas demand. Gas-to-gas competition (spot prices) can be found in North America, the UK and parts of continental Europe and represents one-third of global gas demand. Prices in many other regions are often regulated: they can be set below costs, at cost of service, or be determined politically, reflecting perceived public needs. The IEA follows the reported market prices in Europe, North America and the OECD Pacific region.

Declining US Natural Gas Prices and its effect: Natural gas prices were increasing steadily up to 2008, when they began to drop. The monthly Henry Hub spot price increased by an average of 29% per annum between 2000 and 2008. However, from January 2009 to April 2012, Henry Hub spot prices decreased at an average annual rate of 19%. Over the last decade, spot prices have been volatile, with several spikes resulting in prices as high as $13.80 per million Btu in June 2008. Since the beginning of 2012, natural gas prices have been very low, with the average Henry Hub spot price (the pricing point for natural gas futures contracts) from January 2012, through the end of April 2012 at $2.33 per million Btu and an average price of $1.94 per million Btu for April 2012. These low prices attributed to a warm winter and high production rates have resulted in a buildup of a record 2.1 Tcf of natural gas on March 1, 2012. Low prices can discourage drilling in the shorter term, but this could only be a temporary state that may have little effect on shale gas production rates going forward. But the long term development is subject to whether and to what extent the USA would become an LNG exporter and the implications of such a situation for local natural gas prices. With regard to exporting, the opinions differ: while a few argue for competing in the world market where gas prices are often indexed to oil prices, a few others argue that LNG exports would drive up the domestic gas price because of the fluctuations in the world LNG and oil prices (California Energy Commission 2012).

Although the USA's supplies have increased, domestic demand has remained relatively flat except in the electricity generation sector, which is the main driver of the country's natural gas demand growth. Over the next few years, federal air quality regulations will require major investments in existing coal facilities, many of which are reaching the end of their design life, to substantially reduce emissions. With low natural gas prices, instead of incuring the necessary investments to keep the coal-fired plants running, operators may prefer to shut them down, resulting in an increased demand for natural gas for electric generation in the country (California Energy Commission 2012).

With an expanded shale gas resource base, the natural gas spot prices are projected to be significantly lower than past years as seen from AEO 2009, AEO2010 and AEO 2011 data (Figure 4.8). The energy parity between oil and gas prices seems to already be broken at US markets since January 2009; the oil prices were floating around 100 $/barrel for much of 2010 and 2011 while the US Henry Hub (HH) price had been consistently below 5 $/per million Btu.

The European market has not yet been affected by the rise of the US unconventional gas. Europe currently offers prices in 2010 that are more than double those in the USA. The UK's Natural Balancing Point price (NBP) has been in the 8 to 10 $/per million Btu range for the first nine months of this year, while the Henry Hub price has been consistently below 5 $/per million Btu. Decoupling of price between the USA and EU is observed from January 2010 (Saurez 2012).

There is another point of view regarding the future of gas: the current supply of low-cost natural gas in the US will be for a finite time only, and as the price of

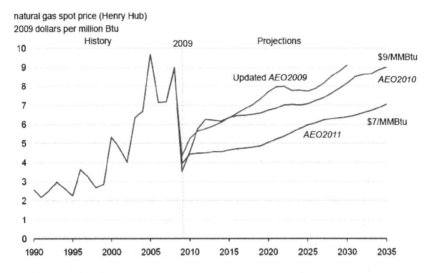

natural gas spot price (Henry Hub)
2009 dollars per million Btu

Figure 4.8 The US Natural gas price projections are significantly lower than past years due to an expanded shale gas resource base (*Sources*: USEIA – AEO 2011; USEIA – AEO 2010; USEIA – An Updated Annual Energy Outlook 2009 Reference Case).

gas increases in the future, coal use will likely expand, as has happened several times in the past in response to increasing gas prices (Shaddix 2012). The National Coal Council (2012) also concludes that in the current US gas markets, the interest in SNG has declined because of the belief that shale gas has permanently institutionalized the expectation of increased natural gas supply at low prices; but the predictions of the future supply and price of gas have a high level of uncertainty.

Levelized costs of electricity: The utilization of natural gas in power generation is projected to increase in the coming years, due to its major advantages over coal-fired power plants, viz, low emission levels when combusted, high volumes of availability due to increased shale gas resources and growth globally, and comparatively low price. The USEIA-AEO 2011 Report has presented the levelized costs (in US$) of electricity produced by plants, utilizing different dispatchable technologies, starting service in 2017. These are given in Annexure. Plants utilizing 'natural gas combined cycle technologies' deliver cheaper power compared to coal-fired power plants.

REFERENCES

Arthur, J.D., Bohm, B. & Layne, M. (2008): Hydraulic Fracturing Considerations for Natural Wells of the Marcellus Shale, presented at *Ground Water Protection Council Annual Forum*, Cincinnati, Ohio. September 2008.

Berman, A. (2011): After the gold rush: A perspective on future U.S. natural gas supply and price, The Oil Drum, 8 Feb. 2011.

BP (2012): *BP Energy Outlook 2030*, London, January 2012.

BP (2012a): *BP Statistical Review of World Energy 2012*, London, UK.

California Energy Commission Staff Report (2012): 2012 Natural gas market trends – In Support of the *2012 Integrated Energy Policy Report Update*, California Energy Commission, May 2012.

ExxonMobil (2012): *The Outlook for Energy:* A View to 2040, at exxonmobil.com

Hughes, D. (2011): *Will Natural Gas Fuel America in the 21st Century?* Post Carbon Institute, May 2011.

Healy, D. (2012): *Hydraulic fracturing or Fracking: A short summary of current knowledge and potential environmental Impacts*; Report prepared for EPA (Ireland) under STRIVE programme 2007–2013, University of Aberdeen, July 2012.

Howarth, R.W., Santoro, R., & Ingraffea, A. (2011): Methane and the greenhouse gas footprint of natural gas from shale formations, *Climatic Change Letters*, doi: 10.1007/s10584-011-0061-5.

IEA/OECD (2013): Natural Gas, available at http://www.iea.org/aboutus/faqs/gas/.

Krauss, C. (2009): *New way to tap gas may expand global supplies*, *New York Times*, 9 October 2009.

National Coal Council (2012): Harnessing Coal's carbon content to Advance the Economy, Environment and Energy Security, June 22, 2012, Study chair: Richard Bajura, National Coal Council, Washington, DC.

Newell, R. (2011): Shale Gas and the Outlook for U.S. Natural Gas Markets and Global Gas Resources, Administrator, OECD, June 21, 2011, Paris, France.

NPC (2007): *Facing the Hard truths about Energy: A comprehensive View to 2030 of Global oil and natural gas*, Washington DC: National Petroleum Council, 2007, pp. 96–97.

Pearson, I., Zeniewski, P., Graccev, F., Zastera, P., McGlade, C., Sorrell, S., Speirs, J., & Thonhauser, G. (2012): *Unconventional gas: Potential energy market Impacts in the European Union, A report by the Energy Security Unit* of the EC's Joint Research Centre, Luxembourg: Publications Office of the European Union, 2012.

Rice University, News and Media Relations (2011): *Shale Gas and U. S. National Security*, 21 July 2011.

Rogner, H.H. (1997): An Assessment of World Hydrocarbon resources, *Annual review of Energy and the Environment*, **22**, 1997.

Schmidt, G.A., Unger, N., & Bauer, S.E. (2009): Improved Attribution of Climate Forcing to Emissions, *Science*, **326** (5953), pp. 716–718.

Shaddix, C.R. (2012): Coal combustion, gasification, and beyond: Developing new technologies for a changing world, *Combustion and Flame*, **159**, 3003–3006.

Stevens, P. (2012): *The Shale Gas Revolution: Developments and Changes*, August 2012, *Chatham House*, at http://www.chathamhouse.org/publications/papers/view/185311.

Suárez, A.A. (2012): The Expansion of Unconventional Production of Natural Gas (Tight Gas, Gas Shale and Coal Bed Methane), *In: Advances in Natural Gas Technology, Hamid Al-Megren (ed.)*, ISBN: 978-953-51-0507-7; In Tech, Available at http://www.intechopen.com/books/advances-in-natural-gastechnology/the-expansion-of-unconventional-production-of-natural-gas-tight-gas-gas-shale-and-coal-bedmethaneShindell, D.T, Faluvegi, G, Koch, D.M.

USEIA (2011): *Annual Energy Outlook 2011 with projections to 2035*, April 2011; Available at www.eia.doe.gov.

USEIA (2012): *Energy in Brief*, US Department of Energy, Washington DC, last updated: December 5, 2012.

USEIA (2011): *Annual Energy Outlook 2011, Prospects for shale gas*, DOE/EIA-0383, Washington, DC, April 2011; available at www.eia.gov/forecasts/aeo/IF_all.cfm#prospect shale.

USEIA (2011a): *EIA – Analysis & Projections: World Shale Gas Resources*: An Initial Assessment of 14 Regions outside the United States, April 5, 2011, USDOE, Washinton DC, available at http://www.eia.gov/analysis/studies/worldshalegas/.

USEIA (2011b): *EIA: Today in Energy*, April 5, 2011, USDOE, Washington, DC; available at www.eia.gov/todayinenergy/detail.cfm?id=811.

USEIA (Nov. 2012): *Annual Energy Outlook 2013 Early Release*, US Department of Energy, Washington DC, November 2012.

White House, Office of the Press Secretary (2011): *Statement on U.S.-China shale gas resource initiative*, 17 November 2009.

Chapter 5

Pollution from coal combustion

Coal is the dirtiest of the fossil fuels and has the potential to pollute the environment as it releases several pollutants when it is combusted. The burning of coal results in relatively higher emissions of CO_2 than oil and gas, due to coal's ratio of hydrogen over carbon, and power generation efficiency is relatively low compared to other fossil fuels. The emissions of particulates, sulfur oxides, nitrogen oxides, mercury and other metals, including some radioactive materials during the combustion of coal are also in a much higher proportion than when other fossils, oil or natural gas, are burned. Burning of coal not only causes local and regional pollution but extends to the global level leading to climate change and its distressing effects. Coal is also responsible for methane emissions, especially from mining. These pollutants cause severe human health problems sometimes proving fatal, besides affecting environment in general (Jayarama Reddy 2010).

Coal's applications are varied, the principal usage being for electricity generation. Steam coal is used in power plants and for process and comfort heat in many industries and in the residential and commercial sectors. Coal is burnt in isolated stoves or industrial boilers for central heating systems. Coking coal is used in the steel industry. Coal plays a relatively low role in transport; for e.g., in old steam locomotives in various developing countries, or as a source for liquid fuels.

The coal-fired power plants produce a few by-products from burning coal, generally referred as *Coal Combustion Products (CCPs)*, some of them having economic value. These by-products include fly ash, bottom ash, boiler slag, flue gas desulfurization gypsum, and other types of substances such as fluidized bed combustion ash, cenospheres, and scrubber residues. Fly ash can be used to replace or supplement cement in concrete. In the USA, for example, more than half of the concrete produced is blended with fly ash. The most significant environmental benefit of using fly ash over conventional cement is that greenhouse gas (GHG) emissions can be significantly reduced. For every tonne of fly ash used for a tonne of Portland cement (the most common type of cement in general use around the world) approximately one tonne of carbon dioxide is prevented from entering the Earth's atmosphere. Fly ash does not require the energy-intensive kilning process required by Portland cement.

Utilizing CCPs reduces GHG emissions, reduces the need for landfill space, and eliminates the need to use primary raw materials. Fly ash produces a concrete that is strong and durable, with resistance to corrosion, alkali-aggregate expansion, sulfate and other forms of chemical attack. Coal combustion products continue to play a major role in the concrete market. Their use in other building products is also expected to grow as sustainable construction becomes more prominent, and more architects and building owners understand the benefits of using Coal.

5.1   ENERGY RELATED CO_2 EMISSIONS

Since the power generation sector around the world utilizes coal as a major source, all countries contribute to the emission of greenhouse gases. The world's top five emitters of carbon dioxide from their power generation plants are China, the USA, India, Russia and Japan. If the 27 member states of the European Union are counted as a single entity, the EU would rank as the third biggest CO_2 polluter, after China and the United States (CARMA 2008).

One of the major emitters, the USA, produces close to 2 billion tons of CO_2 per year from coal-burning power plants. GHG emissions from coal-fired electricity are about one third of total US greenhouse gas emissions and 40% of total CO_2 emissions in 2009 (US EPA, 2011). In 2009, however, the total GHG emissions fell by 6.59% relative to 2008 and CO_2 emissions dropped by 8.76%. Relative to 2007, the drop was about 10% (from a high of 1.58 billion tons of carbon in 2007 to 1.43 billion tons in 2009, the lowest level since 1995) mainly due to economic recession, though other reasons are attributed. Harvard researchers have primarily attributed this decline to a reduced reliance on coal by the industry due to abundant availability and a fall in the price of natural gas in that year. At the same time, emissions in most industrial countries also fell, bringing global CO_2 emissions from fossil fuel use down from a high of 8.5 billion tons of carbon in 2008 to 8.4 billion tons in 2009. Yet this drop follows a decade of rapid growth. Over the previous decade, global CO_2 emissions rose by an average of 2.5% a year – nearly four times as fast as in the 1990s.

The CO_2 emissions in UK fell by over 10% from 2007 to 2009. Emissions in Germany dropped by 8% and in France by 5%. Japan saw its emissions decline nearly 12% over the two-year period. But, the situation in developing countries is different. CO_2 emissions in the world's heavily populated China and India continued to grow rapidly. China's emissions rose to 1.86 billion tons of carbon in 2009, representing nearly a quarter of global emissions from fossil fuel burning. The carbon dioxide emissions in China grew by nearly 9% in that year. With an average annual emissions growth of 8% over the past decade, China overtook the USA in 2007 as the world's leading CO_2 emitter. India's emissions grew by close to 5% a year over the past decade; the country passed Russia in 2007 to become the world's third largest emitter. Although measures are taken whenever possible to improve the efficiency of fossil-fueled power plants, those measures are inherently modest and total global emissions continue to grow rapidly.

In per capita terms, *emissions in developing economies remain far below* those of most of the industrial world. Per capita emissions from the USA are the highest in the world with about 17.3 metric tons of CO_2, 11.6 metric tons in the Russian federation, and 8.6 metric tons in Japan, compared to 5.8 metric tons in China, 1.6 in India, and 0.9 in Brazil in 2009 (IEA Statistics, World Bank 2012). Average per capita emissions from electricity and heat production in the EU is 3.3 tons per year. Per capita emissions in Australia, the USA and Canada are three times that of China and nearly four times the world average. In many developing countries, per capita power consumption is extremely low because millions of people lack access to electricity (CARMA 2008). However, several oil-rich countries, including the tiny nation of Qatar, recorded high per capita emissions. Interestingly, many European countries, the UK, Germany, and France, have comparable standards of living to the United States but emit only half as much carbon dioxide per person.

Emissions projections to 2035: Energy related CO_2 emissions in OECD and non-OECD countries by fuel type projected to 2035 are shown in Figure 5.1.

Figure 5.2 shows projected average annual growth of energy-related CO_2 emissions in non-OECD countries to 2035. The data assumes no change in policy in the use

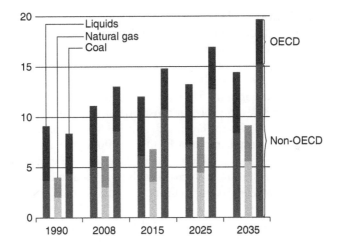

Figure 5.1 OECD and non-OECD energy related CO_2 emissions by fuel type in billion metric tons, 1990–2035 (*Source:* USEIA-IEO 2011).

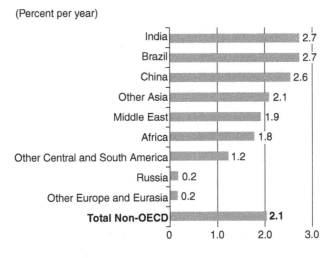

Figure 5.2 Average annual growth of energy-related CO_2 emissions in non-OECD countries (% per year), 2008–2035 (*Source:* USIEA-IEO 2011).

of energy sources. The emissions from OECD countries including the USA are almost flat during the period of projection. In contrast, the non-OECD countries, and especially China, are showing an increasing trend in CO_2 emissions. The top-ten power generating companies in the world in terms of power sector emissions include five in China, two in the USA, one in India, one in South Africa, and one in Germany. The world's biggest corporate carbon emitter is China's Huaneng Power International, whose plants pump out about 285 million tons of CO_2 per year, far more than the 227 million tons produced by all of the power plants in the UK combined and almost as much as the entire continent of Africa (335 million tons). In the USA, the biggest CO_2 emitter is Southern Co. with annual emissions over 200 million tons, followed by American Electric Power Company Inc. at 175 million tons, and Duke Energy Corp. at 112 million tons (CARMA 2008).

A recent study by Stanford University research group found that 22% of Chinese emissions resulted from the production of goods for export. The study also found that the manufacture of goods imported by the USA was responsible for 190 million tons of carbon emissions per year. If emissions totals were adjusted to account for Chinese exports and US imports, the USA would again be the world's leading emitter. While fossil fuel burning is responsible for most carbon dioxide emissions, changes in land use, such as clearing forests for cropland, another serious problem in developing countries, also emit a substantial amount of CO_2. In 2008, global emissions from land use change were estimated at 1.2 billion tons of carbon. The vast majority of these emissions were from deforestation in the tropics; Indonesia and Brazil alone represent over 60% of land use change emissions. This part of emissions is not elaborated in the present context.

More than half of the carbon dioxide emitted annually is absorbed by oceans, soils, and trees. The rapid rate at which carbon dioxide is reaching the atmosphere is overwhelming these natural systems, posing a particular threat to ecosystems in oceans. The large amounts of dissolved CO_2 alter ocean chemistry, making seawater more acidic, which makes it more difficult for organisms such as reef-building corals or shellfish to form their skeletons or shells. The world's oceans are now more acidic than they have been at any time in the past 20 million years. Experts have estimated that if CO_2 emissions continue to rise on their long-term trajectory, coral reefs around the world may be dying off by 2050.

The carbon dioxide that is not absorbed by oceans and other natural sinks remains in the atmosphere, and traps heat. The level of carbon dioxide in the atmosphere, which stood between 260 and 285 parts per million (ppm) until the Industrial Revolution, has risen rapidly in the last two-and-a-half centuries, to over 387 ppm in 2009 (Jayarama Reddy 2010). The last time carbon dioxide levels were this high was roughly 15 million years ago, when the sea level was 25–40 meters (80–130 feet) higher and global temperatures were 3–6°C (5–11°F) warmer.

The increase in atmospheric CO_2 has driven a rapid rise in global temperature, and each decade over the past half-century has been hotter than the previous. This has been established with overwhelming scientific evidence (IPCC-AR4, 2007). The consequences of these rising temperatures have already been witnessed in melting glaciers and ice sheets, shifting weather patterns, and changes in the timing of seasonal events. While much of the global emissions drop in 2009 was thought to be due to declining fossil fuel use associated with the economic recession, the year also saw strong growth in the use of renewable energy. For example, installed wind capacity alone grew by over 30% worldwide. In 2010, global solar photovoltaic production and markets more than doubled compared to 2009, due to government incentives and feed-in-tariffs, and the continued fall in PV module prices. Germany installed more solar photovoltaic in 2010 than the entire world added in 2009. PV markets in Japan and the US almost doubled relative to 2009. Globally, wind power added the most new capacity, followed by hydropower and solar PV, but for the first time ever, Europe added more PV than wind capacity (REN21: 2011). Many countries worldwide have formulated sound environment policies and allocated huge budgets for clean energy and energy efficiency projects, which will boost the growth of renewable energy sources in the years ahead.

However, evidence is mounting that faster, more substantial action is needed to control emissions. The Intergovernmental Panel on Climate Change (IPCC) has modeled a number of scenarios for possible emissions growth in the coming decades (IPCC AR-4, 2007). The likely rise in temperature projected in these scenarios ranges from 1.1° to 6.4°C (2 to 11°F) by the end of the century. Even with the recent drop, carbon dioxide emissions continue to track some of the worst-case IPCC scenarios. Increased scientific evidence shows that atmospheric CO_2 must be stabilized at 350 ppm or less; and in order to achieve this goal, an early fundamental shift in course is needed (for details see Jayarama Reddy 2010). One way to push carbon dioxide emissions onto a rapid downward trend is by shifting to a new sustainable energy economy that relies on carbon-free sources of energy such as wind, solar, hydro, and geothermal instead of climate-threatening fossil fuels. The international community holds unanimous view on this approach, though the initial costs of implementation, efficiency and other aspects is a subject of debate for total introduction. Due to abundant availability of cheap coal in most parts of the world, especially in countries with the highest emissions, and since coal-fired power plants at gigawatt level are already working or are brought on line, utilization of advanced or clean coal technologies on large scale is also very vital for reducing emissions. There are, however, several ways of implementing: coal cleaning (pre-combustion control), incorporating air pollution control devices (post-combustion control), and carbon capture and storage, in addition to improving combustion process technology. These are discussed in more detail later in this book. The intelligent and economical use of coal, which can provide secure and affordable energy, will remain a key challenge until the global energy demand is largely met by renewable sources.

In the USA, the Clean Coal Technology Program has made available carbon control technologies for coal-based electricity generation at low cost while meeting environmental regulations. The short term goal of the program is to meet the current and emerging environmental regulations in the country, which will considerably reduce costs for controlling SO_2, NO_x, particulates, and mercury at both new and existing coal-fired power plants. The midterm goal is to develop low-cost, super clean coal-fired power plants with efficiencies 50% higher than today's average efficiency (NEPD 2001). The higher efficiencies lead to low emissions at minimum cost. In the long term, the goal is to develop low cost, zero-emission power plants with efficiencies nearly double that of today's power plants as well as to coproduce fuels and chemicals, along with electricity, leading to the development of technologies where coal becomes a major source of hydrogen. Meanwhile, the current emission scenario demands all

Table 5.1 Average concentration of hazardous elements in coal (in lbs/billion BTU).

Element	Anthracite	Bituminous	Sub-bituminous	Lignite
Arsenic	NR	0.5	0.1	0.3
Beryllium	NR	0.11	0.03	0.2
Cadmium	NR	0.03	0.01	0.06
Chlorine	NR	35	2.7	24
Chromium	NR	1.1	0.4	2.2
Lead	NR	0.6	0.2	1.0
Manganese	NR	1.8	1.3	20
Mercury	NR	0.007	0.006	0.03
Nickel	NR	0.9	0.4	1.2

NR – Not Reported (Source: US EPA 2010)

major carbon emitters to urgently commit for time bound emission reductions; and the developed countries to provide resources and the new technologies for the developing and poorer countries.

5.2   HAZARDOUS POLLUTANTS FROM COAL

It is known that during the formation of coal, it incorporates substances from the surrounding soil and sediment. These substances include impurities such as sulfur and hazardous elements – lead, mercury, arsenic, nickel, chromium etc. The nature and extent of these impurities may differ from one coal seam to another and depends on the conditions existing during the time of formation. When coal is combusted in a power plant, these impurities are released into the atmosphere through the stack gas, unless controlled/removed by emissions control methods. The average concentrations of these elements present in different types of coals are given in the Table 5.1 (USEPA 2010). Coal-fired power plants emit a large number of hazardous air pollutants posing serious threat to human health and environment (USEPA 2007). These emissions include both 'fuel-based' pollutants (such as metals, HCl, HF, Hg) which depend on the contaminants in the coal, and 'combustion-based' pollutants (like dioxins and formaldehyde) formed during combustion of coal (USEPA 2010). Hazardous air pollutants from coal-fired power plants include: (a) Acid gases, such as hydrogen chloride and hydrogen fluoride; (b) Benzene, toluene and other compounds; (c) Dioxins and furans; (d) Formaldehyde; (e) Lead, arsenic, and other metals; (f) Mercury; (g) Polycyclic Aromatic Hydrocarbons (PAH); and (h) Radioactive materials, like radium and uranium (US EPA 2007; ATSDR 2011).

5.3   IMPACT ON HUMAN HEALTH AND ENVIRONMENT

These toxic emissions cause a dangerous range of health problems to humans as shown in Table 5.2. These emissions can make breathing difficult and can worsen asthma, chronic obstructive pulmonary disease, bronchitis and other lung diseases. They can cause heart attacks and strokes, lung and other cancers, birth defects and premature death. These pollutants also threaten essential life systems. Acid gases are corrosive and can irritate and burn the eyes, skin, and breathing passages. Long term exposure to metals has the potential to harm the kidneys, lungs, and nervous system. Exposure to the metals and dioxins in emissions from coal-fired power plants increases the risk of cancer. Specific forms of arsenic, beryllium, chromium, and nickel have been shown to cause cancer in both human and animal studies. Table 5.2 also identifies those pollutants that have long-term impacts on the environment because they accumulate in soil, water and fish (ATSDR 2011). Coal-fired power plants are also the biggest emitters of airborne mercury among all industrial sources. Mercury is associated with damage to the kidneys, liver, brain, nervous system and can cause birth defects (ATSDR 2011).

The particulate matter from the power plants comes directly from the ash and soot, but smaller particles come from chemical reactions that emitted gases undergo in the atmosphere. The smaller particles produced by fossil fuel combustion worsen asthma and bronchitis, cause heart attacks and strokes, and increase the risk of premature death. This is in part because these fine particles can travel far deeper into the lungs than larger ones that are filtered out by the nose and larger airways. Health problems from power plant emissions can occur when levels are high over a short period or at lower levels over longer time periods (US EPA 2009).

The environment is also affected by these emissions. This includes environmental degradation with the swelling of toxic metals; contamination of rivers, lakes and oceans; degradation of culturally important monuments by acid rain, for e.g., the Statue of Liberty and the Lincoln Memorial. Acid rain reaching soil and water bodies can change their acidity or pH and alter the chemistry and nutrient balance in those environments. As a result, changes happen in the types of plants, animals and microorganisms that inhabit those areas. Hazardous air pollutants also add to pollution in rivers and streams and can cause damage to crops, forests and, finally, to humans (US EPA 2009, US EPA 1997).

Not all power plants are the same. Emissions vary depending on the types of coal used, the types of control equipment in place, and the length of time operated. Effects of the Plant emissions will vary depending on the height of the stacks and their location relative to population centers, topography, and weather patterns.

Hazardous emissions threaten health locally and at great distances. For example, acid gases, such as HCl and HF tend to settle within a day or two, posing high risk to neighborhoods and nearby towns. Mercury and sulfur dioxide emissions also have immediate impact in the local area. Many pollutants also travel much farther and can be carried hundreds or even thousands of miles from their original source. Health effects can only be determined through detailed analyses of relationships between emissions, transport, concentrations, exposure, and effect. Many metals, dioxins and other pollutants adhere themselves to the fine particles, and travel with airborne particles to distant locations. These particles can remain in the air for up to a week or more, travelling long distances, being carried by winds to areas far away from the original source (ALA 2011).

People at greater risk

Everyone faces increased risk of health problems from exposure to these hazardous air pollutants. However, some people – because of their age, general health conditions,

Table 5.2 Toxic emissions from Coal-fired power plants and their impact on human health and environment.

Class of HAPs	Notable HAPs	Impact on Human health	Impact on Environment
Acid gases	HCl, HF	Irritation to skin, eyes, nose, throat and breathing passages	Acid precipitation: damage to forests and crops
Dioxins and Furans	2,3,7,8 Tetra-chlorodioxine	Probable carcinogen: stomach & immune system, Affects reproductive endocrine & immune system	Deposits in rivers, lakes and oceans and is taken up by fish and wild life; Accumulates in food chain
Mercury	Methyl mercury	Damage to brain, kidney, nervous system & liver; causes neurological & developmental birth defects	Consumed by fish and wildlife; accumulates in the food chain
Non-mercury metals and metalloids	Sb, As, Cd, Cr, Be, Ni, Se, Mn	Carcinogens: lung, bladder, skin, kidney, May adversely affect nervous, dermal, cardiovascular, immune & respiratory systems.	Accumulate in soils & sediments; soluble forms may contaminate water bodies; accumulates in soils & sediments.
	Lead	Damages developing nervous system; may adversely affect learning, memory and behavior; may cause cardiovascular and kidney effects; anemia, weakness of ankles, wrists & fingers	Harms plants & wildlife; may adversely affect land and water ecosystems
Polycyclic Aromatic Hydrocarbons (PAH)	Benzo-a-anthracene, Benzo-a-pyrene, Fluoranthene, Chrysene, Dibenzo-a-anthracene	Probable carcinogens. May attach to small particulate matter and deposit in the lungs. May have adverse effects on the kidney, liver & testes. May damage sperm cells and may cause impairment of reproduction	Exists in vapor and particulate phase; Accumulates in soil and sediment
Radioisotopes	Radium	Carcinogen: lung & bone. Broncho-pneumonia, anemia and brain abscess	Deposits into lakes, rivers and oceans; consumed by fish & wildlife; Accumulates in soils & sediments and in the food chain
	Uranium	Carcinogen: lung & lymphatic system, kidney disease	Same as above
Volatile Organic Compounds (VOC)	Aromatic hydrocarbons including benzene, xylene, ethyl-benzene & toluene	Irritation of skin, eyes, throat, nose; difficulty in breathing; impaired function of lungs; delayed response to visual stimulus; impaired memory; stomach discomfort; & effects to liver and kidneys; May also cause adverse effects to the nervous system; benzene is a carcinogen	Accumulates in soil and sediments
	Aldehydes including formaldehyde	Probable carcinogen: lung cancer. Eye, nose, throat irritation; respiratory symptoms	Same as above

[*Source:* American Lung Association (ALA) March 2011]

or direct exposure to the pollutants – face greater risk. They include: children and teenagers, old people, pregnant women, people with asthma and other lung diseases, people with cardiovascular diseases, diabetics, low income groups, people who work outdoors, and others with some sort of health problems (US EPA 2009, 1997). Too often, low income groups or those belonging to ethnic or racial minorities face a greater share of this pollution impact because they live closer to industrial units, including power plants (Levy *et al.* 2002, O'Neill *et al.* 2003).

5.4   CONTROL TECHNOLOGIES

Controlling these pollutants without letting them into the atmosphere is possible, and the technologies available today are shown in Table 5.3. These technologies are discussed in detail in the next chapter. In the USA, the Clean Air Act amendment of 1990 makes it mandatory to incorporate the controlling technologies into the coal-fired power plants.

Table 5.3 Available technologies to control toxic pollutants released by Coal-fired power plants.

Control technology	Pollutants controlled	Principle of the technology
Acid gas Control		
Wet & dry Flue gas Desulfurization (Scrubbers)	*HAPs*: HCl, HF, HCN, Hg *Collateral pollutants*: SO_2, Particulates	Liquid mixed with limestone is sprayed into the emission; and emissions are passed through a stream of liquid mixed with lime or a bed of basic material such as limestone; reactions between sulfur and base compounds produce salts which are removed from the exhaust air stream.
Dry sorbent injection (DSI)	*HAPs*: HCl, HF, HCN *Collateral pollutants*: SO_2	Dry sorbent consisting of sodium bicarbonate, lime or similar material is blown into duct, reacts with acid gases and is captured in downstream PM controls.
Non-Hg metal control		
Electro Static Precipitator (ESP)	*HAPs*: Sb, Be, Cd, Co, Pb, Mn, Ni, Particulate phase organics *Collateral pollutants*: other forms of primary particulates	Particles are charged with electricity and collected on oppositely charged plates; particles are collected for disposal/further treatment
Baghouse Cyclones	Same as above Same as above	Emissions are passed through fabric filters & collected; Use centrifugal force to separate particulates from gas stream
Hg Control		
Activated Carbon Injection (ACI)	Hg, As, Cr, Se, Dioxin and other gas-phase organic carbon-based compounds	Powdered activated carbon (similar to charcoal) is blown into the flue gas after combustion; pollutants are absorbed by carbon and removed by PM controls

(*Source*: ALA March 2011)

REFERENCES

ALA (2011): Toxic Air: The Case for Cleaning Up Coal-fired Power Plants, March 2011, Washington DC: American Lung Association Headquarters office.

ATSDR (2011): Agency for Toxic Substances and Disease Registry, 2011: *Toxic Substances Portal: Toxicological Profiles.* Washington, DC, USA: ATSDR; available at http://www.atsdr.cdc.gov/toxprofiles/index.asp.

CARMA: Global Estimation of CO2 Emissions from the Power Sector – Working Paper 145; *Information Disclosure and Climate: The Thinking Behind CARMA* 12 Working Paper 132, Aug 27, 2008.

IEA Statistics (2012): CO_2 emissions, at http://www.iea.org/states/index.asp; and http://data.worldbank.org/indicator/EN.ATM.CO2E.KT/countries.

IPCC (2007): Climate Change 2007, *Synthesis Report, IPCC 4th Assessment Report,* IPCC, Geneva, Switzerland.

Jayarama Reddy, P. (2010); *Pollution and Global Warming,* BS Publications, Hyderabad, India.

Levy, J.I., Greco, S.L., & Spengler, J.D. (2002): The importance of population susceptibility for air pollution risk assessment: a case study of power plants near Washington, DC, *Environmental Health Perspectives,* **110(12)**, 1253–60.

Miller, B.G. (2011): *Clean Coal engineering Technology,* Elsvier, Amsterdam, ISBN: 978-1-85617-710-8.

O'Neill, M.S., Jerrett, M., Kawachi, I., Levy, J.I., Cohen, A.J., Gouveia, N., *et al.* (2003): Health, Wealth, and Air Pollution: Advancing Theory and Methods. *Environmental Health Perspectives,* **111,** 1861–1870.

REN21: *Renewables 2011 Global Status Report,* available at www.ren21.net/REN21Activities/Publications/ . . .

US EPA (1997): *Mercury Study Report to Congress,* Volumes I–VIII: (EPA-452/R-97-003 through EPA-452/R-97-010), Washington, DC.

USEPA (2002): ALLNEI HAP Annual 01232008; available at http://www.epa.gov/ttn/chief/net/2002inventory.html# inventory data.

USEPA (2007): *National emission Inventory 2002.* Inventory data; Point sector data- ALLNEI HAP Annual 01232008; at www.epa.gov/ttn/chief/net/2002 inventory.htm

US EPA, (2009): *Integrated Science Assessment for Particulate Matter,* EPA 600/R-08/139F; Available at http://cfpub.epa.gov/ncea/cfm/recordisplay.cfm?deid=216546.

USEPA (2009): *Integrated Science Assessment for particulate matter,* EPA/600/R-08/139F), National Center for environment assessment, RTP division, OR&D, EPA

USEPA (2010): *Air toxic standards for Utilities: Utility MACT ICR data.* Part I,II,III: final draft (version3); at www.epa.gov/ttn/atr/utility/utilitypg.html

US EPA (2011): Inventory of U.S. Greenhouse Gas Emissions and Sinks: 1990–2009, Table ES-7, 2011, at http://www.epa.gov/climatechange/emissions/usinventoryreport.html.

USEIA-IEO (2011): Energy Information Administration – Intl. Energy Outlook 2011, DOE, Washington, DC.

Chapter 6

Coal treatment and emissions control technologies

6.1 COAL TREATMENT

6.1.1 Introduction

The nature of the environmental impact that energy generating technologies create is generally dependent on the specific technology used that includes concerns over water resource or pollutant emissions or waste generation or public health and safety concerns or all of these. Coal-fired power generation is not free from these impacts and has been associated with a number of environmental challenges, primarily associated with emissions. The environmental concerns become even more significant with coal being used in larger amounts than ever before. Coal has demonstrated the ability to face such challenges in the past and hopefully it will effectively meet future environmental challenges.

Mined coals are highly heterogeneous varying widely in quality and content from country to country, mine to mine and even from seam to seam. Coal is normally associated with ash-forming minerals and chemical material including sand, rock, sulfur and trace elements. Some materials are intermixed through the coal seam, some – mainly organic sulfur, nitrogen and some mineral salts – are bound organically to the coal, and some are introduced by the mining process. These impurities affect the properties of coal and the combustion process including the stack (flue) gas emissions and byproducts of combustion.

Coal treatment: Coal cleaning and washing at the mine or at the power plant (or at both) removes ash, rock, and moisture from coal to the extent possible. In principle, coal is crushed and washed when it is cleaned at the mine preparation plants which removes the largest amount of sulfur found in coal. The cleaning and washing of coal, also known as the *coal benefaction, coal beneficiation* or *coal preparation* process, improves the quality of coal and provides further benefits that include (a) savings in the transportation, and in capital and operating costs of the power plant, particularly the boiler, coal handling and ash handling systems, and (b) reduction in the cost of power generation if the washed coal increases the plant load factor and the washery rejects are utilized efficiently in fluidized bed boilers (Burnard and Bhattacharya, IEA 2007). It has been recognized that the supply of 'clean coal' of secure quality to utilization purposes can bring sizeable reductions in SO_2, NO_x, particulate, and carbon dioxide emissions (Burnard and Bhattacharya, IEA 2011). Reduced emissions of SO_2 and NO_x mean reduced acid rain. While coal preparation is standard practice in many countries, greater uptake in developing countries is needed as a low-cost way to improve the environmental impact of coal.

In the 1800s, only 5% of the energy in coal was used. Today, using the technology of the 1950–1960s, around 35% of the energy available in coal is used. The new technologies can further improve performance of power plants with increased efficiency and in reducing the emission of pollutants with less cost which ultimately make energy from coal cheaper.

Highly efficient and viable technologies have been developed to control emission of sulfur oxides and nitrogen oxides, particulate matter, and trace elements such as mercury. Recently, the focus has been on developing and deploying technologies to control CO_2 emission, a major greenhouse gas. Many of these modern processes are included in coal-fired power generation technologies to improve thermal efficiency and to reduce the emissions. Efficient plants burn less coal per unit of energy produced and consequently has lower associated environmental impacts. The technology option varies depending on factors such as location, coal source and quality.

The processes utilized to remove the impurities and pollutants are broadly classified as *Pre-combustion,* during *Combustion,* and *Post-combustion* processes, and *Conversion technologies* (IEA 2003, 2011; Balat 2008a, 2008b; Omer 2008; Franco and Diaz 2009).

6.1.2 Pre-combustion processes (coal benefaction)

Coal benefaction/cleaning consists a group of technologies – *physical cleaning* or *washing, chemical cleaning* and *biological methods of cleaning* – to remove sulfur and ash (Satyamurthy 2007). Coal beneficiation also includes drying, briquetting, and blending (Breeze 2005; IEA 2003).

Physical cleaning separates effectively ash content, rocks and *pyritic sulfur* (sulfur combined with iron) and trace elements like mercury from the coal. Often this is done using water. The density of coal and its constituents are not the same and significantly differ in different types of coal. So, these differences are exploited to separate coal from the impurities. Each type of coal has its own washability criteria depending on the chemical composition. When coal is crushed and washed, the heavier impurities separate from the coal, making it cleaner. Since coal can be ground into much smaller sizes similar to powder, it allows for removal, up to 90% of the pyritic sulfur.

The system used in physical cleaning varies among coal cleaning plants but broadly involves the following stages: initial preparation, fine coal processing and coarse coal processing, and final preparation. A process flow diagram for a typical coal cleaning plant is shown in Figure 6.1 (redrawn from RRI 1970; EPA 1988).

In the initial preparation stage, the raw coal is unloaded, stored, conveyed, crushed, and grouped into coarse and fine coal fractions by screening. These fractions are then sent to their respective cleaning processes. In the second stage of fine coal and coarse coal processing, similar operations and facilities, but with different operating parameters, are used for separating impurities. Most of the coal cleaning processes uses upward currents or pulses of water, to fluidize a bed of crushed coal and impurities. The lighter coal particles rise and are removed from the top of the bed and the heavier impurities are removed from the bottom.

In the first stage of preparation processes, moisture from coal is removed to reduce freezing problems, volume and weight; and to enhance the heating capacity. 'Dewatering' is performed by using screens, thickeners, and cyclones followed by thermal drying using either fluidized bed or flash or multilouvered dryer. In the fluidized bed dryer, the coal is suspended and dried above a perforated plate by pushing hot gases. In the flash dryer, coal is fed into a stream of hot gases for instant drying. The dried coal and wet gases are both drawn up a drying column and into a cyclone for separation. In the multilouvered dryer, hot gases are passed through a falling curtain of coal, which is then carried by a specially designed conveyor. Heat from sun can be used for drying, though currently it is carried out by using heat from the plant flue gases.

The emissions from the first phase (Figure 6.1) of wet or dry processes consist primarily of particulate matter as coal dust from roadways, stock piles, refuse areas, rail wagons, conveyor belt pour-offs, crushers, and classifiers. Water wetting is used to reduce/control these emissions. Another technique that applies to unloading, conveying, crushing, and screening operations involves enclosing the process area and

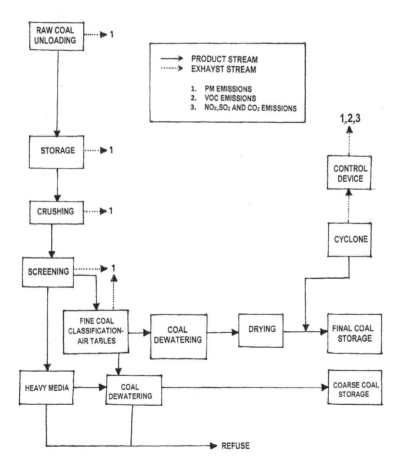

Figure 6.1 Typical Coal Cleaning Plant Process Flow diagram (Redrawn from Refs. RRI 1970; US EPA 1988).

circulating air from the area through fabric filters. The major emission source in the second phase of fine or coarse coal processing is the air exhaust from the air separation processes (air tables). Emissions from wet cleaning processes are very low; for the dry cleaning process, the emissions (particulate matter) are generated when the coal is stratified by pulses of air. These are normally controlled with cyclones followed by fabric filters. The major source of emissions from the final preparation phase is the thermal dryer exhaust containing (a) coal particles carried in the drying gases and volatile organic compounds (VOC) released from coal, and (b) products resulting from coal combustion – the hot gases including CO, CO_2, VOC, SO_2, and nitrogen oxides. The most common technology used to control dryer exhaust is venturi scrubbers and mist eliminators downstream from the product recovery cyclones. The control efficiency of these techniques is over 90% (US EPA 1989).

Dry cleaning methods such as air-heavy medium devices, medium devices, air tables, and air jigs can also be used for cleaning when wet washing is not suitable or when washing plants are considered not economically feasible (Cicek 2008, Burnard and Bhattacharya, IEA 2007).

Briquetting of coal using appropriate binders reduces sulfur dioxide emissions and fixation of some toxic elements. *Blending* of coal helps to improve the quality and combustion and to save the costs (Breeze 2005). For a detailed account on dry benefaction, the reader may refer to an article by Dwari and Rao (2007).

Column flotation is another cleaning method that floats finely ground coal in water. The coal has been chemically conditioned to attach to rising air bubbles. This allows nearly all inorganic matter, such as pyritic sulfur, to sink to the bottom of the flotation column.

Heavy media separators, heavy media cyclones, and jigs are used for coal upgrading. Enhanced gravity separators, selective agglomeration, and froth flotation are the other effective coal cleaning methods of removing mineral matter. The ash content of coal can be reduced by more than half (over 50%) by using these devices (WCI 2005).

Conventional clean-coal technologies and advanced clean-coal processes are summarized and tabulated in Tables 6.1 and 6.2.

Chemical cleaning: Chemical cleaning is used to remove organic sulfur from the coal. In *Molten-caustic leaching*, a chemical cleaning method, coal is immersed in a chemical that actually leaches the sulfur and other minerals from the coal. This method is not yet commercially used.

Biological cleaning involves using microorganisms (bacteria) that literally 'eat' the sulfur out of the coal.

Another advantage of coal cleaning is that several inorganic hazardous pollutants such as antimony, arsenic, cadmium, chromium, copper, cobalt, mercury, lead, manganese, nickel, thorium, uranium, and selenium, generally associated with coal in trace amounts are reduced through the cleaning process. It is likely that many of these are emitted from crushing, grinding, and drying operations in trace quantities.

The physical, chemical and biological cleaning methods can also be used to upgrade the low quality coals. The best option, however, is physical cleaning (using coal washing plants). The removal of mineral matter including pyritic sulfur could reduce sulfur dioxide emissions by around 40% and CO_2 emissions by about 5% (Breeze 2005; IEA 2008b) and improve the efficiency of the power plant.

Table 6.1 Conventional Physical Coal-cleaning Technologies.

Technology	Process
Crushing	Pulverizing coal and sieving into coarse, medium and fine particles. The non-organic material is released during crushing which can be separated due to their dense nature by further processing.
Jigs (G)	Separating medium size particles from coarse ones.
Dense-medium baths (G)	Separating medium size from coarse particles.
Cyclones (G)	Used for coarse to medium particles.
Froth floatation	Separating fine (<0.5 mm) particles by selective attachment of air bubbles to coal particles allowing them to be buoyed up into a froth while leaving the rest of the particles in water; for fine particles, it relies on surface properties of ash vs coal.
Wet concentration (Kawatra & Eisle 2001)	Rapid shaking causes particles of different densities to migrate to different zones on the table's periphery (Specific gravity of pyrite is about 5.0 and that of coal is about 1.8).
Concentration spiral	As the coal pulp is fed from the top of the spiral, it flows downword, and the centrifugal force causes the separation.
Electrokinetics	Physical cleaning is accomplished by using electrophoresis method in which the electrokinetically charged particles in a liquid are migrated toward an electrode of opposite charge in a dc electric field.

G = Gravity/density-based separation (modified from Ghosh and Plares 2009: several references are cited in their book)

Table 6.2 Advanced Cleaning Processes.

Technology	Process
Advanced physical cleaning	Advanced froth floatation (S), Electrostatic (S), & Heavy liquid cycloning (G).
Aqueous phase pretreatment	Bioprocessing, Hydrothermal, Ion exchange, Selective agglomeration, LICADO process, & Spherical oil agglomeration.
Organic phase treatment	Depolymerization, Alkylation, Solvent Swelling, Electrolysis & Rapid pyrolysis.
Other processes	Microwave, Microbial desulfurization, Fluidized bed, Ultrasonic, Liquid-fluidized bed classification, Chemical cleaning processes, Dry benefaction, & High gradient magnetic separation.

G = Gravity-(density)-based separation; S = surface-effect-based separation (Drawn from: Ghosh and Plares 2009, several Refs. on these processes are cited in their book)

Coal benefaction applicability

In principle, the coal cleaning is possible for most bituminous coals and anthracite which account for about two-thirds of global coal production. Around one-third of this quantity is presently cleaned. Most of the coals from the USA, Australia and South Africa are cleaned/washed; and there is prospect for increased use of coal benefaction in China, India, Russia, Poland and other smaller coal producing countries (Ghosh 2007).

Sub-bituminous and lignite coals are often low in ash and sulfur, but generally contain more moisture, 20% to 60%. This causes an array of problems in a coal-fired boiler requiring more energy and maintenance. Benefaction methods are used in drying these coals as efficiently and economically as possible. Another issue with low-rank coal is spontaneous combustion; hence, drying the coal is done instantly prior to combustion.

A number of countries, particularly China, India, Poland, Czech Republic, South Africa, Romania and Turkey use high-ash coals for power generation. Approximate ranges of moisture content, ash content and calorific values of lignite coals for the major countries using high-ash and/or high-moisture coals are given by Burnard and Bhattacharya (IEA 2011). These coals require pre-drying absolutely to improve the efficiency of the power generating plant.

6.1.3 Benefits of pre-drying

Power plants that use high-moisture coals suffer from efficiency decrease. The efficiency is estimated to decline by 4% if the moisture-content in the coal increases from 10% to 40% and by 9% if the moisture-content increases to 60%. Additionally, high-moisture increases coal-handling feed rate and requires extra energy for coal-handling systems leading to an increase in the costs for plant operation and maintenance (Burnard & Bhattacharya, IEA 2007). Coal pre-drying is thus very critical.

Since high-moisture coals are more reactive due to their high oxygen content, there is a risk of spontaneous combustion in drying these coals. Therefore, the pre-drying has to be performed instantly prior to combustion nearer to the plant itself by recirculation of part of the flue gases from the upper portion of the boiler. In turn, this demands a larger size boiler to handle additional volume of water vapor, again leading to extra auxiliary power and reduced efficiency. So, if these coals could be pre-dried using low-grade or waste heat, the boiler size could be smaller and its efficiency could be higher. It is therefore necessary to utilize drying technologies that utilize lower grade energy and recover the exhaust heat from the dryer effluent and/or remove the water without evaporation, avoiding the loss of latent heat in evaporation. If rigorous pre-drying of high-moisture coals could be done, emissions as much as 0.3 billion tonnes CO_2/year could be reduced which is a substantial portion of the country's CO_2 emissions from power production in many countries, e.g., Australia, Germany, Indonesia and Russia.

Coal pre-drying using low grade heat offers many benefits: (a) overall plant efficiency is increased due to increase in boiler efficiency, thereby reducing CO_2 emissions, (b) boiler size and plant's extra power usage are reduced by reducing the flow rates of coal and flue gas, (c) flow rate reduction also allows extra SO_2 capture for high sulfur-content coals by extra scrubber, (d) NO_2 emissions are reduced by increasing coal's heating value and reducing the flow rates of coal and air, (e) Hg oxidation increases during combustion and this oxidized Hg can be removed by wet-lime spray towers, and (f) using low-grade or waste heat avoids the need for alternative heat sources.

Under USDOE's Clean Coal Power Initiative (CCPI Round 1) Program, a promising *low temperature coal-drying process* has been developed by Great River Energy (GRE) (USDOE June 2012). It is a fuel enhancement project utilizing waste heat from the power plant reducing the lignite moisture content in a Fluidized Bed dryer (FBD) at Coal Creek Station in Underwood, North Dakota. Reducing the moisture content of the lignite increases the energy efficiency of the boiler which means the fuel requirement is reduced for a given load. Emissions reductions were achieved as a result of

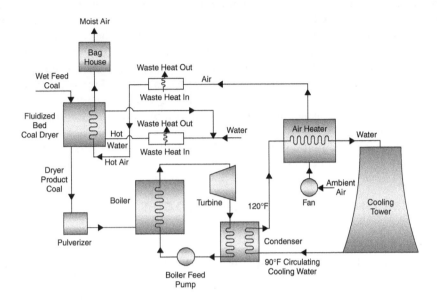

Figure 6.2   Schematic of Lignite Drying Process (USDOE CCPI 2012).

increased fuel quality, segregation of pyrite and mercury in the drying process, and increased oxidation of mercury resulting in greater mercury removal in the FGD system. The moisture content of the lignite coal was reduced by 8.5% resulting in a HHV improvement from 6290 Btu/lb to 7043 Btu/lb. Also, mercury emissions were reduced by 41%, NO_x by 32% and SO_2 by 54%. The schematic of drying process is shown in Figure 6.2. In this process, cooling water leaves the condenser carrying the waste heat rejected by the steam turbine. Before the water reaches the cooling tower, it first passes through an air heater where a fan-driven air stream picks up some of the waste heat from the cooling water. The heated air is then sent to the FBCD, which is configured for two-stage drying to optimize heat transfer. Before arriving at the FBCD, the air stream picks up additional heat from the unit flue gas through another heat exchanger. The twice-heated air stream then enters the FBCD. After picking up moisture from the coal, the moisture-contained air stream passes through a dust collector to remove coal dust liberated during the drying process before being discharged to the atmosphere. Additional heat is added to the FBCD through coils fed with water heated by the unit's flue gas. This additional heat is added to the FBCD to optimize fluidized bed operating characteristics. After leaving the FBCD, dried coal enters a coal storage bunker (not shown in the figure) and then to a pulverizer for size reduction prior to being delivered to the boiler.

The GRE project at Coal Creek station was implemented in two phases. The first phase of the project involved the installation and operation of one prototype dryer, rated at 112.5 tons/hour (225,000 lb/hour) capacity. The prototype dryer was designed to reduce the lignite moisture content from 38% to 29.5% (i.e., 8.5%), which corresponds to an increase in higher heating value from 6,200 Btu/lb to 7,045 Btu/lb. The prototype coal drying system was designed with completely automated control capability, which included startup, shutdown, and emergency shutdown sequences. The heat input to the FBCD is automatically controlled to remove a specified amount of moisture from the lignite feed stream. After operation around the clock and collecting information for about six months, full scale dryers, 135 tons/hour, were installed; and GRE completed the construction of the complete drying system and started operations from late 2009. A potential benefit of the drying system is the reduction in capital costs because a decrease in coal firing rate could result in smaller capacity systems for coal handling and coal processing, as well as for the combustion, flue gas transport, and flue gas cleaning.

Several countries, particularly Australia, Germany, and Russia, have shown interest in this drying process system.

6.2 EMISSIONS CONTROL TECHNOLOGIES

6.2.1 Used during combustion

There are technologies that are used inside the furnace where the coal is actually combusted. These methods remove sulfur dioxide and nitrogen oxides. When coal is burned, its chemical composition is changed and sulfur is released. As we know, sulfur presents in coal both as organic-bound and inorganic form; and inorganic sulfur presents either as pyritic sulfur generally in high fractions or as sulfate compounds in low fractions. The relative proportions of pyritic and organic sulfur vary, and up to around 40%, the sulfur being pyritic. During combustion, majority of the sulfur content (95%) is oxidized into SO_2 and a small amount into SO_3. The approximate SO_2 emissions (Graus and Worrell 2007) in combusting different coals are: anthracite coal: 230 g/GJ$_{NCV}$, bituminous coal: 390 g/GJ$_{NCV}$, and lignite coal: 525 g/GJ$_{NCV}$.

Fluidized-bed combustion is one such technology. In fluidized-bed combustion, finely grounded coal mixed with limestone is injected with hot air into the boiler. This bed of coal and limestone suspends on jets of air and resembles a boiling liquid, a 'fluid'. As the coal burns, the limestone acts as a sponge and captures the sulfur. As in a conventional boiler, water-filled tubes collect the heat generated, creating steam which is used to rotate a generator producing electricity. This technology can reduce the amount of sulfur released by over 90%. Another advantage of this technology is the reduction in the boiler temperature. In conventional boilers, the temperature can reach at least 2,700°F (~1480°C). Because the tumbling motion enhances the burning process, temperatures are usually around 1,400 to 1,600°F (~760° to 880°C) in fluidized-bed combustion. The lower temperature is an advantage because fewer nitrogen pollutants are produced. Several companies utilize this method, for example, Archer Daniels Midland Company (ADM) in Decatur, Illinois and B.F. Goodrich in Henry, Illinois. This technology will be discussed in detail later in the book.

6.2.2 Post-combustion processes

The sulfur oxides, and nitrogen oxides, trace elements and other coal combustion products that are released from burning coal are removed before they reach the smokestack and released into the atmosphere. Methods under use to control these polluting emissions are briefly explained.

SO₂ *control methods*

A number of technologies that are widely used to control the sulfur emissions are collectively referred to as Flue Gas Desulfurization (FGD) or scrubbing technology. A chemical sorbent, generally calcium or sodium-based alkaline reagent, is injected into the flue gas in a spray tower or directly into the duct. The SO₂ in the flue gas is absorbed, neutralized and/or oxidized by the reagent into a solid compound, either calcium or sodium sulfate. This solid is removed from the waste gas stream using downstream equipment.

Most FGD systems (scrubbing systems) employ two stages: one for *fly ash* removal and the other for SO₂ removal. *Scrubbers* are classified as 'once-through' or 'regenerable'. Once-through systems either dispose of the spent sorbent as a waste or use it as a byproduct. Regenerate systems recycle the sorbent back into the system. Both types of systems are further classified as wet or dry or semi-dry scrubbers.

Wet processes: Wet scrubbers are the most common currently in use. They include a variety of processes that use sorbents based on calcium, magnesium, potassium, or sodium, and ammonia, or seawater. The calcium-based scrubbers using limestone or lime are by far the *most popular commercial* technologies. In this wet scrubber, the flue gas is treated with a 5–15% (by weight) slurry containing the sulfite/sulfate salts and the limestone (CaCO₃) or calcium hydroxide, by pumping the slurry into the spray tower absorber (Figure 6.3). The sulfur dioxide in the flue gas is absorbed into the droplets of slurry where a series of complex reactions occur. Around 95% of SO₂ can

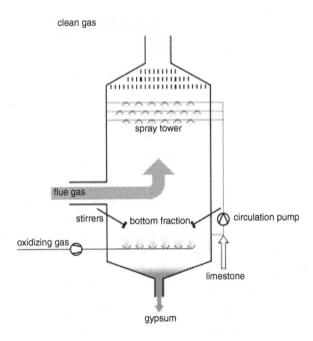

Figure 6.3 Schematic design of an absorber in a FGD System [*Source:* From Wikipedia, Free encyclopedia; Author: Flue_gas_desulferization_unit_DE.svg: Sponk (talk), 2010].

be removed from the flue gas along with about 99.5% fly ash and 100% of hydrogen chloride (HCl). The byproduct, gypsum, has economic value.

$$SO_2 + CaCO_3 + \frac{1}{2}O_2 + H_2O \cdot CaSO_42H_2O \text{ (gypsum)} + CO_2 \qquad (6.1)$$

This chemical reaction represents forced oxidation to obtain gypsum. The inbuilt simplicity, the availability of an inexpensive sorbent (limestone), production of a saleable by-product (gypsum), reliability, and the high removal efficiencies are the advantages of this process. The capital costs are high, but the operating costs are low due to easy accessibility of limestone and the process cost-effectiveness. This method is suited to *medium-to-high sulfur* coals.

Sodium-based wet scrubbers are also in commercial operation. These systems can achieve high SO_2 removal efficiencies from coals having medium- to high-sulfur content. A disadvantage, however, is the production of a waste sludge that requires disposal. In ammonia scrubbing, ammonium sulfate or ammonium is used as a scrubbing agent. About 93% of SO_2 from the flue gas can be removed at commercial-level processes. Ammonium sulfate is the by-product obtained which has commercial value. Seawater scrubbing is a relatively new process wherein flue gas is treated with seawater to neutralize sulfur dioxide. This technology is efficient if the sulfur level in the coal is low, below 2.5 to 3%.

Regenerative FGD processes regenerate the alkaline reagent and convert the SO_2 to a usable by-product. The Wellman-Lord process is the most highly established regenerative technology. The Wellman-Lord process, in principle, uses sodium sulfite to absorb SO_2, which is then regenerated to release a concentrated stream of SO_2. The chemical reactions involved are the following:

$$SO_2 + Na_2SO_3 + H_2O \rightarrow 2NaHSO_3 \qquad (6.2)$$

$$2NaHSO_3 + \text{heat} \rightarrow Na_2SO_3 + H_2O + SO_2 \text{ (concentrated)} \qquad (6.3)$$

Most of the sodium sulfite is converted to sodium bisulfate reacting with SO_2, and part of the sodium sulfite is oxidized to sodium sulfate. The flue gases have to be prescrubbed to saturate and cool to about 130°F which then removes chlorides and any remaining fly ash, and avoids excessive evaporation in the absorber. The sodium sulfite is regenerated in an evaporator-crystallizer by heating, when concentrated SO_2 stream is produced. This concentrated SO_2 stream may be compressed, liquefied, and oxidized to produce sulfuric acid (H_2SO_4) or elemental sulfur. A small portion of collected SO_2 oxidizes to form sulfate and is converted in a crystallizer to sodium sulfate (salt cake) which is useable/saleable (Elliot 1989). The advantages of this process are low consumption of alkaline reagent, least waste production, and use of slurry rather than a solution that helps to prevent scaling and allowing the production of a by-product of commercial value. The disadvantage is the high energy consumption and initial cost due to the complexity of the process.

Other processes used on a limited basis, or currently under development include ammonia-based scrubbing, an aqueous carbonate process, and the Citrate process.

A number of wet scrubber designs have been used, including spray towers, venturis, plate towers, and mobile packed beds to promote maximum gas-liquid surface area and residence time. Since some of these designs have problem with FGD dependability and absorber efficiency, the trend is to use simple scrubbers such as spray towers instead of more complicated ones. The configuration of the tower may be vertical or horizontal, and flue gas can flow cocurrently, counter-currently, or cross-currently with respect to the liquid. The main drawback of spray towers however, is that they require a higher liquid-to-gas ratio requirement for equivalent SO_2 removal than other absorber designs.

Dry processes: These are further divided as semi-dry and dry processes. Semi-dry FGD technology includes lime or limestone spray drying; duct spray-dry, and circulating fluidized bed scrubbers. The dry processes include furnace sorbent injection and dry soda sorbents injection. In these processes, the dry waste products produced are generally easy to dispose than waste products from wet scrubbers. All dry FGD processes are once-through types.

In the spray-dry FGD process, lime or calcium oxide is used as a sorbent. Sometimes, hydrated lime [$Ca(OH)_2$] is also used. The hot flue gas is sent into a reactor vessel and slurry consisting of lime and recycled solids is sprayed into the absorber. The SO_2 and SO_3 in the flue gas is absorbed into the slurry and reacts with the lime and fly ash alkali to form calcium salts. The water that enters with the slurry is evaporated, which lowers the temperature and increases the moisture content of the scrubbed gas. The scrubbed gas then passes through a particulate control device downstream of the spray drier. Nearly 85–90% of SO_2 and total SO_3 and HCl can be removed by this process in coal-fired power plants using low- to medium-sulfur coals. Though the capital costs are low, the operational costs are high due to the continual usage of lime and the disposal of the solid by-product. Combining spray dry scrubbing with other FGD systems such as furnace or duct sorbent injection and particulate control technology such as a pulse-jet baghouse allows the use of limestone as the sorbent instead of the more costly lime (Soud 2000). Sulfur dioxide removal efficiencies can exceed 99% with such a combination.

Duct spray-dry process is similar to spray-dry process except that the slaked lime slurry is directly fed into the ductwork to remove SO_2. The reaction products and fly ash are captured downstream in the particulate removal device (Figure 6.4).

A portion of these solids is recycled and re-injected with the fresh sorbent. Duct spray drying (DSD) is a relatively simple retrofit process capable of moderate level of SO_2 removal. A mixture of calcium compounds in dry powdered form is the resulting by-product. Desulphurization to an extent of 50% to 75% can be achieved with this process which is being tried on pilot-scale now.

In 1990s, the application of FGD technology has become prominent and widely installed in Central and Eastern Europe and Asia. As of 1999, FGD systems are installed globally to control SO_2 emissions from more than 229,000 MW of generating capacity. Of this, around 87% consists of wet FGD technology, 11% of dry FGD technology, and the balance regenerable technology (Srivastava *et al.* 2000). Table 6.3 provides the power generation capacity using FGD installations in and outside the USA.

Circulating fluidized-bed (CFB) uses hydrated lime to remove SO_2, SO_3, and HCl from the flue gas; and CFB systems are both dry and semi-dry types; and the dry type is commercially more common capable of achieving SO_2 removal efficiencies of 93 to 97% at a Ca/S molar ratio of 1.2 to 1.5. This method is suited for complete removal

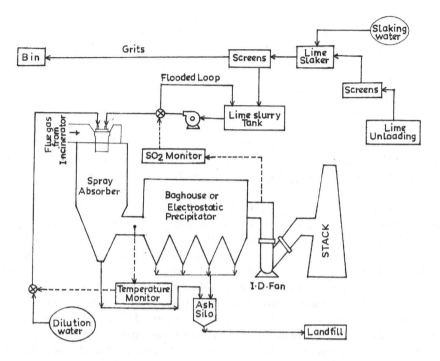

Figure 6.4 Dry scrubber system using slaked lime slurry (Redrawn from Hamon-Research Cottrell, Inc.).

Table 6.3 World's Power generation Capacity (MW) equipped with FGD technology.

Technology	USA	Outside USA	Total (MW)
Wet	82,859	116,374	199,223
Dry	14,386	11,008	25,394
Regenerable	2,798	2,059	4,857
Total FGD	100,043	129,441	229,484

(*Source:* Srivastava, Singer, and Jozewicz 2000)

of SO_3 and HCl. The operation costs are high due to regular lime consumption and disposal costs of the by-product which is a mixture of calcium compounds.

The Furnace sorbent injection (FSI) process consists of the injection of lime into wall- and tangentially-fired boilers to absorb SO_2. The optimum temperature for injecting limestone is approximately 1900 to 2100°F. The sulfur removal efficiency ranges between 30% and 90%. The main advantage (Radcliffe 1991) of this process is its simplicity; the dry reagent is injected directly into the flue gas in the furnace, without requiring a separate absorption vessel. The injection of lime in dry form allows for a less complex reagent handling system. As a result, operating and maintenance costs are low and the problems of plugging, scaling, and corrosion found in slurry handling are eliminated. Power requirements are also reduced since less equipment is used. This process is more suited to small plants where low-sulfur coal is used, though it can be applied to boilers burning low- to high-sulfur coals.

In another dry process, soda sorbents are injected into the flue gas duct downstream of the air heater to react with SO_2, SO_3, and HCl. The compounds that can be used as sorbents are sodium carbonate (Na_2CO_3), sodium bicarbonate ($NaHCO_3$), sodium sesquicarbonate ($NaHCO_3 \cdot Na_2CO_3 \cdot 2H_2O$), and natural materials, $NaHCO_3$ and $NaHCO_3 \cdot Na_2CO_3 \cdot 2H_2O$. Of these, sodium bicarbonate and sodium sesquicarbonate have been the most extensively tested in pilot, demonstration, and full-scale utility applications due to established success and commercial availability. Sodium bicarbonate has demonstrated to have the best sulfur capture in coal-fired boiler applications (Bland & Martin 1990); that is up to 70% efficiency for removal of SO_2, and about 90% efficiency for removal of HCl. The by-product is dry powdered mixture of sodium compound and fly ash. The operational costs are high due to the regular requirement of sodium bicarbonate, and disposal of waste. These dry injection processes have been developed to provide moderate SO_2 removal at low capital costs, and for easy retrofitting to existing power plants.

The selection of a FGD process is dependent on site-specifics, economics, and other criteria. Generally, dry scrubbing systems are considered more economical for power plants using low-sulfur coal, while wet based systems are selected for high-sulfur coal applications.

NO$_x$ control methods

Among the combustion generated atmospheric emissions, NO_x has received the greatest attention in the last three decades. The combustion of coal in the presence of nitrogen, from fuel or air, allow the formation of nitrogen oxides (Miller 2005; Suarez-Ruiz and Ward 2008; Franco & Diaz 2009). The nitrogen oxides consist of three chemical forms: nitric oxide (NO), nitrogen dioxide (NO_2) and nitrous oxide (N_2O). Three primary sources of NO_x formation are identified: (a) thermal NO_x generated by the high-temperature, above 2700°F (1480°C), reaction of oxygen and nitrogen from the combustion air; (b) prompt NO_x which is the fixation of atmospheric nitrogen by hydrocarbon fragments in the reducing atmosphere in the flame zone; and (c) fuel NO_x which is generated due to the oxidation of nitrogen in the coal. Since thermal NO_x is a product of a high-temperature process, it makes a small contribution to the overall NO_x emissions, while fuel NO_x is the primary contributor. Typically, more than 90% of the NO_x is in the form of NO, and approximately 20 to 300 ppm as N_2O, and the balance as NO_2 (Wu 2003).

NO_x formation is very complex and is influenced by several factors related to boiler specifics and coal properties (Mitchell 1998). NO_x emissions formed during the coal combustion are between 90 g/GJ$_{NVC}$ (fluidized-bed) to 540 g/GJ$_{NCV}$ (boiler) (Graus & Worrell 2007). Nitrogen oxides are the worst pollutants responsible for causing acid rain and ozone pollution (also called urban smog).

The nitrogen oxide control technologies are divided into two groups: the first group considers the reduction of NO_x produced in the primary combustion zone (referred to as combustion modification methods or *primary abatement* and control methods), and the second, the reduction of NO_x existing in the flue gas, called *flue gas treatment*

(Miller 2005; Srivastava *et al.* 2005). Several technologies that eliminate NO_x along with other pollutants such as SO_2, mercury, particulates, and air toxics are available (Srivastava *et al.* 2005).

Primary control/combustion modification methods are based on the chemistry of NO_x formation and focus on minimizing peak combustion temperatures and the residence time at peak temperature. These control measures are routinely included in newly built power generation plants and also retrofitted when reductions in NO_x emissions are required. These are briefly explained:

(a) *Low excess air (LEA) combustion*: Technique is based on reducing surplus O_2 at the burner flame where gas temperatures are highest resulting in lower peak flame temperature and less NO_x formation. LEA, if used, reduces the amount of air introduced into the boiler resulting in increased thermal efficiency provided stoichiometric requirements are fulfilled. LEA can be the easiest approach to implement for reducing NO_x emissions since no physical modifications are required except to the combustion controller.

(b) *Low NO_x Burners (LNB)*: NO_x control from these special burners is based on combustion modification. Precise mixing of fuel and air is used to keep the flame temperature low and to dissipate heat quickly through the use of low excess air, non-stoichiometric combustion, and combustion gas recirculation. LNBs are a mature, well proven technology for NO_x control in both wall fired and tangentially fired furnaces. They are commercially available with full performance guarantees and hence significantly used worldwide.

(c) *Staged Combustion (SC)*: Here, combustion take place in two areas; in the first combustion area fuel is fired with less than stoichiometric amount of air, creating a fuel-rich condition near the primary flame; and in the second area, the rest of the combustion air is introduced to complete the fuel consumption. The deficiency of oxygen in the first zone and the low temperature in the second zone both contribute to a reduction in NO_x formation. This is a *well established* technology for NO_x reduction in fossil fuel fired furnaces. The process is equally applicable to both wall and tangentially fired plants; indeed air staging is an inherent part of tangentially fired combustion systems.

(d) *Flue Gas Recirculation (FGR)*: FGR means recycling a portion of the combustion gases from the stack to the boiler wind box. 10 to 30% of the flue gas exhaust is recycled to the main combustion chamber from the stack gas stream and mixed with secondary air entering the wind box. This lowers the flame temperature, diluting the oxygen and reducing NO_x. When utilising FGR for NO_x control it is usual to extract the flue gas from downstream of the particulate control equipment. The recycled flue gas can be introduced to various locations in the furnace, and this can have a considerable impact on the achievable NO_x reduction. The use of FGR for any purpose other than reheat control can have a significant impact on the boiler operation. This technology is primarily for large coal, oil or gas boilers.

(e) *NO_x Reburning*: Reburning is a NO_x staged combustion control technique that suppresses the formation of NO_x in burners and then enables additional NO_x reduction in the furnace area of the boiler. Reburning can be applied to any type of boiler, provided the gas stream residence time is long enough to allow the reburn fuel to burn completely. NO_x reductions of 50% to 70% are possible. Reburning can be used in association with a number of other NO_x control techniques. NO_x reburning is also called reburn, in-furnace NO_x reduction, and staged fuel injection.

The second group, *flue gas treatment technologies*, is post-combustion processes to convert NO_x to molecular nitrogen or nitrates. The two technologies that have been commercially available are Selective catalytic reduction (SCR), and Selective non-catalytic reduction (SNCR). These can be employed separately as well as in combination.

Selective catalytic reduction (SCR): The selective catalytic reduction involves mixing the exhaust gas with a reagent, typically anhydrous ammonia and passing the homogenous mixture over a bed of catalyst designed for the reaction to occur at the gas stream temperature. Ammonia chemisorbs onto the active sites on the catalyst which, in turn, promotes a reaction between NH_3, NO_x and the excess oxygen in the exhaust stream, forming nitrogen and water vapor. In the reactor, the catalyst is placed in the form of parallel plates or honeycomb structures as shown in Figure 6.5. The installed catalyst reacts with the NH_3 blown into it, allowing NO_x to break down. The SCR process is chemically expressed as:

$$4NO + 4NH_3 + O_2 \rightarrow 4N_2 + 6H_2O \qquad (6.4)$$

$$2NO_2 + 4NH_3 + O_2 \rightarrow 3N_2 + 6H_2O \qquad (6.5)$$

$$NO + NO_2 + 2NH_3 \rightarrow 2N_2 + 3H_2O \qquad (6.6)$$

Traditionally, the catalyst is shaped as a monolithic honeycomb structure with cell density varying from 30 to 200 cells per sq in. The monoliths provide for a low pressure drop at linear velocity in excess of 10 ft per second, which translates into a compact catalyst bed design. The turbine exhaust gas must contain a minimum amount of oxygen and be within a particular temperature range for a proper operation of the SCR system. The temperature range is influenced by the catalyst, which is

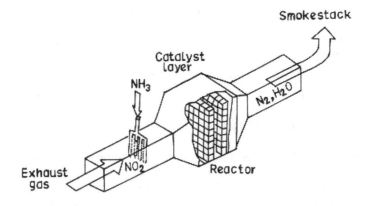

Figure 6.5 Schematic of Selective Contact Reduction (SCR) Method.

Table 6.4 Matching the catalyst to the optimum temperatures.

Temperature	Catalyst type
Low: 250–660°F (120–350°C)	Porous extrudiates in packed bed reactor.
Medium: 500–800°F (265–425°C)	Vanadia/titania catalyst on high-density honeycomb structure.
High: 650–1100°F (345–590°C)	Zeolite catalyst on ceramic substrate.

typically made of noble metals or base metal oxides, or zeolite-based material. For many industrial applications, vanadia/titania catalysts are used for which the typical operating temperatures range between 500 and 800°F (265–425°C).

Low temperature gases can be treated using heat recovery systems, but advances in catalyst technology have created catalysts that can directly treat operating temperatures as low as 250°F (120°C). For high temperature applications up to 1100°F (595°C), zeolite catalysts provide for high performance and stability. Proper matching of the catalyst to the process temperature is essential for an optimized performance and low operation costs (Table 6.4).

For example, in the case of vanadia/titania catalyst, it is important to keep the exhaust gas temperature within the ranges specified in the Table. If it drops, the reaction efficiency gets too low and increased amounts of NO_x and ammonia will be released out of the stack. If the reaction temperature gets too high, the catalyst may begin to decompose.

Since turbine exhaust gas is generally above 1000°F (540°C), the heat recovery steam generators cool the exhaust gases before they reach the catalyst. This extracted heat energy is used to generate steam for use in industrial processes or to turn a steam turbine. In simple-cycle power plants where no heat recovery is performed, high temperature catalysts that can operate at temperatures up to 1050°F (565°C) are the alternative.

Oxidation of SO_2 to sulfur trioxide (SO_3) also occurs on the catalyst, i.e.,

$$SO_2 + \frac{1}{2}O_2 \rightarrow SO_3 \tag{6.7}$$

At a temperature lower than the optimum temperature (350°C), SO_3 in the exhaust gas reacts with ammonia (NH_3), producing ammonium hydrogen sulfate (NH_4HSO_4) that covers the surface of the catalyst, thereby reducing the ability to remove NO_x. The chemical reactions forming ammonium sulfate ($(NH_4)_2SO_4$) and bisulfate ($(NH_4)HSO_4$):

$$2NH_3 + SO_3 + H_2O \rightarrow (NH_4)_2SO_4 \tag{6.8}$$

$$NH_3 + SO_3 + H_2O \rightarrow NH_4HSO_4 \tag{6.9}$$

The formation of these salts is highly dependent upon the concentration of each constituent; therefore, each component is a key design parameter for the system.

At a temperature higher than 350°C, the NH_4HSO_4 decomposes, improving the removal of NO_x despite the SO_3 concentration. At a temperature above 400°C, NH_3 is oxidized decreasing its volume, thereby reducing its ability to remove NO_x.

The process is also designed to limit ammonia leaks from the reactor to 5 ppm or less. If a significant quantity should leak, it will react with the SO_3 in the exhaust gas, producing NH_4HSO_4, which clogs the piping when separated out by an air pre-heater. The NO_x removal efficiency is around 80–90% for pulverized coal-fired power plants.

Measures to equally disperse and mix the NH_3 with the exhaust gas as well as to create greater uniformity of the exhaust gas flows to cope with growing boiler sizes have been developed. These measures which ensure best possible SCR performance include placing a current plate, called a 'guide vane' at the gas inlet, or dividing the gas inlet into grids, each to be equipped with an NH_3 injection nozzle. Anhydrous ammonia, though toxic and hazardous, has been commonly used as reagent in most currently-working SCR systems worldwide (Wu 2002). Recently, urea-based reagents such as dry urea, molten urea, or urea solution, are being developed as alternatives to anhydrous ammonia that needs very careful storage and handling.

Three types of configurations for SCR are known for coal-fired power plants: high-dust, low-dust and tail-end systems, of which the high-dust configuration is more popular. Each configuration has its advantages and disadvantages in terms of operation and equipment needs. Several issues including coal characteristics, catalyst and reagent selections, process conditions, ammonia injection, catalyst cleaning and regeneration, low-load operation, and process optimization must be considered in the design and operation of SCR systems (Wu 2002). It is more suitable for low sulfur coals due to the corrosive effects of sulfuric acid formed by side reactions.

The efficiency of NO_x removal depends on its volume, density, operating temperature and type of SCR catalyst. Several highly active and selective NO_x reduction catalysts with proven performance and longevity in different industrial applications are commercially available. For high dust applications, catalysts have been developed that are supported on stacked ceramic plates, providing large channel size and low dust retention. The cleaning service for these catalysts is very infrequent. The reagent injection systems need to be designed to ensure complete mixing of the reagent with the flue gas for aiding the conversion process. Excess residual ammonia (ammonia slip) is environmentally harmful and its level should be minimized. For many sources, compliance must be proven at very low NO_x and ammonia emission rates which mean that the methods selected for compliance are as important as the selection of the abatement technology. SCR was patented by Englehard Corporation, USA in 1957 (DOE 1997). This technology can achieve NO_x reductions around 85 to 95% and has been widely used for over 30 years in commercial applications in Western Europe especially Germany, Japan, USA and China (McIlvaine *et al.* 2003).

Progress: Selective catalytic reduction provides for the highest NO_x removal efficiency for many industrial applications, including ceramic and glass manufacturing. However, there are concerns related to secondary emissions of ammonia reagent (ammonia slip), the handling and storage of reagents, process control at variable flow rates, NO_x concentration and temperature, and high costs of equipment and control system. Significant advances have been made during the past years to address these concerns and improve the catalyst, the reagent delivery and control system, and the operational costs associated with the SCR technology.

Selective Non-catalytic Reduction (SNCR) is another post-combustion technology to remove nitrogen oxides from the boilers by the injection of ammonia or urea reagents into the flue gas without a catalyst. The principle of reacting NO_x with ammonia in

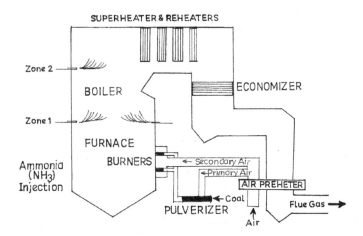

Figure 6.6 The process works by injecting a reagent (ammonia or urea) into the radiant and convection regions of a coal-fired boiler (Redrawn from Hamon Research-Cottrell, Inc.).

the presence of oxygen, utilizing gas-phase free-radical reactions instead of a catalyst promoting reactions is used. The critical factor to the SNCR process is optimization of reagent injection with the flue gas within a specific temperature range to achieve high NO_x reduction. For urea, this range is around 1800–2100°F (982–1149°C), and for ammonia, it is slightly lower at 1600–1800°F (871–982°C). The temperature range can be dropped to a lower range by the co-injection of hydrogen with ammonia. Temperature mapping using high velocity temperature (HVT) probes, at different operating loads is critical to determine the locations of the desired temperature windows (Figure 6.6).

The HVT mapping and system geometry is used along with computer combustion CFD modeling to characterize the system. The temperatures and gas streamlines are then predicted for a range of loads. In addition to the temperature mapping and boiler/furnace modeling, a computer chemical kinetic reaction model is also utilized. The latter provides a solution for the competing gas phase reactions and predicts NO_x removal. To achieve proper reduction levels, spray patterns and dispersion are also crucial. Typically, there are two types of reagent injectors placed in the system to achieve reagent coverage: Retractable multiple port lances such as steam, water or glycol cooled are used to inject the reagent over the width of the system. The SNCR reaction requires ammonia as the primary reducing agent. The ammonia may be provided based on anhydrous, aqueous or U_2A product gas ammonia or from urea solution injection. The reagent is piped to distribution and mixing modules for metering to each of the injectors. The advantages of SNCR are: not requiring a catalyst, lower installation costs, and no-waste generation. SNCR performance is exclusive to each application and NO_x reduction levels ranging from 30 to 70% have been reported. This technology is mainly used at small commercial boilers and waste incinerators. More NH_3 is also leaked than with the SCR method, requiring measures to cope with NH_4HSO_4 precipitation in the event of high SO_3 concentrations in the exhaust gas. SNCR can be combined with other upstream and downstream NO_x control technologies. Recently, for more effective control, hybrid systems have been introduced (Nordstrand et al. 2008). By combining SCR and SCNR technologies, improvements can be seen in chemical and catalyst use, making the hybrid combination often more flexible and effective. A hybrid SNCR/SCR system, though uses a highly expensive reagent, balances the costs over the life cycle for a specific NO_x reduction level, and provides improvements in reagent utilization, and increases overall NO_x reduction (Wu 2002). However, these hybrid systems have only been in demonstration stage so far showing NO_x reductions as high as 60 to 70%.

Zero ammonia technology (ZAT): Also known as 'NO$_x$ trap', it is the latest technology that does not require the injection of ammonia or urea. ZAT is a catalytic-based system that converts all of the NO_x into NO_2 (i.e., it oxidizes the NO) and adsorbs the NO_2 onto the catalyst. Portions of the catalyst are isolated from the exhaust stream, and the adsorbed NO_2 is reduced to N_2 using diluted hydrogen, or some sort of hydrogen reagent gas. The N_2 is then desorbed from the catalyst.

Particulate matter control methods

Several combustion products are formed during the coal combustion due to the presence of mineral impurities in the coal stock. Fly ash is the main potential source of contamination that requires control (Akar et al. 2009) as it contains fine particles (PM_{10}) which may pass into the atmosphere by the dust collection devices. Particulate pollution is due to high concentration and surface association of a few trace elements in the composition. The different trace elements originating from various minerals and macerals in coal behave differently during coal combustion (Suarez-Ruiz & Ward 2008), and can be classified as low volatile, volatile, and highly volatile. During combustion, the low volatile trace elements tend to distribute between fly ash and bottom ash or stay in bottom ash. The volatile and highly volatile trace elements, on the other hand, are vaporized in the furnace and may be incorporated in any slagging deposit (vitreous mass residue left by the smelting) or fouling (mineral matter). They are mostly found to condense onto the existing fine fly ash or are emitted in the vapour phase (Xu et al. 2003; Suarez-Ruiz & Ward 2008; Vejahati et al. 2009).

The particulate emissions generated in coal combustion are grouped as PM_1, $PM_{2.5}$ and PM_{10} depending on the particle size. The first two ultra-fine particulates may remain air-suspended for longer periods and adversely affect environment and human health (Bhanarkar et al. 2008; Breeze 2005; Miller 2005). The finest fraction of the particulates originates mostly from the ash-forming group vaporized during combustion; the rest, larger than one micron, generally formed by the mineral impurities in coal, remain as the residual ash (Senior et al. 2000b; Ohlstrom et al. 2006).

A number of technologies have been developed to control particulate emissions from coal-fired power plants which are widely deployed in both developed and developing countries. They include electrostatic precipitators (ESPs), fabric filters or baghouses, wet particulate scrubbers (wet flue gas desulfurization systems), cyclones and hot gas filtration systems (Soud & Mitchell 1997). Of these, ESPs and fabric filters are currently the technologies of choice, capable of cleaning large volumes of flue gas with very high collection efficiencies, and in removing fine particles.

Except volatile elements like mercury and selenium, most of the trace metals are captured with primary particles (Sondreal 2006). The fine particulates ($PM_{1.0}$ and $PM_{2.5}$) are difficult to control due to their small size and small fraction in the total mass; but it is vital to capture them because they are the most unsafe pollutants from the point of view of human health (Ohlstrom *et al.* 2006). The technologies used to control emissions of particulate matter in power plants include electrostatic precipitators (ESP), cyclones, fabric filters (baghouses), Wet scrubbers, and hot gas filtration systems. The design aspects and details can be found in many books.

Mercury emissions control

Mercury (Hg), a dangerous toxin pollutant, is present in sulfide minerals of coal; it is also present as organically bound to coal macerals (Pavlish *et al.* 2003; Sondreal *et al.* 2004). During combustion of coal, mercury passes to the flue gas as a mixture of its three different chemical states: Hg^0, Hg^{2+}, and particulate-bound mercury, Hg^p (Lee *et al.* 2006). The concentration of Hg pollution in the flue gas varies in the range, 1–$20\,\mu g/m^3$ (Yang *et al.* 2007). Of the three Hg pollutants, Hg^0 is hard to capture by emission control equipment, as it is highly volatile and insoluble. Hence, it is completely emitted into the atmosphere and travels over long distances due to long life-time in the atmosphere creating pollution globally (Wang *et al.* 2008; Pavlish *et al.* 2003, 2009; Sondreal *et al.* 2004). Hg^{2+} is soluble and tends to form surface associations with particulate matter. Both Hg^{2+} and Hg^p have short lifetimes of few days in the atmosphere. It is possible to efficiently control both the pollutants using conventional emission control equipment (Senior *et al.* 2000a; Pavlish *et al.* 2003; Sondreal *et al.* 2004; Wang *et al.* 2008). There is no single technology for reducing Hg in the flue gas, and combination of existing emission control devices can help to remove Hg to certain extent. The type of coal strongly influences the rate of Hg removal. The electrostatic precipitator (ESP), fabric filter baghouse and wet flue gas desulfurization systems can remove part of particulate-bound and oxidized forms of Hg from the flue gas with varied range of efficiency. The scrubber systems need additional expensive equipment. Fabric filters help to achieve highest reduction rate. Cold-side electrostatic precipitators (ESP) are more effective than hot-side ESPs; and the cost of these systems is also moderate to high and requires additional space for installation. This air pollution control equipment is not uniformly efficient for use in combustion of all types of coal; they are highly efficient in the control of particulate emissions in the combustion of bituminous coals, less efficient in sub-bituminous coal and almost inefficient in lignite coal (Pavlish *et al.* 2003; Kolker *et al.* 2006).

Sorbent injection technology is considered as highly efficient in removing Hg^0 and Hg^{2+} from the flue gas (Yang *et al.* 2007). Sorbents such as activated carbon, petroleum coke, zeolites, fly ash, chemically treated carbons and carbon substitutes so on are injected into the upstream of either an ESP or a fabric filter baghouse to control Hg emissions.

Waste control

The combustion of coal generates waste consisting primarily of non-combustible mineral matter along with a small amount of unreacted carbon. The production of this waste can be minimized by coal cleaning prior to combustion. This represents a cost-effective method of providing high quality coal, while helping to reduce power station waste and increasing efficiencies. Waste can be further minimized through the use of high efficiency coal combustion technologies. There is increasing awareness of the opportunities to reprocess power station waste into valuable materials for use primarily in the construction and civil engineering industry. A wide variety of uses have been developed for coal waste including boiler slag for road surfacing, fluidized bed combustion waste as an agricultural lime and the addition of fly ash to cement manufacturing (www.worldcoal.org/coal-the-environment/coal-use-the-environment/).

Multi-Pollutant control: Several control methods for removal of SO_2, NO_x, particulates and mercury individually applicable to coal-fired power plants have been seen so far. In the context of mandated reductions in the pollutants (including CO_2), an integrated reduction approach might prove to be the best one to implement. There have been several studies to demonstrate a multi-pollutant control process that can achieve more stringent pollution controls by reducing emissions of NO_x, SO_2, mercury, acid gases, and particulate matter cost-effectively. The USEPA has released a report, 'Multi-pollutant Emission Control Technology Options for Coal-Fired Power Plants' (EPA-600/R-05/034, 2005) which is a critical analysis of various existing and novel control technologies designed to achieve multi-emission reductions. The report is confined to technologies with a certain level of maturity; nonetheless, rapid technological progress is expected in the development and commercialization of several more multi-emission control processes in future. Some of this research activity will now be briefly outlined.

Electro-catalytic oxidation technology (ECO), a promising multi-pollutant removal process, consists of three steps: first, the effluent is led through a gas reactor, which oxidizes the pollutants to higher oxides. Then the effluent goes through a wet electrostatic precipitator system for collecting products of the oxidation process and other fine particles. Finally, this effluent is treated to recover valuable by-products, concentrated sulfuric and nitric acids. Pilot tests have shown strong reductions (90%) in NO_x, SO_2, $PM_{2.5}$, and mercury. Emissions of $PM_{2.5}$ were reduced by over 96% in the tests, and the mercury was undetectable in the waste stream (EPA 2005, & http://www.physics.ohio-state.edu/~wilkins /energy/ Companion/E14.1.pdf.xpdf).

Fan and co-workers (Gupta *et al.* 2007) have demonstrated the OSCAR (Ohio State Carbonation Ash Reactivation) process that involves the reactivity of two novel calcium-based sorbents toward the capture of sulfur and trace heavy metals such as arsenic, selenium, and mercury in the furnace sorbent injection (FSI) mode on a slipstream of a bituminous coal-fired stoker boiler. The sorbents are synthesized by bubbling CO_2 to precipitate calcium carbonate from both the unreacted calcium present in the lime spray dryer ash, and calcium hydroxide slurry that contained a negatively charged dispersant. The heterogeneous reaction between these sorbents and SO_2 occurred under entrained flow conditions by injecting fine sorbent powders into the flue gas slipstream. The reacted sorbents were captured either in a hot cyclone ($\sim$650°C) or in the relatively cooler downstream baghouse ($\sim$230°C). The baghouse samples indicated $\sim$90% toward sulfation and captured arsenic, selenium and mercury to 800 ppmw, 175 ppmw and 3.6 ppmw, respectively.

LoTOx is a gas phase 'low temperature oxidation' process, which involves injection of ozone in the flue gas upstream of a wet FGD to oxidize NO_x to higher oxides of nitrogen (N_2O_5) and mercury to HgO. Subsequently these compounds are removed in a wet FGD as they are water-soluble. Studies have confirmed NO_x reduction efficiencies of 70–95%, and mercury reduction up to 90%, depending on the rank of coal, residence time, and operating temperature. This is demonstrated up to a capacity of 25 MW (Gross *et al.* 2001; Gross 2002).

Levendis and co-workers evaluated an integrated approach for reducing NO_x, SO_2, and particulate emissions concurrently from power plants (Ergut *et al.* 2003) using a dry-sorbent injection for SO_2 reduction, coal injection for NO_x reduction and a ceramic honeycomb filter for particulate capture. Mounted in an elevated temperature region, the filter retains sorbent particles for extended periods of time allowing their utilization until it is regenerated. The chosen low-cost sorbents – calcium carbonate ($CaCO_3$); calcium hydroxide [$Ca(OH)_2$]; calcium oxide (CaO); and sodium bicarbonate ($NaHCO_3$) – for performance evaluation were powdered and blended with the three main types of pulverized coal to achieve NO_x reduction. The sorbents were injected in a simulated effluent gas at a gas temperature of 1150°C upstream of the ceramic filter, kept at 600°C or 800°C. The molar Ca/S ratio was in the range of 0.5–5, and the fuel-to-air equivalence ratio was around 2 for all the studies. Numerical simulations using the modified 'pore tree' mathematical model were done. For comparison, studies using the costly porous sorbent, calcium formate, Ca $(COOH)_2$ were also performed. At Ca/S ratio of 2, 80% SO_2 reduction (i.e., 40% calcium utilization) with calcium formate, and about 40% SO_2 reduction (that is, 20% Ca utilization) with the less porous and cheaper sorbents was achieved. Performance of sodium bicarbonate was better, 50% SO_2 reduction at Na/S ratio of 2, with sodium utilization as high as 50% when injected alone. Coal-mixed sodium bicarbonate achieved much higher reduction, > 70%. With higher Ca/S and Na/S values, SO_2 reduction efficiencies improved but not the sorbent utilization. NO_x removal efficiencies were around 45–55% at fuel-air equivalence ratio of ~2. Particulate removal efficiencies by the filter were in the range of 97–99%.

Another technology (airborne process) that combines the use of dry sodium bicarbonate injection, coupled with enhanced wet sodium bicarbonate scrubbing was also developed to provide for the control of SO_x, NO_x, mercury, and other heavy metals. The experimental and analytical results of a lab and pilot scale 0.3 MW coal-fired combustion test facility and the progression to an integrated 5 MW facility are discussed (Mortson and Owens II 2012; Johnson *et al.* 2012).

Levendis and his group also studied calcium magnesium acetate (CMA) sorbent, for the concurrent removal of SO_2 and NO_x in oxygen-lean atmospheres (Steciak *et al.* 1995; Shukerow *et al.* 1996). When injected into a stimulated hot environment in the boiler, the salts calcined and formed highly porous 'popcorn'-like cenospheres. Dry-injected CMA particles at a ratio of 2, residence time of 1 s and bulk equivalence ratio of 1.3 removed over 90% of SO_2 and NO_x at gas temperatures ≥ 950°C. Fine mists of CMA sprayed in the furnace at temperatures of 850–1050°C removed 90% of SO_2 at a molar ratio of 1, about half of the amount used in the dry injection experiments to achieve a similar SO_2 reduction. As for NO_x removal, the same reduction efficiency was achieved as with dry injection (25–30%). While SO_2 and NO_x emissions could be substantially controlled, the vapor phase emissions comprising mercury, arsenic, and selenium, and particulate emissions within the respirable range that cover heavy metals, polyaromatic hydrocarbons, and volatile organics could not be controlled. The utility of CMA sorbent on mercury vapor capture was studied using CMA ash after combustion with the gases (SO_2 and NO_x). The effect of calcium magnesium carbonate (CMC) currently used in industrial furnaces was also investigated utilizing the same combustion conditions as CMA. Results showed that CMA ash combusted with the gases performed better at higher mercury removal (40%) while CMC yielded only 4% removal. Further studies using a model particulate ($FeSO_4$) showed CMA ash was able to capture more of the model air particulates than the CMC ash after combustion, signifying CMA's efficacy in particulate toxin capture. The economics of the integrated approach to control SO_2, NO_x, HCl and particulate emissions was also studied by Levendis and co-workers (Shemwell *et al.* 2002).

REFERENCES

Akar, G., Arslan, V., Ertem, M.E., & Ipekoglu, U. (2009): Relationship between ash fusion temperatures and coal mineral matter in some Turkish coal ashes, *Asian J. Chem.*, 21, 2105–2109.

Balat, M. (2008a): The future of clean coal, *In: Future Energy: Improved, Sustainable and Clean Options for our Planet*, Letcher, T.M. (ed.), Amsterdam: Elsevier.

Balat, M. (2008b): Coal-fired Power Generation: Proven technologies and pollution control systems, *Energ. Source*, Part A, 30, 132–140.

Bhanarkar, A.D., Gavane, A.G., Tajne, D.S., Tamhane, S.M., & Nema, P. (2008): Composition and size distribution of particulates emissions from a coal-fired power plant in India, *Fuel*, 87, 2095–2101.

Bland, V.V., & Martin, C.E. (1990): *Full-Scale Demonstration of Additives for NO_2 Reduction with Dry Sodium Desulfurization*, Electric Power Research Institute, EPRI GS-6852, June 1990.

Breeze, P. (2005): *Power Generation Technologies*, Oxford: Elsevier

Burnard, K. & Bhattacharya, S. (2011): *Power generation from Coal – Ongoing Developments and Outlook, Information paper*, OECD/IEA, Paris, October 2011.

Dwari, R.K. & Rao, K.H. (2007): Dry benefaction of Coal – A Review, *Miner. Process Extract Metallurgy Rev.*, 28, 177–234.

EPA, Environmental Protection Agency (1988): *Second Review of New Source Performance Standards for Coal Preparation Plants*, EPA-450/3-88-001, February 1988, US EPA, Research Triangle Park, NC.

EPA (1989): *Estimating Air Toxic Emissions from Coal and Oil Combustion Sources*, EPA-450/2-89-001, April 1989, US EPA, Research Triangle Park, NC.

EPA (2009): *Latest Findings on National Air Quality Status and Trends through 2006*, Office of Air Quality Planning and Standards, U.S. Government Printing Office, 2009.

EPA (2008): *Acid Rain and Related Programs: 2008 Emission, Compliance, and Market Analysis*, Office of Air Quality Planning and Standards, U.S. Government Printing Office, January 2008.

Environmental Engineering, Inc. (1971): *Background Information for Establishment of National Standards of Performance for New Sources*: Coal Cleaning Industry, EPA Contract CPA-70-142, July 1971, EEI, Gainesville, FL.

Elliot, T.C. (1989): (ed.) *Standard Handbook of Power Plant Engineering*, McGraw-Hill.

Environmental Controls: Understanding and Controlling NO_x Emissions, Feb. 1, 2002 at www.ceramicindutry.com/articles/environmental-controls-understanding-and-controlling-nox-emissions, retrieved April 4, 2012.

Ergut, A., Levendis, Y.A., & Simons, G.A. (2003): High temperature injection of sorbent-coal blends upstream of a ceramic filter for SO_2, NO_x, and particulate pollutant reductions, Combustion Science and Technology, 175, 597–617.

Franco, A., & Diaz, A.R. (2009): The future challenges for 'Clean Coal technologies': Joining efficiency increase and pollutant emission control, *Energy*, 34, 348–354.

Ghosh, S.R. (2007): Global Coal benefaction scenario and Economics of using Washed Coal, Workshop on Coal benefaction and Utilization of Rejects: Initiatives, Policies and Practice, Ranchi, India, 22–24 August 2007.

Ghosh, T.K. & Prelas, M.A. (2009): *Energy Resources & Systems, Vol-1: Fundamentals & Non-renewable Resources*, Springer 2009, ISBN:978-90-481-2382-7.

Gupta, H., Thomas, T.J., Park, A.A., Iyer, M.V., Gupta, P., Agnihotri, R., Jadhav, R.A., Walker, H.W., Weavers, L.K., Butalia, T. & Fan, L.S. (2007): Pilot scale demonstration of OSCAR process for high temperature multi-pollutant control of coal combustion flue gas, using carbonated fly-ash and mesoporous calcium carbonate, *Ind. Eng. Chem. Res.*, 46, 5051–5060.

Graus, W.H.J., & Worrell, E. (2007): Effects of SO_2 and NO_x control on energy-efficiency power generation, *Energy Policy*, 35, 3898–3908.

Gross, W.L. (2002): Multi-pollutant control system installation at Medical college of Ohio, technical transfer paper, June 28, 2002.

Gross, W.L., Lutwen, R.C., Ferre, R., Suchak, N., & Hwang, S.C. (2001): report of the stand-up of a multi-pollutant removal system for NO_x, SO_x, and particulate control using $LoTO_x$ on a 25 MW coal-fired boiler, Presented at Power-Gen 2001, Las Vegas, NV, December 12, 2001.

IEA (International Energy Agency) (2008): *Clean coal technologies – Accelarating Commercial and Policy drivers for Deployment*, Paris: OECD/IEA.

IEA (2010a): World Energy Outlook 2010, OECD/IEA, Paris

IEA (2010b): Energy technology Perspectives 2010, OECD/IEA, Paris.

Johnson, D.W., Ehrenschwender, M.S., & Seidman, L. (2012): The Airborne process – Advancement in multi-pollutant emissions control technology and by-product utilization, paper#131, Power plant Air pollutant Control Mega Symposium, Baltimore, MD, Aug. 20–23, 2012.

Kolker, A., Senior, C.L., & Quick, J.C. (2006): Mercury in coal and the impact of coal quality on mercury emissions from combustion systems, *Appl. Geochem.* 21, 1821–1836.

Lee, S.H., Rhim, Y.J., Cho, S.P., & Baek, J.I. (2006): Carbon-based novel sorbent for removing gas-phase mercury, *Fuel*, 85, 219–226.

McIlvaine, R.W., Weiler, H., & Ellison, W. (2003): SCR Operating Experience of German Power plant Owners as Applied to Challenging, U.S., High-Sulfur Service, in: *Proc. of the EPRI-DOE-EPA Combined Power Plant Air Pollution Control MEGA Symposium*, 2003.

Miller, B.G. (2005): *Coal Energy Systems*, London: Elsevier.

Mitchell, S.C. (1998): NO_x in Pulverized Coal Combustion, IEA Coal Research

Mortson, M.E. & Owens II, S.C. (2012): Multi-component control with the Airborne process, paper #59, Power plant Air pollutant Control Mega Symposium, Baltimore, MD, August 20–23, 2012; at www.aitbornecleanenergy.com/papers/59.pdf (accessed 05/22/2013).

Nordstrand, D., Duong, D.N.B., & Miller, B.G. (2008): Post-combustion Emissions control, In: *Combustion Engineering Issues for Solid Fuel systems*, Miller, B.G., & Tillman, D. (eds), London: Elsevier.

Ohlstrom, M., Jokiniemi, J., Hokkinen, J., Makkonen, P., & Tissari, J. (2006): *Combating Particulate Emissions in Energy Generation and Industry*, Herring, P. (ed.) Helsinki: TEKES.

Omer, A.M. (2008): Energy, Environment and Sustainable development, *Renew. Sust. Energ. Rev.*, 12, 2265–2300.

Pavlish, J.H., Sondreal, E.A., Mann, M.D., Olsen, E.S., Galbreath, K.C., Laudal, D.L., & Benson, S.A. (2003): Status review of mercury control options for coal-fired power plants, *Fuel Process. Technol.* 82, 89–165.

Radcliffe, P.T., (1991): Economic Evaluation of Flue Gas Desulfurization Systems, Electric Power Research Institute.

Resources Research Inc. (April 1970): *Air Pollutant Emissions Factors*, Contract CPA-22-69-119, RRI, Reston, VA.

Satyamurthy, M. (2007): Coal benefaction Technology 2007: Initiatives, Policies and Practices, *Workshop on Coal Benefaction: ...,* Ranchi, India, 22–24, August 2007; available at http://fossil.energy.gov/international/International_Partner/August_2007_CWG_Meeting.html.

Sen, S. (2010): An Overview of Clean Coal technologies I: Pre-combustion and Post-combustion Emission Control, *Energy Sources, Part B,* 5, 261–271.

Senior, C.L., Bool, III, L.E., Srinivasachar, S., Pease, B.R., & Porle, K. (2000a): Pilot scale study of trace element vaporization and condensation during combustion of a pulverized sub-bituminous coal, *Fuel Process. Technol.* 63, 143–165.

Senior, C.L., Sarofim, A.F., Zeng, T., Helble, J.J., & Mamani-Paco, R. (2000b): Gas phase transformations of mercury in coal-fired power plants, *Fuel Process. Technol.* 63, 197–213.

Shemwell, B., Ergut, A., & Levendis, Y.A. (2002): Economics of an Integrated Approach to Control SO₂, NOₓ, HCl and Particulate Emissions from Power plants, Journal of the Air & Waste Management Association, 52, 521–534.

Shuckerow, J.I., Steciak, J.A., Wise, D.L., Levendis, Y.A., Simons, G.A., Gresser, J.D., Gutoff, E.B., & Livengood, C.D. (1996): Control of air toxins particulates and vapor emissions after coal combustion utilizing calcium magnesium acetate, Resources, Conservation and Recycling, 16, 15–69.

Sondreal, E.A., Benson, S.A., & Pavlish, J.H. (2006): Status of research on Air quality: mercury, trace elements and particulate matter, *Fuel process. Technol.* 65–66, 5–19.

Sondreal, E.A., Benson, S.A., Pavlish, J.H., & Ralston, N.V.C. (2004): An overview of air quality III: Mercury, trace elements and particulate matter, *Fuel Process. Technol.* 85, 425–440.

Soud, H.N., & Mitchell, S.C. (1997): *Particulate Control Handbook for Coal-Fired Plants,* IEA Coal Research, 1997.

Soud, H.N. (2000): *Developments in FGD,* IEA Coal Research, 2000.

Srivastava, R.K., Hall, R.E., Khan, S., Culligan, K., & Lani, B.W. (2005): Nitrogen oxide emission control options for coal-fired electric boilers, *J. Air Waste Manage.* 55, 1367–1388.

Srivastava, R.K., Singer, C., & Jozewicz, W. (2000): SO₂ Scrubbing Technologies: A Review, in: *Proc. of AWMA 2000 Annual Conference and Exhibition,* 2000.

Steciak, J., Levendis, Y.A., Wise, D.L., & Simons, G.A. (1995a): Dual SO₂-NOₓ concentration reduction by calcium salts and carboxylic acids, J. Environmental Engineering, 121(8), 595–604.

Steciak, J., Levendis, Y.A., & Wise, D.L. (1995b): The effectiveness of calcium magnesium acetate as a dual SO₂-NOₓ emission control agent, AIChE Journal, 41(3), 712–722.

Suarez-Ruiz, I., & Ward, C.R. (2009): Coal combustion, *In: Applied Coal Petrology: The Role of Retrology in Coal utilization,* Suarez-Ruiz, I and Crelling, J.C. (eds). London: Elsevier.

USDOE (1997): Clean Coal Technology, Control of Nitrogen Oxide Emissions: Selective Catalytic Reduction (SCR), Topical Report No. 9, USDOE, July 1997.

USDOE CCPI (2012): Clean Coal power Initiative Round 1 Demonstration projects, Clean Coal Technology, Topical Report No. 27, DOE: OFE, NETL, June 2012.

USEIA (2008): *Electric Power Annual 2007,* US Department of Energy, Office of Coal, Nuclear, Electric and Alternate Fuels, US Government Printing Office, January 21, 2008.

USEIA (2009): International Energy Outlook 2009, US DOE, Washington, D.C.

Vejahati, F, Xu, Z, & Gupta, R (2009): Trace elements in coal: Associations with coal and minerals and their behavior during coal utilization – A review, *Fuel,* 89, 904–911.

Wang, Y., Duan, Y., Yang, L., Jiang, Y., Wu, C., Wang, Q., & Yang, X. (2008): Comparison of mercury removal characteristic between fabric filter and electrostatic precipitators of coal-fired power plants, *J. Fuel Chem. Technol.* 36, 23–29.

Wikipedia, Free encyclopedia: Flue gas desulfurization at http://en.wikipedia/wiki /fuel_gas_desulfurization.

World Coal Institute (WCI) (2007): *The Coal Resource – A Comprehensive Overview of Coal*, London: WCI.

Wu, Z. (2002): NO_x Control for Pulverized Coal Fired Power Stations, *IEA Coal Research*, London.

Wu, Z. (2003): Understanding fluidized combustion, *IEA Coal Research*, London.

Xu, M., Yan, R., Zheng, C., Qiao, Y., Han, J., & Sheng, C. (2003): Status of Trace element emission in a coal combustion process: A Review, *Fuel process. Technol.* 85, 215–237.

Yang, H., Xua, Z., Fan, M., Bland, A.E., & Judkins, R.R. (2007): Adsorbents for capturing mercury in coal-fired boiler flue gas. *J. Hazard. Materials*, 146, 1–11.

NO_x Removal: www.durrenvironmental.com

Chapter 7

Coal-based electricity generation

Electricity generation using coal started towards the end of the 19th century. The earliest power stations had an efficiency of around 1%, and needed 12.3 kg of coal for the generation of 1 kWh. This amounted to 37 kg CO_2 emissions per kWh. With research and development and increasing experience, these low efficiency levels improved rapidly. In addition, technical advancements with coal processing and combustion technology which enabled a steady increase in the steam parameters 'pressure' and 'temperature', resulted in a further rise in efficiency. By 1910, efficiency had increased to 5%, reaching 20% by 1920. In the 1950s, power plants achieved 30% efficiency, but the average efficiency of all operating power plants was still a modest 17%. In the next stage, the use of cooling towers for the removal of heat that could not be converted to electricity became a necessity, in addition to the removal of pollutants, SO_x and NO_x, from exhaust gases; this resulted in a reduction of efficiency because these technologies consume energy. However, continuous development resulted, around the mid 1980s, in an average efficiency of 38% for all power plants, and best values of 43%. In the second half of the 1990s, a Danish power plant set a world record at 47%. Today, the average efficiency of coal-fired power plants in the world is around 33% on LHV basis.

Coal-based power generation has come to stay as a more stable technology for power production worldwide. Although coal usage significantly enhances global warming, the general perception is that coal will remain, by necessity, a key component in our electricity generating choice for the foreseeable future; coal and a low carbon economy need not be incompatible; emissions cost does not necessarily invalidate coal's other advantages; and coal can be carbon-free (or nearly so) with some added cost (e.g., MIT 2007; Burnard 2009).

Coal fuels more than 40% of the world's electricity, though in some countries it is much higher. For instance, South Africa – 93%, Poland – 92%, China – 79%, Australia – 77%, Kazakhstan – 70%, India – 69%, Israel – 63%, Czech Republic – 60%, Morocco – 55%, Greece – 52%, USA – 49%, Germany – 46% (IEA 2010).

Coal-fired power plants worldwide produced 8698 TWh in 2010. Country-wise, China tops the list with 3273 TWh, followed by the USA – 1994 TWh, India – 653 TWh, Japan – 304 Twh, Germany – 274 TWh, South Africa – 242 TWh, and so on (IEA – Key World Energy Statistics 2012).The importance of coal to electricity generation is projected to continue in the developing world due to their growing energy needs, with coal fuelling 44% of global electricity in 2035 (IEA 2011).

To maximize the value of coal use in power generation, plant efficiency is an important performance parameter. Efficiency improvements offer several benefits: (i) the life of coal reserves will be extended by reducing utilization; (ii) emissions of CO_2 and conventional pollutants may be reduced (a 1% improvement in efficiency can result in up to 3% reduction in CO_2 emissions); (iii) the power output is increased; and (iv) operating costs are reduced.

7.1 CONVENTIONAL POWER GENERATION PLANT

A traditional coal-fired power plant based on coal combustion is schematically shown in Figure 7.1. The 'cleaned' coal is ground into a fine powder in a pulverizer to increase the surface area which allows coal to burn more quickly. The powder then enters into the boiler furnace along with sufficient air. The mixture ignites immediately and burns to release a substantial amount of heat. The carbon in the coal and oxygen in the air combine to produce carbon dioxide and heat. The heat from combustion of the coal boils water in the boiler to produce high pressure and high temperature steam. The high pressure steam is passed through a series of turbines containing thousands of propeller-like blades. The impulse and thrust created by steam pushes these blades causing the turbine shaft to which a generator is mounted to rotate. The rapid rotation in a strong magnetic field generates electricity based on the principle of Faraday's electromagnetic induction. The exiting steam from turbine is condensed and returned to

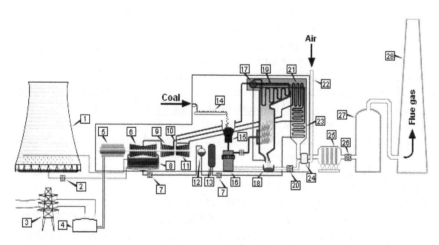

[1 Cooling tower; 2 Cooling water pump; 3 Three-phase transmission line; 4 step-up transformer, 5 Electrical generator, 6 Low pressur steam turbine, 7 Condensate and feedwater pump, 8 surface condenser, 9 Intermediate pressure steam turbine, 10 Steam control valve, 11 High pressure steam turbine, 12 Deaerator, 13 Feedwater heater, 14 Coal conveyor, 15 Coal hopper, 16 Coal pulverizer, 17 Boiler steam drum, 18 Bottom ash hopper, 19 Superheater, 20 Fan, 21 Reheater, 22 Combustion air intake, 23 Economiser, 24 Airpreheater, 25 ESP, 26 Fan, 27 Flue gas desulfurization scrubber, 28 Flue gas stack]

Figure 7.1 Coal-fired Power plant (*Source:* Citizendium, Author & Copyright © Milton Beychock, 2010) at http://en.citizendium.org/wiki/conventional_coal-fired_power_plant).

the boiler to be heated again and the cycle is repeated. Steam coal, also known as thermal coal, is normally used to generate electricity. Heavy ash from the burned coal in the boiler collects in an ash hopper while the lighter 'fly ash' leaves along with exhaust gases. The fly ash is then removed along with other pollutants such as particulates by mechanical and electrostatic precipitators before going into environment. In practice, many systems and sub systems and several technologies such as combustion, aerodynamics, heat transfer, thermodynamics, pollution control and logistics are involved in the energy conversion to electricity. Conventional coal-fired power plants are highly complex and custom designed, and they provide most of the electrical energy used in many countries. Most power plants, 500 MW size, were built in the 1980s and early 1990s. The types of steam generator, steam turbine, and steam condenser used in these Plants are explained in the following pages. See Annex for coal, air, water requirements of a typical power plant of capacity 500 MW. The conventional power plants use relatively cheap coal with low ash content, water at subcritical condition and steam turbines as the only source to generate electrical power; and the operating efficiencies of these old boilers are typically 30–32%.

Generating efficiency

Estimating power plant efficiency is not as simple as it may appear. There is no definitive methodology. Plant efficiency calculations of different plants in different regions are often calculated using different assumptions and expressed on different bases. For example, the heat rate of European power plants can appear to be 8–10% lower than their US counterparts (and so appear 3–4% points more efficient). This may be partly due to real plant differences, but differences in calculation methodologies for identical plants can also be of this magnitude (IEA-CIAB 2010). Efficiency is expressed as the ratio of electric energy output to the fuel energy input of a thermal power plant, expressed in percentages. Heat Rate (HR) – the fuel energy input required for the generation of unit of electricity (Btu/kWh), or (kJ/kWh) – is another parameter used to determine efficiency. The fuel energy input can be taken either as the higher (gross) or as the lower (net) heating value of the fuel (HHV or LHV); but when comparing the efficiency of different energy conversion systems, it is to be ensured that the same type of heating value is used. Generation efficiency is expressed as:

$$[3600(kJ/kWh)/HR(kJ/kWh)] \times 100, \quad \text{or} \quad [3414\,Btu/kWh/HR(Btu/kWh)] \times 100$$

The directly determined heating value by calorimetric measurement in the laboratory is HHV. In this measurement, the fuel is combusted in a closed vessel, and water that surrounds the calorimeter receives the heat of combustion. The combustion products are cooled to 15°C (60°F) and hence, the heat of condensation of the water vapor originating from the combustion of hydrogen, and from the evaporation of the coal moisture, is included in the measured heating value. For determining the lower heating value, LHV, the heat of condensation has to be deducted from the HHV. In US engineering practice, efficiency calculations for steam plants are based on HHV, while in the European practice, the calculations are based on LHV. For Gas turbine cycles, however, LHV is used both in the USA and Europe. Perhaps one reason for this difference in the method of calculating steam power plant efficiencies is that US electric utilities purchase coal on a $/MBtu (HHV) basis and want to know their efficiency also on that basis, while the European practice is based on the understanding that the heat of condensation is not a recoverable part of the fuel's energy, because it is not practicable to cool sulfur containing flue gas to below its dew point in the boiler (Beer 2009, web.mit.edu).

Termuchlen and Emsperger (2003) calculated LHV using the International Energy Agency (IEA) formula:

$$LHV = HHV - [91.14 \times H + 10.32 \times H_2O + 0.35 \times O],$$

where LHV and HHV are in Btu/lb, and H, H_2O and O are in %, on 'as received' basis, and in SI units as:

$$LHV = HHV - [0.2121 \times H + 0.02442 \times H_2O + 0.0008 \times O],$$

where LHV and HHV are in MJ/kg, and H, H_2O and O are in %. The difference in efficiency between HHV and LHV for bituminous coal is about two percentage points absolute (i.e., 5% relative), but for high moisture sub-bituminous coals and lignite coals, the difference is about 3–4 percentage points (>8% relative). The average efficiency of coal based electricity generating plant in the USA is about 34% (LHV). Typical modern coal units range in thermal efficiency from 33% to 43% (HHV).

Enhancing Efficiency: The focus today is how to enhance the efficiency and minimize the pollution from coal-fired power plants. Modern combustion technologies have to satisfy several severe requirements, coming mainly from the use of low quality coals, and unconventional fuels. Energy technologies have to be: highly efficient (combustion efficiency >99%, boiler efficiency >85%), with high fuel flexibility (burning at the same time or consecutively different fuels), wide range of load following (1:5) and low emissions, SO_2 < 200 ppm, NO_x < 200 ppm (Simeon Oka 2001). Techniques for removal of SO_2, NO_x, mercury, and particulate matter from the stack gas before it is released are well established. But carbon dioxide is the major problem, and it is recognized that CCS technology is the ultimate technology for capturing CO_2. All these technologies generally referred as clean coal technologies may be grouped into: (a) Pollution control technologies, explained in an earlier chapter, and (b) Power plant efficiency improvement technologies.

Efficiency is influenced by a number of factors and operating parameters that include coal type and quality (e.g., high ash/moisture content), steam temperature and pressure (steam cycle severity), Boiler design, and condenser cooling water temperature, and so on.

Extensive research and development related to coal-fired power plants over the years, has resulted in an increase in the steam 'temperature' and 'pressure', and materials development for boilers and so on leading to high-temperature and high pressure Pulverized Coal technologies (referred as supercritical and ultra-supercritical). In addition, new concepts such as Circulating Fluidized Bed Combustion (CFBC), Pressurized Circulating Fluidized Bed Combustion (PCFBC), and Coal Gasification technologies (e.g., IGCC) have come into operation. These are discussed in detail in the next chapters. The achievable efficiencies of these technologies are plotted in Figure 7.2. A one percentage point improvement in the efficiency of a conventional pulverized coal combustion plant results in a 2–3% reduction in CO_2 emissions (IEA 2006).

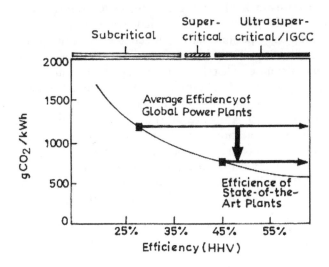

Figure 7.2 Efficiency ranges for subcritical, super-critical, and ultra super-critical technologies (Redrawn from IEA – Focus on Clean Coal 2006).

More efficient coal-fired combustion technologies reduce emissions of CO_2, and pollutants such as NO_x, SO_x and particulates. This is particularly necessary in countries where existing plant efficiencies are generally low. The average global efficiency of coal-fired plants is 28% (in 2005) compared to 45% for today's most efficient plants (Figure 7.2). Retrofitting existing coal-fired plants to improve their efficiency as well as installing newer and more efficient plants will result in significant CO_2 reductions. Local site specifics and ambient environmental conditions and the available *coal quality* by and large are the factors to be considered for the deployment of new, highly efficient plants. The designs of the newly installed plants must be 'ready' for retrofitting with CCS in the future. Further, the efficiency of the plants has to be high, as inefficient operating plants will weaken the capacity to deploy CCS technologies.

7.2 STEAM PROPERTIES

Steam characteristics are very critical for the operation of coal-fired power plants. What happens when water is heated at normal atmospheric pressure? (i) As the heating of the water goes on, the temperature of water increases till it reaches 100°C (212°F). This is the Sensible Heat addition, and the heat required is the enthalpy of water, measured in KJ/Kg or Btu/lb. When water boils, both water and steam are at the same temperature called the saturation temperature related to the particular boiling pressure; (ii) Further heating does not increase the temperature; instead small bubbles of steam start to form within it and rise to break through the surface. The space immediately above the water surface thus becomes filled with less dense steam molecules. When the number of molecules leaving the liquid surface is more than those re-entering, the water freely evaporates, and at this point it has reached boiling point, as it is saturated with heat energy. If the pressure remains constant, adding more heat does not cause the temperature to rise any further but causes the water to form saturated steam. The temperature of the boiling water and saturated steam within the same system is the same, but the heat energy per unit mass is much greater in the steam. If the pressure is increased, this will allow the addition of more heat and an increase in temperature without a change of phase. Therefore, increasing the pressure effectively increases both the enthalpy of water, and the saturation temperature. The relationship between the saturation temperature and the pressure is known as the steam saturation curve. Water and steam can coexist at any pressure on this curve, both being at the saturation temperature. Steam at a condition above the saturation curve is a superheated steam. Enthalpy of evaporation involves no change in the temperature of the steam/water mixture, and all the energy is used to change the state from liquid (water) to vapour (saturated steam). Enthalpy of saturated steam, or total heat of saturated steam is the total energy in saturated steam, and is the sum of the enthalpy of water and the enthalpy of evaporation:

$$h_g = h_f + h_{fg} \tag{7.1}$$

where h_g = total enthalpy of saturated steam (kJ/kg)
h_f = liquid enthalpy (sensible heat) (kJ/kg)
h_{fg} = enthalpy of evaporation (kJ/kg)
The enthalpy (and other properties) of saturated steam can easily be referenced using the steam tables.

Steam phase diagram

The relation among temperature, pressure, and enthalpy of saturated and superheated steam, called 'steam phase diagram' is illustrated in Figure 7.3.

As water is heated from 0°C to its saturation temperature, its condition follows the saturated water line until it has received all of its liquid enthalpy, h_f, (A–B). If further

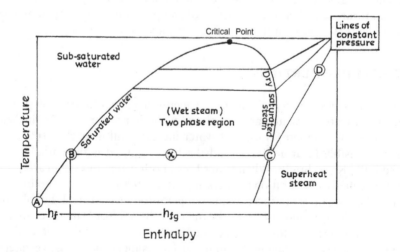

Figure 7.3 Temperature – Enthalpy diagrams of saturated and superheated steam (*Source:* Spiraxsarco website).

heated, the water changes phase to a water/vapor mixture and continues to increase in enthalpy while remaining at saturation temperature, h_{fg}, (B–C). As the water/vapor mixture increases in dryness, its condition moves from the saturated liquid line to the saturated vapor line. At a point exactly halfway between these two states, the dryness fraction (χ) is 0.5; similarly steam is 100% dry on the saturated steam line. Once it has received all of its enthalpy of evaporation, it reaches the saturated steam line. Further continuous heating after this point, the pressure remains constant but the temperature of the steam will begin to rise as superheat is imparted (C–D). The saturated water and saturated steam lines enclose a region in which wet steam (a water/vapor mixture) exists. In the region to the left of the saturated water line only water exists, and in the region to the right of the saturated steam line only superheated steam exists. The point at which the saturated water and saturated steam lines meet is known as the critical point – the highest temperature at which water can exist. Above the critical point the steam may be considered as a gas. Any compression at constant temperature above the critical point will not produce a phase change whereas compression at constant temperature below the critical point will result in liquefaction of the vapor as it passes from the superheated region into the wet steam region. For steam, this occurs at 374°C (705°F) and 220.6 bar (22.06 MPa). Above this pressure, the steam is termed 'supercritical'.

Conventional steam power plants operate at steam pressures of around 170 bar (17 MPa), and are called 'subcritical' power plants. The new generation of power plants operates at pressures higher than the critical pressure, in the range of 230 to 265 bar (23.5 to 26.5 MPa) which are called Supercritical power plants. In the pursuit for higher efficiency the trend is to go for still higher operating pressures in the range of 300 bar (30 MPa), and the plants operate with such steam conditions are Ultra supercritical Power plants. These units operate at temperatures of 615 to 630°C.

7.3 STEAM GENERATORS/BOILERS

7.3.1 Early steam boilers

The steam generator/boiler is an important component in the power plant where water is converted into steam at the desired temperature and pressure by thermal energy produced from combustion of coal. A steam power plant has the following components: (a) a furnace to burn the fuel, (b) steam generator or boiler containing water; heat generated in the furnace is utilized to convert water into steam, (c) main power unit such as a turbine to use the heat energy of steam and perform work, and (d) piping system to convey steam and water. In addition, the plant requires various auxiliaries and accessories depending upon the availability of water, fuel and the service (to run a turbine to generate electricity) for which the plant is intended. A steam power plant using steam as working substance works basically on Rankine cycle, discussed later.

Figure 7.4 shows a schematic arrangement of a steam power plant showing different systems and components. The coal storage area receives coal and is transferred into the furnace by coal handling unit. The coal is burned and the heat produced is utilized to convert water contained in boiler drum into steam at suitable pressure and temperature. The steam generated is passed through the super heater, and the superheated steam then flows through the turbine. After performing work in the turbine, the steam

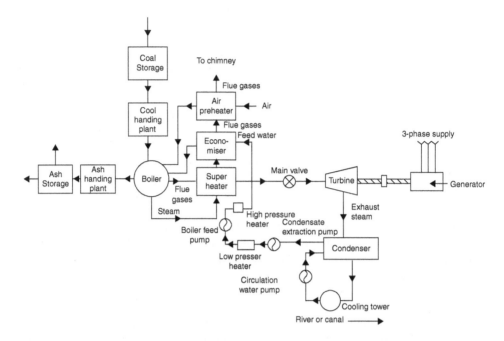

Figure 7.4 Schematic of a Steam Power plant (Redrawn from Unit-2-58SteamPowerPlant.pdf).

pressure is reduced. Steam leaving the turbine passes through the condenser which is maintained by the low pressure of steam at the exhaust of the turbine. Steam pressure in the condenser depends upon flow rate and cooling water temperature and on the efficiency of air removal equipment. If sufficient quantity of water is not available the hot water coming out of the condenser may be cooled in cooling towers and circulated again through the condenser. Bled steam taken from the turbine at suitable withdrawal points is sent to low pressure and high pressure water heaters.

The development of boilers for steam generation in power plants has a long history beginning in the late 17th century. The first popular design, water-tube boilers, used in power generating systems was developed by Bob & Wilcox in the 19th century. In Water-tube boilers, the water is circulated inside the tubes, with the heat source surrounding them. The tube diameter is significantly smaller resulting in much higher pressures being tolerated for the same stress. Many water-tube boilers operate on the principle of natural water circulation (also known as 'thermo-siphoning'). Figure 7.5a illustrates the principle. Higher density cooler feedwater is introduced into the steam drum behind a baffle where it descends in the 'downcomer' towards the lower or 'mud' drum, displacing the warmer water up into the front tubes. Continued heating creates steam bubbles in the front tubes, which are naturally separated from the hot water in the steam drum, and are taken off. However, when the pressure in the water-tube boiler is increased, the difference between the densities of the water and saturated steam falls, consequently less circulation occurs. To keep the same level of steam output at higher design pressures, the distance between the lower drum and the steam drum must be increased, or some means of forced circulation must be introduced.

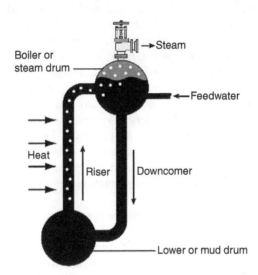

Figure 7.5a Schematic of Water tube Boiler with natural water circulation (*Source:* Spirax Sarco at www.spiraxsarco.com/resources/...).

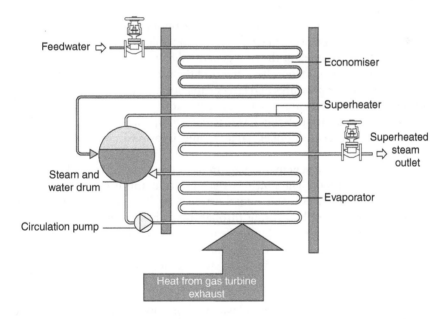

Figure 7.5b Forced circulation Water tube boiler used in a CHP system (*Source:* Spirax Sarco website).

Water-tube boilers are used in power station applications that require, (a) high steam output up to 500 kg/s, or (b) high pressure steam up to 160 bar, or (c) super-heated steam up to 550°C. A forced circulation water-tube boiler used on a CHP plant is shown in Figure 7.5b. The water-tube boiler, over years, went through several stages of design and development. They are available as 'horizontal straight tube boilers', 'bent tube boilers' and 'cyclone fired boilers'. Fire tube boilers are classified as 'external furnace' and 'internal furnace' types of boilers. The boilers are also classified according to (a) the position of furnace (internally-fired or externally-fired), or (b) the position of principal axis (vertical or horizontal or inclined), or (c) the type of circulation of water (natural or forced circulation) or (d) steam pressure (low, high, medium), or (e) their application (stationary or mobile). In *fire tube boiler*, on the other hand, the hot products of combustion pass through the tubes, which are surrounded by water. Fire tube boilers have low initial cost, and are more compact and are competitive for steam rates up to 12,000 kg/hour and pressures up to 18 kg/cm^2. But they are more prone to explosion; water volume is large and cannot meet quickly the change in steam demand due to poor circulation. For the same output the outer shell of fire tube boilers is much larger than the shell of water-tube boiler.

With the optimization of boilers for the purpose of power generation, the steam temperature and pressure got increased – nearly doubled. The key achievements in R & D included (a) using pulverized coal-firing (using coal particle of small sizes) in place of stoker-fired boilers (large coal particle size) to take advantage of the pulverized coal's higher volumetric heat release rates, (b) increasing system efficiencies by using super heaters (c) economizers (d) and combustion air pre-heaters and (e) improving construction materials, making it possible for steam generators to reach steam pressures in excess of 1,200 psig (Kuehn 1996).

The development of superheaters, reheaters, economizers and air preheaters, and soot blowers and cooling towers played a significant role in improving overall system efficiency by utilizing as much of the heat generated from burning the coal as possible. The separation of the steam from the water and the use of superheaters and reheaters allowed for higher boiler pressures and larger capacities. A few details on superheaters, air preheaters, and reheaters are given in the Annexes.

The pulverized coal firing design has reached widespread usage by the mid-1920s, and has resulted in increased boiler capacity and improved combustion and boiler efficiencies.

Figure 7.6 illustrates a schematic diagram of a large, coal-fired subcritical steam generator in a power plant. The unit has a steam drum and uses water tubes rooted in the walls of the generator's *furnace* combustion zone. The *saturated steam* from the steam drum is superheated by passing through tubes heated by the hot combustion gases. These hot gases are also used to preheat the boiler feed water entering the steam drum and the combustion air entering the combustion zone.

There are three configurations for such steam generators (Figure 7.7).

(a)   *Natural* circulation in which liquid water flows downward from the steam drum via the down comer (see Figure 7.6) and a mixture of steam and water returns to the steam drum by flowing upward via the tubes embedded in the furnace wall. The difference in density between the downward flowing water and the upward flowing mixture of steam and water provides sufficient driving force to induce the circulating flow;

(b)   In *Forced* circulation, a pump in the downcomer provides additional driving force for the circulating flow. The pump is usually provided when generating steam pressure is above around 170 bar because, at these pressures, the density

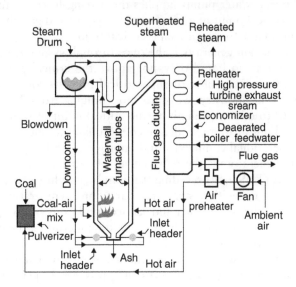

Figure 7.6 Simplified diagram of a modern, large, coal-fired subcritical steam generator in a Power plant [*Source:* Boiler (Power generation) in Wikipedia, the free encyclopedia at http://en.wikipedia.org/wiki/Boiler (Steam generator); Author: Milton Beychok, improved version in 2012].

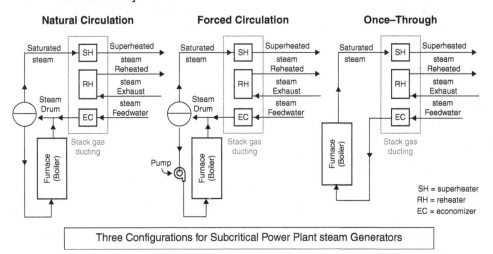

Figure 7.7 Three configurations of steam generators for subcritical power plant (*Source:* Steam Generator, at http://en.citizendium.org/wiki/Steam_generator; Author: Milton Beychok, 2008).

difference between the downcomer liquid and the liquid-steam mixture in the furnace wall tubes is reduced sufficiently to limit the circulating flow rate;

(c)   In *once-through system*, no steam drum is included; and the boiler feed water goes through the economizer, the furnace wall tubes and the superheater section in one continuous pass and there is no recirculation.

In essence, the feed water pump supplies the driving force for the flow through the system. At the critical point of a substance, the liquid and gas phases do not exist distinctly and result in a homogenous supercritical fluid.

For supercritical steam generators, the once-through system is the configuration of choice, since there is no liquid or vapor above the critical point and hence no need for a steam drum. A number of supercritical pressure 'once-through' systems were built for the utility industry, with pressures ranging from 310 to 340 bar and temperatures of 620 to 650°C (well above the critical point of water). Subsequently, these systems were built at more moderate conditions of about 240 bar and 540 to 565°C to reduce operational complexity and improve reliability. The main issue with supercritical steam generators is the requirement of extremely pure feed water with total dissolved solids (TDS) not exceeding 0.1 ppm by weight (Nag 2008; Elliot *et al.* 1997). Boiler designs, capacities, steam pressures, and temperatures, among other parameters, vary with the type of fuel and service conditions. Although steam generators and their auxiliary components have many design criteria, the important issues are efficiency, reliability/availability, and cost.

Advances in materials, system designs, and fuel firing have led to increasing capacity and higher steam-operating temperatures and pressures. Depending upon whether the pressure of the steam being generated is below or above the critical pressure of water (221 bar), a power plant steam generator is known as either *subcritical* (below 221 bar) or *supercritical* (above 221 bar) steam generator. The output superheated steam from subcritical steam generators, in power plants using fuel combustion, usually range in pressure from 130 to 190 bar, in temperature from 540 to 560°C and at steam rates from about 400,000 to about 5,000,000 kg/hour. Some of these high pressure boilers are briefly outlined.

7.3.2 High pressure boilers

High pressure boilers (steam capacities of 30 to 650 tons/hour and above, pressure above 60 bars up to 160 bars, and maximum steam temperature of about 540°C) are universally used in all modern power plants, as they offer several advantages:

(a)   Efficiency and the capacity of the plant can be increased because amount of steam required is reduced if high pressure steam is used for the same power generation.
(b)   Forced circulation of water through boiler tubes provides alternative in the arrangement of furnace and water walls, in addition to the reduction in the heat exchange area.
(c)   Tendency for scale formation is reduced due to high velocity of water through the tubes.
(d)   Risk of overheating is reduced and thermal stress problem is simplified due to the uniform heating of all the parts.
(e)   Differential expansion is reduced due to uniform temperature which in turn reduces the prospect of gas and air leakages.
(f)   Steam can be raised quickly to meet the variable load requirements without the use of complicated control devices.

In order to achieve efficient operation and high capacity, forced circulation of water through boiler tubes appears to be helpful.

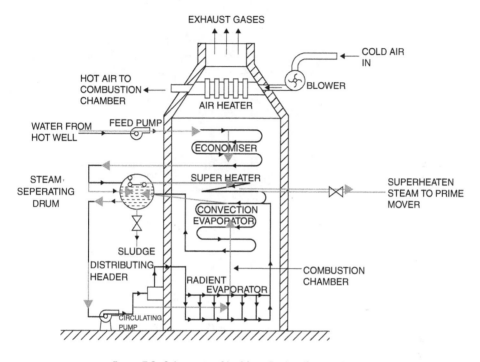

Figure 7.8 Schematic of La Mont Boiler (Source: Joshi).

Some special types of boilers operating at super critical pressures and using forced circulations are the following:

(a)   La Mont Boiler: This was introduced in 1925 by La Mont (Figure 7.8). These boilers have been built to generate 45 to 50 tonnes of superheated steam at a pressure of 120 bars and temperature of 500°C. Recently forced circulation has been introduced in large capacity power. The main issue is the formation and attachment of bubbles on the inner surfaces of the heating tubes which reduce the heat flow and steam generation as it offers high thermal resistance than water film.

(b)   Benson Boiler: If the boiler pressure is raised to critical pressure (225 atm), the steam and water will have the same density and hence the danger of bubble formation as it happens in the La Mont boiler can be completely avoided. Benson developed such a high pressure boiler and erected in 1927. The maximum working pressure obtained so far from commercial Benson boiler is 500 bar. Boilers of 150 tones/hr generating capacity and having as high as 650°C temperature of steam are in use. During starting, the water is passed through the economiser, evaporator, super heater and back to the feed line via starting valve A while the valve B is closed. As the steam generation starts and it becomes superheated, the valve A is closed and the valve B is opened. See Figure 7.9.

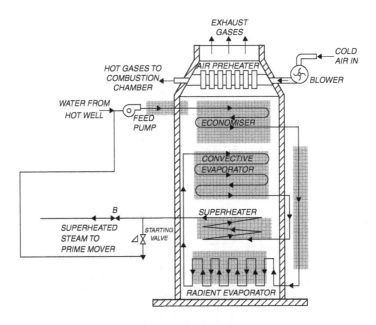

Figure 7.9 Benson Boiler (*Source:* Joshi).

The erection of Benson boiler is easier and quicker as all the parts are carried to the site and welded. The furnace walls of the boiler can be more efficiently protected by using small diameter and close pitched tubes. The super heater in the Benson boiler is an integral part of forced circulation system; therefore no special starting arrangement for super heater is required. The Benson boiler can be operated most economically by varying the temperature and pressure at partial loads and overloads. As there are no drums, the total weight of Benson boiler is 20% less than other boilers. This also reduces the cost of boiler. The desired temperature can also be maintained constant at any pressure. The blow-down losses of Benson boiler are hardly 4% of natural circulation boilers of same capacity. Explosion hazards are not severe as it consists of only tubes of small diameter and has very little storage capacity compared to drum type boiler.

(c)   Loffler Boiler: In a Benson boiler, there is a problem of deposition of salt and sediment on the inner surfaces of the water tubes which reduce the heat transfer and ultimately the generating capacity. Further, the danger of overheating the tubes due to salt deposition as it has high thermal resistance may be increased. The difficulty was solved in Loffler boiler by preventing the flow of water into the boiler tubes. Most of the steam is generated outside from the feed water using part of the superheated steam coming-out from the boiler (Figure 7.10).

This boiler can carry higher salt concentration than any other type and is more compact than indirectly heated boilers having natural circulation. These features allow for use in land or sea transport power generation. Loffler boilers with generating capacity of 94.5 tones/hr and operating at 140 bar have been under operation.

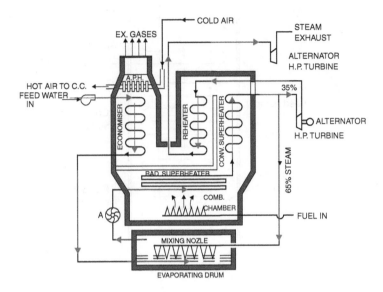

Figure 7.10 Loffler Boiler (*Source:* Joshi).

(d)   *Supercritical Boiler:* As the pressure of water or steam is raised, the enthalpy of evaporation is reduced, and at critical pressure (221.05 bar) the enthalpy of evaporation becomes zero. When water is heated at constant supercritical pressure it is suddenly converted into steam. The high pressure (above critical point) water enters the tube inlets and leaves at the outlet as the superheated steam. Since there is no drum, there should be a transition section where the water is likely to flash in order to accommodate the large increase in volume.

The merits are: heat transfer rates are considerably large; no two-phase mixture and hence the problem of erosion and corrosion are minimized; easy operation and comparative simplicity and flexibility facilitate adaptability to load fluctuations.

Several other boiler designs such as Schmidt-Hartmann Boiler and Velox-Boiler are also available. For technical details, the reader may refer to literature on high pressure steam boilers.

7.3.3 Performance evaluation of boilers

The performance parameters of a boiler – efficiency and evaporation ratio – diminish with the age due to poor combustion, heat transfer surface fouling and poor operation and maintenance. For a new boiler also, deteriorating fuel quality and water quality can result in poor boiler performance. A heat balance helps to identify avoidable and unavoidable heat losses, and Boiler efficiency tests help to find out the deviation of boiler efficiency from the best efficiency and the problem area for corrective action. Typical heat losses from a coal-fired boiler are due to (a) dry flue gas, (b) steam in flue gas, (c) moisture in fuel, (d) moisture in air, (e) unburnt fuel in residue, and (f) radiation & other unaccounted loss. If 100% fuel energy enters the boiler, after

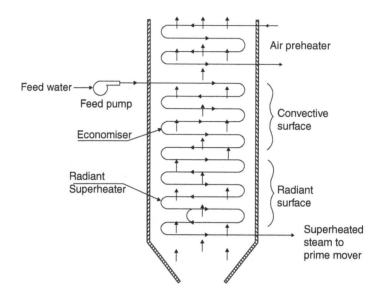

Feed water
Feed pump
Economiser
Radiant Superheater

Air preheater
Convective surface
Radiant surface
Superheated steam to prime mover

Figure 7.11 Schematic of a Supercritical Boiler (Joshi).

accounting for above losses, the net heat in the steam is 73.8%. To enhance the efficiency, some of the avoidable losses can be removed by corrective actions:

Stack gas losses: By reducing excess air and stack gas temperature;
Losses by unburnt fuel in stack and ash: By optimizing operation & maintenance, and with better burner technology;
Blow down losses: By treating fresh feed water and recycling condensate; Condensate losses: By recovering the largest possible amount of condensate; and Convection and radiation losses: By better insulating the boiler.

Thermal efficiency of a boiler, normally, is defined as 'the percentage of (heat) energy input that is effectively useful in the generated steam'. There are two methods of assessing boiler efficiency (UNEP 2006).

(i)   The Direct Method in which the energy gain of the working fluid (water and steam) is compared with the energy content of the boiler fuel. The parameters to be monitored for the calculation of boiler efficiency by this method are (a) quantity of steam generated per hour in kg/hr, (b) quantity of fuel used per hour in kg/hr, (c) working pressure in kg/cm^2 and superheat temperature (°C), if any, (d) temperature of feed water (°C), and (e) type of fuel and gross calorific value of the fuel (GCV) in kcal/kg of fuel. This method has the advantage of using fewer instruments and parameters, and offers easy estimation by plant staff. However, information as to why efficiency of the system is lower and the calculation of various losses accountable for various efficiency levels cannot be derived by this method.

(ii) The Indirect Method: The efficiency is the difference between the energy input and the heat losses; that is why it is also called the heat loss method.

The reference standards for Boiler Testing at Site using the indirect method are the British Standard, BS 845:1987 and the USA Standard ASME PTC-4-1 Power Test Code Steam Generating Units.

The data required for calculation of boiler efficiency using the indirect method are: (a) Ultimate analysis of fuel – H_2, O_2, S, C, moisture content, ash content, (b) O_2 or CO_2 content in the flue gas in %, (c) Flue gas temperature (T_f) in °C, (d) Ambient temperature (T_a) in °C and humidity of air in kg/kg of dry air, (e) GCV of fuel in kcal/kg, (f) Combustible in ash in % (in case of solid fuels), and (g) GCV of ash in kcal/kg (in case of solid fuels). The calculation of efficiency by indirect method is laborious, though it enables to obtain a complete mass and energy balance for each individual stream, making it easier to identify options to improve boiler efficiency. But the main issues are the calculations which are time-consuming, and require laboratory facilities for analysis.

Brief details of 'furnaces' used in coal-fired power plants are given in the Annexes. Cyclone furnaces, introduced in 1946, provide the same benefits as pulverized coal firing, with the additional advantage of utilizing slagging coals, reducing costs due to less fuel preparation (i.e., the coal can be coarser and no need to be pulverized), and reducing the furnace size.

7.4 STEAM TURBINE AND POWER CYCLES

Steam turbines, in a power plant, convert the thermal energy of the steam into mechanical energy, or shaft torque, which is then converted to electrical energy. The conversion of thermal energy into mechanical energy is accomplished by controlled expansion of the steam through stationary nozzles and rotating blades. The shaft torque drives electric generators to produce electricity.

Superheated steam is directed by nozzles onto a rotor causing the rotor to turn. In this process, the steam loses some energy, and if the steam was at saturation temperature, this loss of energy would cause some of the steam to condense. Turbines have a number of stages; the exhaust steam from the first rotor will be directed to a second rotor on the same shaft. This means that saturated steam would get wetter and wetter as it went through the successive stages. Not only would this promote water-hammer, but the water particles would cause severe erosion within the turbine. The solution is to supply the turbine with superheated steam at the inlet, and use the energy in the superheated portion to drive the rotor until the temperature/pressure conditions are close to saturation and then exhaust the steam.

The process of transferring heat to pressurized water to produce high potential energy steam, and translation of the energy to work through interaction of steam and turbine blades, is performed through a Rankine cycle (a thermodynamic cycle). Rankine cycle is one of the ways of classifying a steam turbine (Elliot 1989); it is a way of describing the process by which steam-operated heat engines generate power in a power plant. In this cycle, the turbine exhaust steam which is at low pressure and temperature is condensed and recycled back to the boiler. The generation of heat is through combustion of coal in the present case.

When an efficient turbine is used, the Temperature-Entropy (T-S) diagram of Rankine cycle resembles that of Carnot cycle, the main difference being that heat

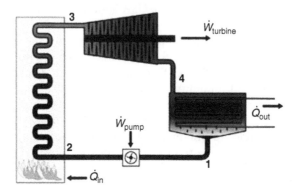

Figure 7.12 Illustration of the four main components in the Rankine cycle (*Source:* English Wikipedia, Author & Copyright © Andrew Ainsworth).

addition (in the boiler) and rejection (in the condenser) are isobaric in the Rankine cycle and isothermal in the theoretical Carnot cycle. A pump is used to pressurize the working fluid received from the condenser in liquid-phase instead of in a gas-phase. All of the energy in pumping the working fluid through the complete cycle is lost, as is most of the energy of vaporization of the working fluid in the boiler. This energy is lost to the cycle because the condensation that can take place in the turbine is limited to about 10% in order to minimize blade erosion; and the condenser rejects vaporization energy from the cycle. But pumping the working fluid through the cycle as liquid requires a very small energy. The efficiency of a Rankine cycle is usually limited by the working fluid. Without the pressure reaching super critical levels for the working fluid, the temperature range over which the cycle can operate is quite small: turbine entry temperatures are typically 565°C (the creep limit of stainless steel) and condenser temperatures are around 30°C. This gives a theoretical Carnot efficiency of about 63% compared with an actual efficiency of 42% for a modern coal-fired power plant.

The working fluid, water, in a Rankine cycle follows a closed loop and is reused constantly. The water *vapor* with accompanied droplets often seen surging from power plants is generated by the cooling systems (not from the closed-loop Rankine power cycle) and represents the waste heat energy (pumping and condensing) that could not be converted to useful work in the turbine. The cooling towers operate using the latent heat of vaporization of the cooling fluid. Water is used as a working fluid, having favorable thermodynamic properties.

One of the merits of the Rankine cycle is that during the compression stage relatively little work is required to drive the pump, thus contributing to a much higher efficiency for a real cycle. The benefit of this is lost to some extent due to the lower heat addition temperature. There are four processes in the Rankine cycle (Figure 7.12), identified by numbers in the Ts-1 diagram shown in Figure 7.13.

Process 1–2: The working fluid entering at state 1 is pumped from low to high pressure. As the fluid is a liquid at this stage, the pump requires little input energy; this is an isentropic compression in pump; *Process 2–3:* At state 2, the high pressure liquid enters a boiler where it is heated at constant pressure by heat from combustion

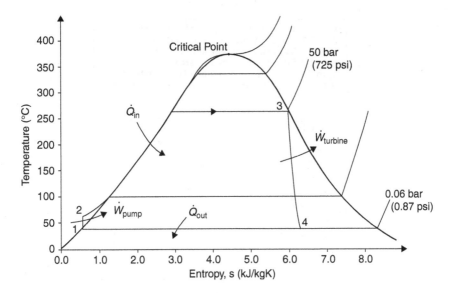

Figure 7.13 T-s diagram of a typical Rankine cycle (*Source:* w.en: File:Rankine cycle Ts.png; author & copyright © Andrew Ainsworth).

gases to become a dry saturated vapor. It is constant pressure heat addition in boiler which acts as a heat exchanger. The input energy required can be easily calculated using Mollier diagram or h-s chart or enthalpy-entropy chart also known as steam tables; Process 3–4: The dry saturated vapor expands through a turbine, generating power; the temperature and pressure of the vapor are decreased resulting in some condensation. It is isentropic expansion in the turbine. The output in this process can be easily calculated using the enthalpy-entropy chart or the steam tables; Process 4–1: The wet vapor (water-steam mixture) then enters a condenser where at a constant temperature it condenses to a saturated liquid. It is constant pressure heat rejection in the compressor to a cooling medium. So, steam leaves the condenser as saturated liquid and enters the pump, completing the cycle.

Hence, in an ideal Rankine cycle, the pump and turbine would generate no entropy and hence maximize the net work output. Processes 1–2 and 3–4 represented by vertical lines on the T-s diagram, more closely resemble that of the Carnot cycle. The Rankine cycle shown here prevents the vapor ending up in the superheat region after the expansion in the turbine, which reduces the energy removed by the condensers.

Figure 7.14 represents a Rankine cycle using 'superheated' steam. All four components, the pump, boiler, turbine and condenser, are steady-flow devices, and thus all four processes that formulate the Rankine cycle can be analyzed as steady-flow process. Starting with steady-flow equation, the 'thermodynamic efficiency' of a simple Rankine cycle can be arrived at:

$$\eta_{\text{therm}} = \frac{\dot{W}_{\text{turbine}} - \dot{W}_{\text{pump}}}{\dot{Q}_{\text{in}}} \approx \frac{\dot{W}_{\text{turbine}}}{\dot{Q}_{\text{in}}}$$

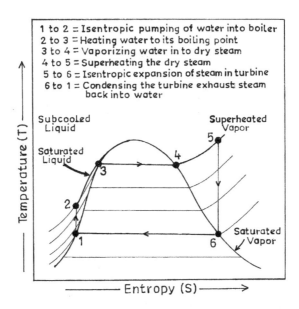

Figure 7.14 T-s diagram (with superheated steam) (*Source:* Redrawn from en.citizendium.org; Author: Milton Beychok 2009).

As the work required by the pump is often around 1% of the turbine work output, the equation is simplified by neglecting W_{pump}. Here Q is the heat flow rate to or from the System (energy per unit time), and W is mechanical power consumed by or provided to the System (energy per unit time).

Actual Cycle Vs Ideal Cycle: The actual vapor power cycle differs from the ideal Rankine cycle because of irreversibilites in various components. 'Fluid friction' and 'heat loss' to the surroundings are the two common sources of irreversibilites.

Fluid friction causes pressure drop in the boiler, the condenser and the piping between various components. Moreover the pressure at the turbine inlet is somewhat lower than that at the boiler exit due to the pressure drop in the connecting pipes; and to compensate for this drop, the water must be pumped to a sufficiently higher pressure than the ideal cycle. This requires a large pump and larger work input to the pump. The other major source of irreversibility is the heat loss from the steam to the surrounding as the steam flows through various components.

As a result of the irreversibilites occurring within the pump and the turbine, a pump requires a greater work input, and a turbine produces a smaller work output. Under the ideal condition the flow through these devices is isentropic.

Improving efficiency of Rankine cycle

Three approaches are available to increase efficiency of the Rankine cycle: (a) by lowering the condenser pressure which lowers steam temperature, and thus the temperature at which heat is rejected; but there is a problem of increasing the moisture content of the steam which affects the turbine efficiency; (b) by superheating the steam to

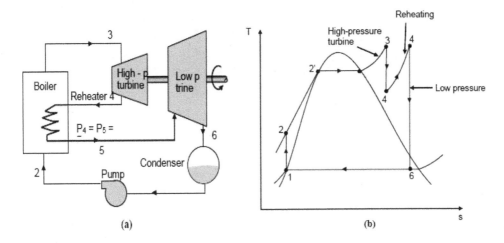

Figure 7.15 (a), (b): Schematics of Ideal Reheat Rankine cycle.

high temperatures which helps to decrease the moisture content of the steam at the turbine exit, but the 'high' temperature is limited by materials considerations; and (c) by increasing the boiler pressure, which automatically raises the temperature at which boiling happens leading to a raise in the average temperature at which heat is added to the steam, leading to increase in the thermal efficiency of the cycle.

The reheat Rankine cycle

The efficiency of the Rankine cycle can be increased by expanding the steam in the turbine in two stages, and reheating it in between. Reheating is a practical solution to the excessive moisture problem in turbines, and is commonly used currently in steam power plants.

The schematic and T-*s* diagram of the ideal reheat Rankine cycle is shown in Figures 7.15(a) and (b). In the ideal reheat Rankine cycle, the expansion process happens in two stages. In the first stage (the high-pressure turbine), steam is expanded isentropically to an intermediate pressure and sent back to the boiler where it is reheated at constant pressure, usually to the inlet temperature of the first turbine stage. Steam then expands isentropically in the second stage (low-pressure turbine) to the condenser pressure.

The work output for a reheat Cycle is therefore equal to

$$\dot{W}_{\text{turbine, out}} = \dot{W}_{\text{turbine, I}} + \dot{W}_{\text{turbine, II}}$$

The regenerative Rankine cycle

T-s diagram for the Rankine cycle in the Figure 7.16 shows that the heat is transferred to the working fluid during process 2–2′ at a relatively low temperature. This lowers the average heat-addition temperature and thus the cycle efficiency. To overcome this problem, the temperature of the liquid (called feedwater) exiting the pump has to be increased before it enters the boiler, which is called regeneration, and the thermal efficiency of the Rankine cycle increases as a result of regeneration. Regeneration raises

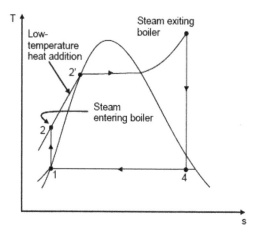

Figure 7.16 Ideal regenerative Rankine cycle.

the average temperature at which heat is transferred to the steam in the boiler by raising the temperature of the water before it enters the boiler.

During a regeneration process, feed water leaving the pump is, thus, heated by steam bled off the turbine at some intermediate pressure in a device called the feed water heater. Two types of feed water heaters – Open feed water heater, and Closed feed water heater – are in use. An open (or direct-contact) feed water heater is basically a mixing chamber, where the steam extracted from the turbine mixes with the feed water exiting the pump. Ideally, the mixture leaves the heater as a saturated liquid at the heater pressure. In the closed feedwater heater, heat is transferred from the extracted steam to the feedwater without any mixing taking place. The two streams can now be at different pressures, since they do not mix.

7.5   STEAM CONDENSER

Steam Condenser is one of the four main components in a Rankine cycle. The main purpose of the condenser is to condense the exhaust steam from the turbine for reuse in the cycle and to maximize turbine efficiency by maintaining proper vacuum. With the lowering of the operating pressure of the condenser (i.e., as the vacuum is increased), the enthalpy drop of the expanding steam in the turbine will also increase. This will increase the amount of available work from the turbine (electrical output). By lowering the condenser pressure and operating at the lowest possible pressure (i.e., highest vacuum), the gains such as 'increased turbine output', 'increased plant efficiency', and 'reduced steam flow for a given plant output' can be achieved.

Condenser types: There are two primary types of condensers that can be used in a power plant: *Direct Contact* condenser and *Surface* condenser (HEI 2005). Direct contact condensers condense the turbine exhaust steam by mixing it directly with cooling water. The older type Barometric and Jet-Type condensers operate on similar principles.

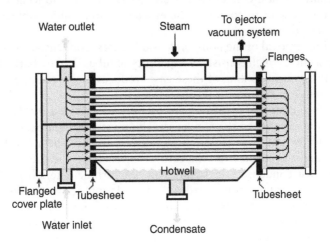

Figure 7.17 Typical Water-cooled Surface Condenser (*Source:* Citizendium, Author: Milton Beychock, 2008).

The second type, Steam surface condensers are the most commonly used condensers in modern power plants. The exhaust steam from the turbine flows on the shell side (under vacuum) of the condenser, while the plant's circulating water flows in the tube side. The main heat transfer mechanisms, therefore, are the condensing of saturated steam on the outside of the tubes and the heating of the circulating water inside the tubes. Thus for a given circulating water flow rate, the water inlet temperature to the condenser determines the operating pressure of the condenser. As this temperature is decreased, the condenser pressure will also decrease, increasing the plant output and efficiency. The source of the circulating water can be either a closed-loop from a cooling tower or spray pond, or once-through water from a lake, ocean, or river. The condensed steam from the turbine (condensate) is collected in the bottom of the condenser, which is called a hotwell. The condensate is then pumped back to the steam generator to repeat the cycle. Schematic of a typical water-cooled surface condenser is shown in Figure 7.17. Because the surface condenser operates under vacuum, non-condensable gases will migrate towards the condenser. These gases consist of mostly air leaked into the cycle from components that are operating below atmospheric pressure (like the condenser). The gases can also result from the decomposition of water into oxygen and hydrogen by thermal or chemical reactions. These gases will increase the operating pressure of the condenser, and as more gas is leaked into the system, the condenser pressure will rise resulting in decrease of the turbine output and efficiency. These gases must be vented from the condenser to avoid covering the outer surface of the tubes, severely lessening the heat transfer of the steam to the circulating water, and thus increasing the pressure in the condenser. Further, with increase in oxygen content, the corrosiveness of the condensate in the condenser increases, affecting the life of the components. Oxygen causes corrosion, mostly in the steam generator.

To vent the non-condensable gases, Steam Jet Air Ejectors and Liquid Ring Vacuum Pumps are utilized. While Steam Jet Air Ejectors (SJAE) use high-pressure motive steam to evacuate the non-condensable gases (Jet Pump), Liquid Ring Vacuum Pumps use a liquid compressant to compress the evacuated non-condensable gases and then discharge them to the atmosphere.

To aid in the removal of the non-condensable gases, condensers are equipped with an Air-Cooler section which consists of a quantity of tubes that are baffled to collect the non-condensable gases. Cooling of these gases reduces their volume and the required size of the air removal equipment (HEI 2005).

Steam surface condensers (SSC's) are broadly categorized *by the orientation of the steam turbine exhaust* to the condenser; the most common being *side* and *down* exhaust. In a side exhaust condenser, the condenser and turbine are installed adjacent to each other, and the steam from the turbine enters from the side of the condenser. In a down exhaust condenser, the steam from the turbine enters from the top of the condenser and the turbine is mounted on a foundation above the condenser. These condensers are also described *by the configuration of the shell and tube sides*. The tubeside of a steam surface condenser can be classified by the number of tubeside passes or the configuration of the tube bundle and waterboxes, and most SSC's have either one or more multiple tubeside passes. The number of tubeside passes is defined as how many times circulating water travels the length of the condenser inside the tubes. Condensers with a once-through circulating water system are often one pass. Multiple pass condensers are typically used with closed-loop systems. The tubeside may also be classified as divided or non-divided. In a divided condenser, the tube bundle and waterboxes are divided into sections. One or more sections of the tube bundle may be in operation while others are not. This allows maintenance of sections of the tubeside while the condenser is operating. In a non divided tubeside, all the tubes are in operation at all times.

The shellside of a steam surface condenser can be classified by its geometry as Cylindrical or Rectangular. The choice of the configuration is determined by the size of the condenser, plant layout, and manufacturer preference. Steam surface condensers can also be multiple shell and multiple pressure configurations (HEI 2005).

Main components in a condenser

Shell, Hotwell, Vacuum system, Tube sheets, Tubes, Water boxes and condensate pumps are the main components in a steam condenser.

Shell is the condenser's outermost body and contains the heat exchanger tubes. Carbon steel plates are used to fabricate the shell which is stiffened to provide rigidity. If the selected design so requires, intermediate plates are installed to serve as baffle plates that provide the desired flow path of the condensing steam. The plates also provide support to help prevent sagging of long tube lengths. For most water-cooled surface condensers, the shell is under vacuum during normal operating conditions.

Hotwell (sump) is provided at the bottom of the shell for the condensate to collect, which is pumped for reuse as boiler feedwater.

Vacuum system: For water-cooled surface condensers, the shell's *internal vacuum* is commonly supplied and maintained by an external steam jet ejector system. Such an ejector system uses steam as the motive fluid to remove non-condensable gases that are present in the surface condenser. As already mentioned, motor driven mechanical vacuum pumps, such as the liquid ring type, are popularly used for this purpose.

Tube sheets: At each end of the shell, a sheet of sufficient thickness usually made of stainless steel is provided, with holes for the tubes to be inserted and rolled. The inlet end of each tube is also bell mouthed for streamlined entry of water. This is to avoid eddies at the inlet of each tube giving rise to erosion, and to reduce flow friction. Plastic inserts are also suggested at the entry of tubes to avoid eddies eroding the inlet end. To take care of length-wise expansion of tubes some designs have expansion joint between the shell and the tube sheet allowing the latter to move longitudinally. In smaller units some sag is given to the tubes to take care of tube expansion with both end water boxes fixed rigidly to the shell.

Tubes: The tubes are generally made of stainless steel, copper alloys, cupro nickel, or titanium depending on several selection criteria. Copper bearing alloys such as brass or cupro nickel being toxic, they are rare in new plants. Titanium condenser tubes are usually the best technical choice; however due to sharp increase in costs, these tubes have been virtually eliminated. The tube lengths extend to about 17m for modern power plants, depending on the size of the condenser. Transportation from the manufacturer's site and ease of installation at the Plant site basically decide the size of the tube.

Waterbox: The tube sheet with tube ends rolled, for each end of the condenser is closed by a box cover known as a waterbox, with flanged connection to the tube sheet or condenser shell. The waterbox is usually provided with man-holes on hinged covers to allow inspection and cleaning. Both at the inlet and outlet sides of the waterbox, flanges, butterfly valves, vent and drain connections are arranged. Thermometers are inserted at the inlet and outlet to measure the temperature of cooling water.

Condensate pumps: These pumps are used to collect and transport condensate back into a steam system for reheating and reuse, or to remove unwanted condensate. Boiler-feed pump and Sump pump are two types of condensate pumps in use.

The condensate pump is normally located adjacent to the main condenser hotwell, actually below it. This pump sends the water to a make-up tank closer to the boiler. Often, the tank is also designed to remove dissolved oxygen from the condensate, and is called De-aerating feed tank (DFT). The output of the DFT supplies the feed booster pump which, in turn, supplies the feed pump (feedwater pump) which returns the feedwater to the boiler allowing the cycle to start over. Two pumps in succession are used to supply sufficient Net Positive Suction Head to prevent cavitations and possible damage associated with it.

A surface condenser used in power plants must satisfy the following requirements: (a) the steam entering the condenser should be evenly distributed over the whole cooling surface of the condenser vessel with minimum pressure loss, (b) the cooling water circulated in the condenser should be so regulated that its temperature when leaving the condenser is equivalent to saturation temperature of steam corresponding to steam pressure in the condenser; this will help in preventing under cooling of condensate, (c) the deposition of dirt on the outer surface of tubes needs to be prevented; this can be achieved by sending the cooling water through the tubes and the steam to flow over the tubes, and (d) there should be no air leakage into the condenser to avoid breaking of vacuum in the condenser which results in the reduction of the work obtained per kg of steam. If there is leakage of air into the condenser air extraction pump should be used to remove air as rapidly as possible.

'Cooling towers' and 'Plant Control Systems' are two other important component systems associated with a Power Plant. The reader may refer to literature for details.

REFERENCES

Andrew Ainsworth (Author) (2007): Rankine Cycle, English Wikipedia, at http://en.wikipedia.org/wiki/rankine_cycle.

Citizendium (2009): Steam Generator, at http://en.citizendium.org/wiki/ Steam _ generator; Figure author: Milton Beychok 2009.

Bengtson, H. (2010): *Steam Power Plant Condenser Cooling-Part 3: The Air Cooled Condenser, Bright Hub Engineering*, Stonecypher, L. (ed.); at http://www.brighthubengineering.com/power-plants/64903-steam-power-plant-condenser-cooling-peart-three-the-air-cooled-condenser/#).

Booras, G., & N. Holt (2004): *Pulverized Coal and IGCC Plant Cost and Performance Estimates*, in *Gasification Technologies 2004*, Washington, DC.

Citizendium (2012): Steam Generator, at http://en.citizendium.org/wiki/Steam_ generator, May 2012.

Elliot, T.C., Chen, K. & Swanekamp, R.C. (1997): *Standard Handbook of Power plant Engineering*, 2nd Edition. McGraw-Hill, ISBN 0-07-019435-1.

Energy/Power Plant Engineering, Unit-2-58SteamPowerPlant.pdf.

HEI (Heat Exchange Inst.) (2005): Condenser Basics, Tech Sheet #113, February 2005, available at Condenser Basics – Tech Sheet 113.pdf.

Johnzactruba: Coal Fired Thermal Power Plant: The Basic Steps and Facts, Lamar Stonecypher (ed.), available at www.brighthub.com › . . . ›

Joshi, K.M.: Power Plant Engineering – High Pressure Boilers and FBC, SSAS Inst. of Technology, Surat, at powerplantengineering.pdf.

IEA-CIAB (2010): *Power Generation from Coal – Measuring and Reporting Efficiency, Performance and CO$_2$ Emissions*, IEA, Paris, France, 2010.

Kuehn, S.K. (1996): Power for the Industrial Age: A brief history of boilers, *Power Engineering*, 100 (2), 5–19.

Nag, P.K. (2008): *Power Plant Engineering*, 3rd edition, Tata McGraw-Hill, ISBN 0-07-064815-8.

Rubin, E. (2005): *Integrated Environmental Control Model 5.0*, 2005, Carnegie Mellon University, Pittsburgh.

Schilling, Hans-Dieter (2005): How did the efficiency of coal-fired power stations evolve, and what can be expected in the future? *In: The efficiency of Coal-fired Power stations*, at www.sealnet.org in cooperation with www.energie-fakten.de; January 11, 2005, pp. 1–8.

UNEP (2006): *Thermal Energy Equipment: Boilers & Thermic Fluid Heaters, Energy Efficiency Guide for Industry in Asia*, at www.energyefficiencyasia.org.

Wikipedia, Free Encyclopedia: Rankine Cycle, at http://en.wikipedia.org/wiki/rankine_cycle, retrieved on May 2012.

Chapter 8

Advanced coal-based power generation

Section A: Coal combustion technologies

Pulverized coal (PC) combustion based on several decades of practice is the major technology used in coal-fired power plants worldwide as explained earlier.

PC combustion technology, during this period, continued to undergo technological improvements that have increased efficiency and reduced emissions. The technologies preferred for today's conditions may not be most favorable in future, if incompatible for further modifications. For example, Carbon dioxide capture and sequestration in coal-based power generation is an important emerging option for controlling carbon dioxide emissions; and this additional feature would add further complexity to the generating technology that is opted today.

8.1 SUBCRITICAL PC TECHNOLOGY

In a pulverized coal unit, the coal is ground to talcum-powder fineness, and injected through burners into the furnace with combustion air (Field *et al.* 1967; Smoot & Smith 1985; Beer 2000). The fine coal particles heat up rapidly and undergo pyrolysis before ignition. The bulk of the combustion air is then mixed into the flame to completely burn the coal char. The flue gas from the boiler then passes through the 'cleaning units' to remove SO_x, particulates and NO_x from the flue gas. The flue gas leaving the clean-up units is at atmospheric pressure and contains about 10–15% carbon dioxide. Dry, saturated steam is generated in the furnace boiler tubes and is heated further in the super heater section of the furnace. This high pressure, superheated steam drives the steam turbine coupled to an electric generator. The low-pressure steam leaving the steam turbine is condensed, and the condensate is pumped back to the boiler for conversion into steam again and the process continues. A schematic of advanced pulverized coal-fired forced circulation boiler equipped with scrubbers for flue gas desulfurization (FGD) and deNO$_x$ unit for significant reduction of NO_x is shown in Figure 8.1. Subcritical operation refers to steam pressure *below* 22.0 MPa (~3200 psi) and steam temperature around 550°C (1025°F). Subcritical PC units have generating efficiencies between 33 and 37% (HHV), depending on coal quality, operations and design parameters, and location.

Key material flows and conditions for a 500 MWe subcritical PC unit is shown schematically in Figure 8.2 (Rubin 2005; Parsons *et al.* 2002; MIT Report 2007).

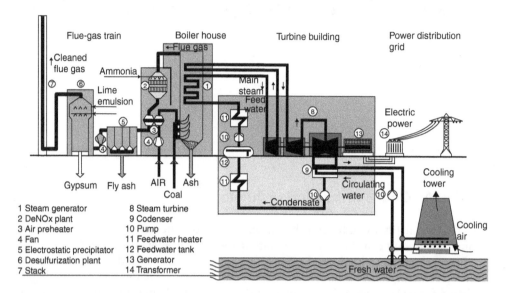

Figure 8.1 Advanced pulverized coal-fired power plant (*Source:* Termuehlen & Empsberger 2003, Janos Beer 2009; reproduced with the permission of Professor Janos Beer © Elsevier).

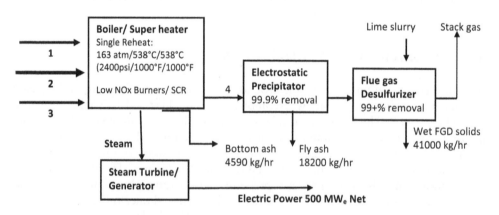

Figure 8.2 Sub critical PC 500 MW$_e$ unit without CO_2 capture showing stream flows and conditions: 1 – Feed air (2,110,000 kg/hr); 2 – Coal feed(208,000 kg/hr); 3 – Infiltration air (335,000 kg/hr); 4 – exhaust gas (159°C); Lime slurry (22600 kg/hr – lime, 145000 kg/hr – water); The stack gas is released at 2,770.000 kg/hour at 55°C, 0.10 MPa (*Source:* MIT 2007, Redrawn).

This unit burns 208,000 kg/hour of coal and requires about 2.5 million kg/hour of combustion air. Air infiltrates into the boiler because it operates at below-atmospheric pressure so that hot, untreated combustion gases do not escape into the environment. The particulate material is removed by an electrostatic precipitator to an extent of 99.9%, most of it as fly ash. Particulate emissions to the air are 11 kg/hr. NO_x emissions is reduced to 114 kg/hr by a combination of low-NO_x burners and SCR. The flue gas desulfurization unit removes 99+% of the SO_2 reducing to 136 kg/hr. For Illinois #6 coal, the mercury removal with the fly ash and in the FGD unit should be 70–80% or higher.

For these operating conditions, the Carnegie Mellon Integrated Environmental Control Model (IECM) projects a generating efficiency of 34.3% for Illinois #6 coal. For Pittsburgh #8 (bituminous coal) at comparable SO_x and NO_x emissions, IECM projects a generating efficiency of 35.4% (Rubin 2005). For Powder River Basin sub bituminous coal and North Dakota lignite at comparable emissions, IECM projects generating efficiencies of 33.1% and 31.9% respectively.

The by-products, namely, coal mineral matter, fly ash and bottom ash can be used in cement and/or brick manufacture. The wet solids formed by desulfurization of the flue gas may be used in wallboard manufacture or disposed by environmental-safe method.

Booras and Holt (2004), using an EPRI electricity generating unit design model, project 35.6% generating efficiency for Illinois #6 coal, at 95% sulfur removal and <0.1 lb NO_x/million Btu. Under the same operating and emissions control conditions, they calculated a generating efficiency of 36.7% for Pittsburgh #8 coal, similar to the efficiency reported by the NCC study (NCC 2004). The efficiency difference between Illinois #6 and Pittsburgh #8 is due to the difference in coal quality and is the same for both models, about 1%.

The IECM and EPRI model differences are attributed to the higher levels of SO_x and NO_x removal that were used and to differences in model parameter assumptions. For Illinois #6 coal, increasing SO_x and NO_x removal from the levels used by Booras and Holt to those used in the MIT study (2007) reduces the generating efficiency by about 0.5 percentage points. The rest of the difference is almost certainly due to model parameter assumptions, say, cooling water temperature, which has a large effect.

8.2 SUPERCRITICAL AND ULTRA SUPERCRITICAL PC TECHNOLOGIES

If the system can be designed to operate at a higher steam temperature and pressure (i.e., at higher steam cycle severity), the thermal efficiency of the sub-critical system can be enhanced. This represents an advance from subcritical to supercritical (SC) to ultra-supercritical (USC) PC power technologies.

Thermodynamically, supercritical is a state where there is no distinction between liquid and gaseous phase. At about 221 bar (22.1 MPa), water/steam reaches supercritical state, and above this operating pressure, the cycle is supercritical with a single-phase fluid contrary to the situation in the case of subcritical boiler. Table 8.1 shows higher steam pressure and temperature resulting in higher efficiencies of the PC plant. Switching from sub-critical to USC steam conditions would raise efficiency by around 4–6%. A potential 50% efficiency (LHV basis) is foreseen for ultra supercritical technology with the availability of proper boiler materials.

In addition to increasing steam conditions, Schilling (1993) has illustrated how measures such as decreasing air ratio (excess air factor) and stack gas temperature,

Table 8.1 Approximate Steam Pressure, Temperature and Efficiency ranges for sub-critical, super critical and ultra super critical PC power plant.

PC Plant	Main steam pressure (MPa)	Main steam temperature (°C)	Reheat steam temperature (°C)	*Efficiency, % net, (HHV basis)
Sub-critical	<22.1	Up to 565	Up to 565	33–39
Super critical	22.1–25	540–580	540–580	38–42
Ultra super critical	>25	>580	>580	>42

*Bituminous coal, (*Source:* Nalbandian 2009)

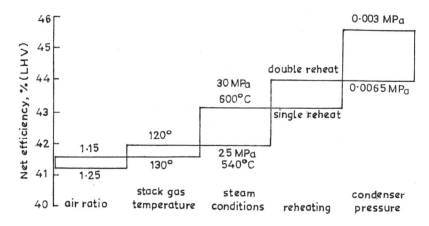

Figure 8.3 Effect of various measures for improving the efficiency (LHV) of pulverized coal-fired power generating plant (Redrawn from Schilling 1993).

reheating, and reducing condenser pressure achieve higher efficiency of PC SC power plants (Figure 8.3).

The first two steps in Figure 8.3 concern the *waste gas heat loss*, about 6–8%, the highest among the boiler's heat losses. The air ratio, called *excess air factor*, represents the mass flow rate of the combustion air as a multiple of the notionally required air for complete combustion. The excess air increases the boiler exit-gas mass flow and, hence, the waste gas heat loss. Improved combustion technology (e.g., finer coal grinding and improved burner design) would allow achieving complete combustion with lower excess air. Some of these steps – finer coal grinding and increasing the momentum flux of the combustion air through the burners – require small additional energy; but the efficiency gain due to the reduced excess air would more than compensate this energy penalty.

The boiler *exit gas temperature* can be reduced by suitable boiler design. The excess air of combustion and the lower limit of boiler exit gas temperature fired by a sulfur containing fuel are directly related. Higher excess air leads to an increase in the oxidation of SO_2 to SO_3 which promotes sulfuric acid formation. In turn, sulfuric acid vapor increases the dew point of the flue gas and hence raises the allowable minimum exit gas temperature. At an exit gas temperature of 130°C (266°F), a reduction of every 10°C (18°F) in boiler exit temperature increases the plant efficiency by about 0.3%.

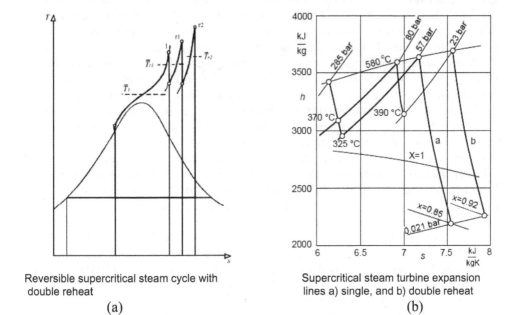

Reversible supercritical steam cycle with
double reheat

(a)

Supercritical steam turbine expansion
lines a) single, and b) double reheat

(b)

Figure 8.4 a&b: Temperature-Entropy (T-s) & Enthalpy-Entropy (h-s) diagrams of supercritical steam
cycle with reheat (*Source:* Buki 1998; Janos Beer 2009; Reproduced with the permission of
Professor Janos Beer © Elsevier).

The Rankine cycle efficiency is directly proportional to the pressure and temperature of heat addition to the cycle, and inversely to the condenser pressure, and hence to
the temperature of the cooling medium. The usual design basis for *condenser pressure*
in the USA is 67 mbar, whereas in Northern Europe with access to lower temperature
cooling water use, 30 mbar pressure. This difference can produce *an efficiency gain* of
more than 2%.

As steam pressure and temperature are increased, the steam becomes supercritical
beyond 225 atm (3308 psi), and increases the Rankine cycle efficiency due to the
higher pressure and higher mean temperature of heat addition (seen earlier). This is
illustrated by the T-s and h-s diagrams of supercritical steam cycles with reheat in
Figures 8.4 a&b. The supercritical fluid is expanded through the high-pressure stages
of a steam turbine, generating electricity. After expansion through the high-pressure
turbine stages, the steam is returned to the boiler to be reheated to boost the steam
properties and thereby increase the amount of electric power generated. Reheating,
single or double, raises the mean temperature of heat addition to the Rankine cycle
resulting in increase of the cycle efficiency. Figure 8.4b illustrates the expanding steam
returning once (shown as 'a') or twice (shown as 'b') from the turbine to the boiler for
reheating to the same initial temperature of 580°C.

It should be noted that the definition of super-critical and ultra super-critical boiler
pressure and temperature profiles differs from country to country. Also, the usage of
the term ultra super-critical varies, but the ranges given in the table are frequently used.

Efficiencies expressed in LHV (in Europe) are generally somewhat higher than HHV (in the USA) because the energy used to vaporize the water is not taken into account in LHV. As a result, for identical plant performance, the US efficiency is reported to be 2–4% lower than European efficiency (Nalbandian 2009).

Once-through boilers are favored in super critical plants because they do not require a boiler blow down. This helps the water balance of the plant with less condensate needing to be fed into water-steam cycle. These plants, however, require a condensate polisher and facility to maintain steam purity. Once-through boilers can be used with steam pressures more than 30 MPa without any changes in process engineering, though the tubes and headers need to be designed to withstand the pressure levels. The convective section of the plant contains the super heater and reheater and economizer parts. In the final stage, there is an air heater to recover heat which brings down the flue gas temperature to 120–150°C from economizer exit temperature of 350–400°C. Rotary air heaters are conveniently used in many large plants.

Supercritical electricity generating efficiency ranges from 37 to 42% (HHV), depending on the design, operating parameters, and coal type. Current state-of-the-art supercritical PC generation involves 24.3 MPa (3530 psi) and 565°C (105°F), giving a generating efficiency of about 38% (HHV) for Illinois #6 coal (MIT 2007). The values given in the Table 8.1 are for bituminous coal.

Supercritical PC power plants were commercialized in the late 1960s. A number of supercritical units were built in the USA during 1970s and early 1980s, but they encountered some materials and fabrication problems. Having overcome these problems, the current supercritical PC combustion is a proven and reliable technology and is currently operating in several countries. Several supercritical boilers are installed in Europe, Asia, Russia, Taiwan, and the USA. India has begun to introduce supercritical PC technology for *high ash-coals* and its operating experience at Sipat and Barh plants and the new units to be built is expected to help India as well as other countries that depend on high ash coals such as Bulgaria, China, Poland, Rumania and South Africa for power generation. The SC's share in coal-fired power generation has risen from 18% in 2004 to 25% in 2009 and continues to increase with capacity addition in China, India, and other countries. It is expected that about 50% of the incremental coal fired capacity addition between 2012 and 2017 will be supercritical units (Burnard and Bhattacharya IEA 2011).

The major units under construction/planning globally are the following (Burnard & Bhattacharya IEA 2011).

Major Supercritical Units: Installed, under construction/planned
Australia: Kogan Creek, 750 MW$_e$ 2007
Canada: Genesee Unit 3, 450 MW$_e$ 2005
China: Waigaoqio, 2 × 1000 MW$_e$ 2008; Yuhuan, 4 × 1000 MW$_e$ 2007–08;
   Under construction – 50,000 MW$_e$
   Planned by 2015 – >110,000 MW$_e$
India: Sipat, 3 × 660MW$_e$ 2007–09; Barh, 3 × 660 MW$_e$, 2009
   Jhajjar project (CLPIndia) 2 × 660 MW$_e$, 2012;
   UltraMega Projects – 2012, 5 × 4000 MW$_e$; unit size – 660 MW$_e$ or 800 MW$_e$

USA:	545 MW$_e$ and 890 MW$_e$ 2008; Oak Grove, Texas, 800 MW$_e$ 2009, 800 MW$_e$ 2010; Under construction – 6500 MW$_e$ 2009–12
Italy:	Torrevaldaliga Nord, 3 × 660 MW$_e$ 2010; Planned by 2015 – 3 × 660 MW$_e$
Mexico:	Pacifico, 700 MW$_e$ 2010
Netherlands:	Eemshaven, under construction, 2 × 800 MW$_e$ 2013
South Africa:	6 × 800 MW$_e$ 2011–15
Russia:	Berezovskya, 800 MW$_e$ 2011; Novocherkasskaya, 330 MW$_e$ (CFB), 2012; Petrovskaya, 3 × 800 MW$_e$ 2012–14
Germany:	Niederaussem, 1000 MW$_e$ (lignite), 2003; Walsum, 750 MW$_e$ 2010; Neurath – under construction, 2 × 1100 MW$_e$ (largest lignite-fired USC plants) 2011 Hamm, under construction, 2 × 800 MW$_e$ 2012
Poland:	Lagisza, 460 MW$_e$ (CFB) 2009; Belchatow, 833 MW$_e$ 2010
Korea:	Tangjin, 2 × 519 MW$_e$ 2006; 5 × 500 MW$_e$ & 2 × 870 MW$_e$ 2008–10

*All are PC combustion plants unless otherwise mentioned

Current materials based on ferrite/martensite alloys allow steam temperatures up to 600°C and pressures of 300 bar in the state-of-the-art super critical boilers. It was however recognized that nickel-based superalloys would have better scope for progress of the technology; i.e. the peak operating conditions could be 720°C and 370 bar which should allow the coal power plants with nearly 50% thermal efficiency (Shaddix 2012).

The development of new materials and capabilities has further expanded the potential operating range; and the coal-fired power generating industry in Europe and Japan (Blum 2003), has moved to higher steam pressure and temperature, mainly higher temperatures, referred to as *ultra supercritical* (USC) conditions.

USC parameters of 300 bar and 600/600°C (4350 psi, 1112/1112°F) are realized today, resulting in efficiencies of 45% (LHV) and higher, for bituminous coal-fired power plants. There are several years of experience with these high temperature (600°C) plants in service, with excellent availability (Blum & Hald 2002); and the operational availability of these units to date has been comparable to that of subcritical plants. USC steam plants in service or under construction in Europe and in Japan during the last decade listed by Blum and Hald (2002) are given by Janos Beer (2009). Further improvement in efficiency is explained in the last chapter on Outlook for clean coal technologies.

The USC plants in operation in Denmark, Germany, Italy and Japan account for less than 1% of global power generation. China is also building USC plants, e.g., Yuhuan Plant in Zhejiang province with two 1000 MW$_e$ units and steam parameters of 26.25 MPa/600°C/600°C (Minchener 2010). The improved efficiency represents a reduction of about 15% in the CO_2 emission compared to the emission from installed capacity. Combined heat-power generationg plant operating in Denmark since 2001 is shown below. This is the world's most energy-efficient plant using *ultrasupercritical* steam conditions.

Avadore Power Station, Denmark; Unit 2: CHP: 585 MW of electricity and 570 MW of heat; USC boiler; Coal primary fuel, gas and biomass other fuels; Operating since 2001; World's most energy-efficient plant with 49% efficiency (while producing power); (*Source*: Wikipedia, Avedorevaerket, Copenhagan, Denmark, By G®iffen).

The timeline of materials development and its relationship with advanced steam parameters are discussed by Henry *et al.* (2004). The plant efficiency *increases* by about 1% for every *20°C rise* in superheat temperature. Figure 8.5 illustrates the percentage reduction of CO_2 emissions with the improvement of efficiency (Booras and Holt 2004).

Reduction of CO_2 emissions is the main issue to be successfully determined in the supercritical and ultra supercritical technology. This technology combined with the best available air pollution-control equipment will reduce current pollution levels by burning less coal per mega watt-hour produced, thus reducing total emissions including carbon dioxide.

A block diagram of a 500 MWe ultra-supercritical PC power plant showing key flows is given in Figure 8.6 (MIT Report 2007). The coal combustion side of the boiler and the flue gas treatment are the same as for a subcritical boiler. Coal required to generate a given amount of electricity is about 21% lower than for subcritical generation, which means that CO_2 emissions per MW_e-h are reduced by 21%. The efficiency projected for these design-operating conditions is 43.3% (HHV) as against 34.3% for subcritical conditions. In this case, 2,200,000 kg/hr of stack gas at 55°C, 1 atmos, is released containing $N_2 = 66.6\%$, $H_2O = 16.7\%$, $CO_2 = 11\%$, $O_2 = 4.9\%$, $Ar = 0.8\%$, $SO_2 = 22$ ppm, $NO_x = 38$ ppm, $Hg < 1$ ppb (vol %).

Benefits of supercritical and ultra supercritical technologies include increased efficiency, reduced fuel costs and plant emissions including CO_2, and reduced effect of plant load operation. The costs are comparable to subcritical technology and can be fully integrated with the appropriate new or retrofitted CO_2 capture technology (Nalbandian 2009). However, if the operating steam conditions reach around 720°C

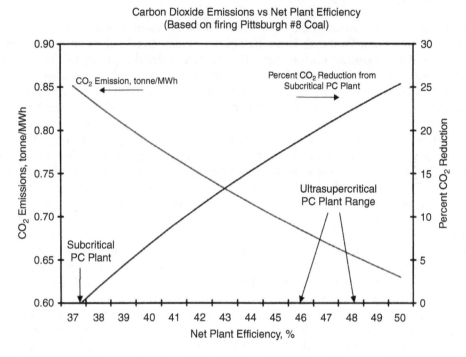

Figure 8.5 CO_2 Emission as a function of Plant Efficiency (HHV) (*Source:* Booras and Holt 2004, Janos Beer 2009, reproduced with the permission of Professor Janos Beer © Elsevier).

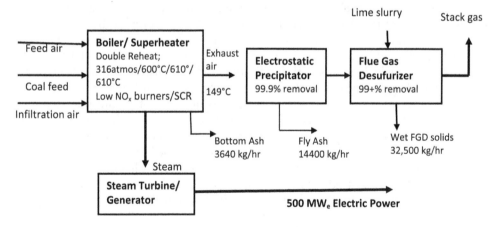

Figure 8.6 Ultra Supercritical 500 MW_e PC Power plant without CO_2 capture – Feed air: 1,680,000 kg/hr; coal feed: 164,000 kg/hr; Infiltration air: 265000 kg/hr; lime slurry: 17900 kg/hr lime and 115,000 kg/hr water (*Source:* MIT report 2007, Redrawn).

and 370 bar (with nickel-based alloys) the issues that arise would be high temperature corrosion and an increased slagging tendency at the superheater. Issues related to large scale implementation of SC/USC power plants are briefly mentioned in the last chapter.

8.3 FLUIDIZED BED COMBUSTION

Fluidized bed combustion (FBC) is a variation of PC combustion in which coal is burned with air in a fluid bed. It is a flexible combustion technology for power generation and has significant advantages: (i) compact boiler design, (ii) fuel flexibility, can burn a variety of fuels, coal, biomass, wood, petro-coke and coal cleaning waste, (iii) higher combustion efficiency, (iv) environment-friendly, reduced emission of SO_x and NO_x (>90%), and (v) decreased capital costs by 8 to 15% compared to PC combustion plants.

When evenly distributed high pressure air or gas is passed upward through a finely divided bed of solid particles such as sand (inert material) supported on a fine mesh, at a particular velocity of air, the individual particles appear suspended in the air stream – the bed is then called 'fluidized'. With further increase in air velocity, there is bubble formation, vigorous turbulence, rapid mixing and formation of dense defined bed surface. The bed of solid particles appears as a fluid – 'bubbling fluidized bed' that provides more efficient chemical reactions and heat transfer. If sand particles in a fluidized state are heated to the ignition temperature of coal, while injecting coal continuously into the bed, the coal will burn rapidly and bed attains a uniform temperature; and the combustion takes place at about 840° to 950°C which is much below the ash fusion temperature. Hence, melting of ash and associated problems are avoided, as well as NO_x formation is also minimized. The high coefficient of heat transfer due to rapid mixing in the fluidized bed and effective removal of heat from the bed through heat transfer tubes fixed in the bed as well as walls of the bed enables to achieve lower combustion temperature. The gas velocity is maintained between minimum fluidization velocity and particle entrainment velocity. This ensures stable operation of the bed and avoids particle entrainment in the gas stream. Most operational boilers of this type are of the Atmospheric Fluidized Bed Combustion (AFBC), and have a wide capacity range, 0.5 T/hr to over 100 T/hr. In practice, coal is crushed to a size of 1–10 mm depending on the rank of coal and fed to the combustion chamber. The atmospheric air, which acts as both fluidization and combustion air, is delivered at a pressure after being preheated by the exhaust fuel gases. The in-bed heat transfer tubes carrying water act as the evaporator. The gaseous products of combustion pass over the superheater sections of the boiler, the economizer, the dust collectors and the air preheater before being released to atmosphere. With the addition of a sorbent such as limestone or dolomite into the bed, much of the SO_2 formed can be captured. Due to the mixing action of the fluidized bed, the flue gases contact with the sorbent, undergo a thermal decomposition (known as calcinations) which is an endothermic reaction. With limestone as sorbent, the calcination reaction occurs above 774°C (1400°F):

$$CaCO_3 + heat \rightarrow CaO + CO_2$$

(limestone + heat → lime + carbon dioxide) (8.1)

Calcination is considered to be necessary before the limestone can absorb and react with gaseous sulfur dioxide. The gaseous SO_2 is captured via the following chemical reaction to produce calcium sulfate, a solid substance:

$$CaO + SO_2 + \tfrac{1}{2}O_2 \rightarrow CaSO_4$$

(lime + sulfur dioxide + oxygen → calcium sulfate) (8.2)

The sulfation reaction always requires an excess amount of limestone which depends on the sulfur content in the fuel, temperature of the bed, and the physical and chemical nature of the limestone. The sorbent's main task in FBC process is to maintain air quality, and also the bed stock that influences the heat transfer characteristics as well as the quality and treatment of ash. Depending on the sulfur content of the fuel, the limestone can comprise up to 50% of the bed stock, with the remaining being fuel ash or any inert material. This is especially true of FBCs firing waste from bituminous coal cleaning plants that contains high sulfur content. When the bed is comprised of a large quantity of calcium (oxide or carbonate), the ash disposal is not easy because the pH of the ash can become very high. More than 95% of the sulfur pollutants in the fuel can be captured inside the boiler by the sorbent; also captures some heavy metals, though not as effectively as the cooler wet scrubbers on conventional units.

The technology is more suited to fuels that are difficult to ignite, like petroleum coke and anthracite, low quality fuels like high ash coals and coal mine wastes, and fuels with highly variable heat content, including biomass and mixtures of fuels.

In the early 1980s, FBC technology was used in the USA to burn petroleum coke and coal mining waste for power generation. In Germany, FBC evolved while trying to control emissions from burning sulfate ores without using emission controls.

FBC systems fit into two groups, non-pressurized systems (AFBC) and pressurized systems (PFBC), and two subgroups, bubbling or circulating fluidized bed: (i) Non-pressurized FBC systems operate at atmospheric pressure and are the most widely applied type of FBC. They have efficiencies similar to PC combustion plants, around 30–40%, (ii) Pressurized FBC systems operate at elevated pressures and produce a high-pressure gas stream that can drive a gas turbine, creating a more efficient combined cycle system, over 40%, (iii) Bubbling fluidized-bed unit uses a low fluidizing velocity so that the particles are held mainly in a bed and is generally used with small plants offering a non-pressurized efficiency of around 30%, and (iv) Circulating bed uses a higher fluidizing velocity so that the particles are constantly held in the flue gases; used for much larger and high efficiency (over 40%) plants.

8.3.1 Bubbling fluidized-bed units

These systems characteristically operate with a mean particle size of between 1,000 and 1,200 mm and fluidizing velocities between the minimum fluidizing velocity and the entraining velocity of the fluidized solid particles (4 to 12 ft/s).

Under these conditions, a defined bed surface separates the high solids loaded bed and the low solids loaded freeboard region. Most bubbling-bed units, however, utilize reinjection of the solids escaping the bed to obtain satisfactory performance. Some bubbling-bed units have the fuel and air distribution so arranged that a high degree of internal circulation occurs within the bed (Virr 2000). A generalized schematic of a bubbling fluidized-bed boiler is shown in Figure 8.7. For fuels with high moisture content and low heating value such as biomass, municipal wastes, paper and pulp industry wastes, sludge etc. and small capacities, bubbling fluidized bed technology is recommended. Also, in situations where these types of fuels need to be used to substitute expensive imported liquid fuels or natural gas for energy generation in industry and district heating, and where distributive energy generation based on local and low quality fuels become attractive, BFBC is desirable.

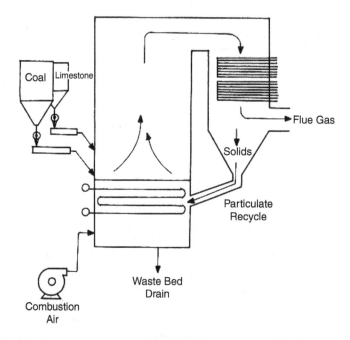

Figure 8.7 Generalized schematic diagram of a bubbling fluidized-bed boiler (Redrawn from Gaglia & Hall 1987).

BFBC boilers and furnaces are economical to use in the power range, 0.5 MW$_{th}$ to 500 MW$_{th}$, which is suitable unit size for distributive energy generation. Also, it is possible to burn fuels of different origin and quality simultaneously or alternatively in BFBC boilers.

The BFBC technology is considered the best solution for distributive energy generation, and hence highly appropriate for developing countries; and they can produce and maintain BFBC boilers and furnaces locally (Oka 2001a).

8.3.2   Circulating fluidized-bed (CFB) units:

CFB technology utilizes fluidization process to mix and circulate the fuel particles with limestone as they burn in a low temperature (800°–900°C) combustion process. The combustion air is fed in two stages – primary air, direct through the combustor, and secondary air, way up the combustor above the fuel feed point. The limestone captures the sulfur oxide pollutants as they are formed during the burning process, while the low burning temperature minimizes the creation of NO$_x$ pollutants.

The fuel and limestone particles are recycled over and over back to the process which results in high efficient burning of the fuel, capturing pollutants, and for transferring the heat energy into high quality steam used to produce power. Due to the vigorous mixing, long burning time, and low temperature combustion process, CFBs can cleanly burn virtually any combustible material. Waste fuels with a high percentage of non-combustibles (heating value 5–35 MJ/kg) are considered highly

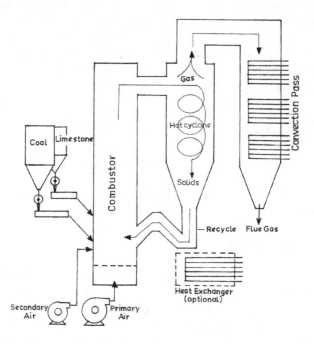

Figure 8.8 Generalized schematic diagram of a circulating fluidized-bed boiler (Redrawn from Gaglia & Hall 1987).

suitable. Unlike conventional steam generators, CFBs capture and control unwanted pollutants during the burning process without relying on additional pollution control equipment.

The atmospheric CFB units operate with a mean particle size, 100 and 300 mm, and fluidizing velocities up to about 30 ft/s. A generalized schematic of a CFB boiler is shown in Figure 8.8. The bed parameters are so maintained that solids elutriation from the bed is promoted. They are lifted in a relatively dilute phase in a solids riser, and a down-comer with a cyclone provides a return path for the solids. There are no steam generation tubes immersed in the bed. The heat transfer surfaces are usually embedded in the fluidized bed and steam generated is passed through the conventional steam cycle operating on Rankine Cycle. i.e., generation and super heating of steam takes place in the convection section, water walls, and at the exit of the riser.

Advantages of CFB technology (Jalkote– EA-0366)

(a)   Fuel flexibility: As the furnace temperatures are lower than the ash fusion temperatures for all fuels, the furnace design is independent of ash characteristics which allow a given furnace to handle a wide variety of fuels. Low grade coals, high-ash coals and low volatile coals, biomass, wastes and other difficult fuels can be burnt easily in the CFBC boiler;

(b)   Improved combustion efficiency: The long residence time in the furnace resulting from collection/recirculation of solid particles via the cyclone, plus the vigorous solids/gas contact in the furnace caused by the fluidization airflow lead to high combustion efficiency, even with fuels tough to burn. Around 98–99% carbon burnout has been achieved. The very high internal and external re-circulating rates of solids result in uniform temperatures throughout the combustor;

(c)   Control of pollutants: By feeding limestone sorbent into the bed, sulfur removal is accomplished in the combustion zone itself; SO_2 removal efficiency of over 95% has been demonstrated using a good sorbent. Low furnace temperature plus staging of air feed to the furnace produce very low NO_x emissions. Chlorine and fluorine are largely retained in ash;

(d)   Operating flexibility: Possible to design for cyclic or base load operation; Part loads down to 25% of MCR and load change rates of up to 7% per minute are possible;

(e)   Simplified fuel feeding: Crushing is sufficient for fuel feed; fuel pulverization is not required;

(f)   Proven technology: The commercial availability of most of these units exceeds 98%; e.g., Foster Wheeler has more than 150 CFB steam generators in operation;

(g)   More economical: CFBC boilers are more economical than AFBC boilers for industrial application requiring more than 75–100 T/hr of steam. For large units, the taller furnace characteristics of CFBC boilers offers better space utilization, greater fuel particle and sorbent residence time for efficient combustion and SO_2 capture, and easier application of techniques for NO_x control than AFBC steam generators. Several new lignite-burning CFB units have been recently constructed (Wilson & Monea 2004).

CFBC system components: The following major components constitute CFBC unit: Fluidized Bed Combustor and associated systems, Fluid bed Heat exchanger, Solids separation system – Recycling Cyclone for external recirculation and U-beam particle separators for internal particle recirculation, Conventional steam turbine systems, Fuel preparation and feeding System, and Ash Removal System.

Fuel feed system and sorbent feed system are very crucial. Fuel feed system is either pneumatic or wet type, and coal is fed as coal-water mixture as it offers more even burning. Ash and sulfur content in the coal-feed decides the optimum system design. For fuels with low ash contents, coal-water mixture (slurry) is favored; for high ash coals, large quantities of water are needed, which affects the efficiency. The fuel coal-water paste contains 25% water by weight, and the fuel feed size is lower than 0.75 in.

Sorbents are not combustibles and are generally fed either continuously or intermittently. In the latter case, lock hoppers are used. The sorbent is crushed to about 3 mm top size, dries and fed in lock hoppers.

Global status of CFB: Since early 1990s, the focus on research and development activity in developed countries has shifted from bubbling FBC to circulating FBC. Low prices of the liquid fuel and gas and more stringent emission standards make burning of coal in bubbling FBC boilers and furnaces non competitive compared to the boilers burning liquid fuels or natural gas in the range of unit power from small to medium.

Hundreds of CFBC boilers are operating worldwide. Foster Wheeler has more than 150 CFB steam generators in operation. The commercial availability of most of these units exceeds 98%. Several of M/s Lurgi Lentjes Babcock Energietechnik's CFBC steam generators (>8700 MW) and M/s Babcock & Wilcox's CFBC units are operating worldwide. These are the major manufacturers of CFBC systems (Jalkote).

In the USA at Morgentown Energy Center, North Dakota and Texas, lignite coals were burned in various fluidized beds with combustion efficiency up to 99% (Rice et al. 1980). The most recent coal fired plants tended to be circulating fluidized bed (CFB) plants such as Warrior Run, Red Hills (250 MW), and Puerto Rico with the selective catalytic reduction (SCR), which has some of the lowest emissions for a coal-fired plant in the world. The largest operating subcritical CFBC unit is of 320 MW$_e$ at Jacksonville, Florida.

In Australia, fluidized bed combustor burning lignite is used for power generation. The lignite was dried previously to reduce the moisture content from 70% to 20% (McKenzie 1978).

In Taiwan, CFB units (2 × 150 MW) are currently being supplied, targeting very stringent emission levels (NO$_x$ < 96 ppm and SO$_2$ < 235 ppm corresponding to a 98% desulphurization level). Two thermal power plants (80 MW$_e$) with CFB technology have been running since 1989 and 1993 in Japan. China has developed a number of fluidized bed combustors of varying scale to burn low quality coals including lignite. China is estimated to have about 2000 fluidized bed hot water and steam generator systems having capacity as high as 10,000 tons/h (Cao and Feng 1984; Bazzuto et al. 2001). Two 125 MW CFB plants have been in progress in India (Eskin & Hepbasli 2006).

Fluidized bed technology in Europe is wide-spread; France, Germany, Sweden, Spain, Turkey and Eastern Europe (especially in Poland) utilize for burning poor quality fuels. Emile Hucket and Gardanne in France are the two biggest applications; Gardanne is designed to burn bituminous coal and heavy petroleum residues. The furnace with 'pant-leg' bottom is large enough for a 300 MWe unit firing regular coal, thus offering additional fuel flexibility (Bazzuto et al. 2001). There has been considerable increase in the capacity of CFBC units recently with the total plant capacity globally at around 20 GW, which is projected to grow.

CFBC technology previously available in sub-critical designs, has reached an economic scale of SC conditions. Due to its fuel flexibility, a large size CFBC plant, >300 MW$_e$ can be built to operate on coal and biomass of different types which can offer less CO$_2$ emissions while providing power at affordable cost. It is estimated that a supercritical CFBC plant of size, 660 MW$_e$, burning 20% of biomass will emit 32% less CO$_2$ emissions than a conventional plant (Giglio and Wehrenberg 2009). The first largest and first super critical CFBC unit, 460 MW$_e$, 282 bar, 563°C/582°C, designed by Foster-Wheeler is operating in Lagisza, Poland (2009) using Polish lignite coal with as-received heating value ranging 4300–5500 kcal/kg and has a design efficiency of 43.3% (LHV, net). In addition, the design allows provision for introduction of coal slurry through lances up to 30% total fuel heat input (Utt & Giglio 2012). A second unit of 330 MW$_e$ will be installed in Russia at the Novocherkasskaya GRES facility (Jantti et al. 2009).

460 MWe Łagisza CFB OTSC Unit in Poland (*Source:* Utt & Giglio, Foster-Wheeler 2012; Reproduced with the permission of Robert Giglio).

Table 8.2 CO$_2$ Emissions from a 600 MW$_e$ Plant (millions tons/year).

Plant type	Emissions (million tons/year)	Reduction with respect to PC coal plant
PC Coal plant	5407	–
CFB plant with 20% Biomass	4325	20%
SC CFB plant with 20% Biomass	3644	32%
CFB Flexi-Burn plant	54	50%
CFB Flexi-Burn Plant with 20% Biomass	−54	110%

(Data taken from Giglio & Wehrenberg 2009)

Designs for SC units of 600 MW$_e$ and 800 MW$_e$ are now available. Since USCs require much higher than 600°C superheat or reheat temperatures, the designs need to be improved considerably, as the CFBC technology operates at comparatively low temperatures. If that could happen along with achieving steam pressure of 28 MP$_a$, efficiency of over 45% (LHV, net) or 43% (HHV, net) for hard coal can be achieved (Burnard & Bhattacharya IEA 2011).

Foster Wheeler is developing Flexi-Burn CFB technology which allows the CFB to produce a CO$_2$ rich-flue gas contributing to global CCS efforts (Giglio and Wehrenberg 2009). The technology has the potential to reduce coal plant CO$_2$ emissions to the atmosphere by over 90% while minimizing the cost impact and technology risk to consumers. Further, by combining its carbon capture capability with its biomass co-firing capability, a net reduction in atmospheric CO$_2$ as shown in Table 8.2 can be achieved. This Foster-Wheeler technology uses a mixture of oxygen and recycled CFB flue gas instead of air. By doing this, the flue gas becomes rich in CO$_2$ (containing over 90% CO$_2$ on a dry basis), rather than rich in nitrogen, and hence doesn't require expensive and energy intensive equipment to remove the CO$_2$ from its flue gas. Further, it has the potential to produce carbon free electricity at a very low cost as compared to other technologies.

Biomass Co-firing: To reduce the greenhouse gas intensity, one option is to use biomass cofiring, which is possible for both PC power plants and Gasification plants. Industrial practice shows that up to about 15% by thermal content biomass cofiring in a coal-based power plant does not have negative effect on efficiency or availability of the coal plant. If large biomass cofiring (upto 30–50%) is desired, the biomass could be gasified in a separate, atmospheric pressure CFB gasifier, and the gas so generated could be piped without cleanup to and cofired with coal in the pulverized coal boiler. Because of the lower operating temperature of the CFB gasifier, the biomass ash is generated in a form that makes it easily removed from the gasifier.

8.3.3 Pressurized fluidized bed combustion (PFBC)

Pressurized fluidized bed combustion is an advanced coal combustion system with higher thermal efficiency and lower pollution than typical modern systems. The PFBC systems utilize a combination of Rankine Cycle and Brayton Cycle with the objective of achieving high cycle efficiency and lower emissions. In the first-generation PFBC system, a mixture of coal and limestone is fluidized and burned efficiently under *high air pressure* (forced draft air) from a compressor (860°C, and 16–18 bars). The heat release rate in the bed is proportional to the bed pressure and hence the heat release in the combustor is high at high air pressures. Therefore, a deep bed is used to extract large amount of heat which will allow improving the combustion efficiency and sulfur dioxide absorption in the bed. Depending upon pressure, full load bed depths range between 3.5 and 4.5 meters. The in-bed tubes recover the generated heat as steam which drives a steam turbine generator. The hot exhaust gas from the combustor which is still under pressure is recovered, cleaned off all the suspended particulate using high efficiency cyclones and used to drive a gas-turbine generator. The steam turbine is the major source of power in PFBC, delivering about 80% of the total power output; the remaining 20% is produced in gas turbines. The first large-scale demonstration of PFBC technology in the USA (Tidd project, 70 MW capacity) is shown in Figure 8.9. At Tidd Power Station, gas turbine is a two-shaft machine. On one shaft, the variable speed, low-pressure turbine is coupled to low pressure compressor; and on the other shaft, the high-pressure turbine drives both the compressor and generator. There is an intercooler between low and high pressure compressors. The advantage of two-shaft design is that the free spinning low-pressure turbine can hold reduced gas temperature and consequent reduction in airflow, as load is reduced while keeping constant speed at generator. Fuel feed system is either pneumatic or wet type. The fuel is fed in the form of coal-water paste with 25% water by weight. This has been typically followed at Tidd. The fuel feed size is lower than 0.75 in.

This TIDD was shut down in 1994 after 8-year demonstration in which considerable useful data recovered and experience obtained (DOE 1999):

(a)   Sorbent size had the greatest effect on SO_2 removal efficiency and the stabilization and heat transfer characteristics of the fluidized-bed;

(b)   Tests showed that SO_2 removal efficiencies of 90 & 95% were achievable at full load and at 1580°F with Ca/S molar ratios of 1.14 and 1.5, respectively;

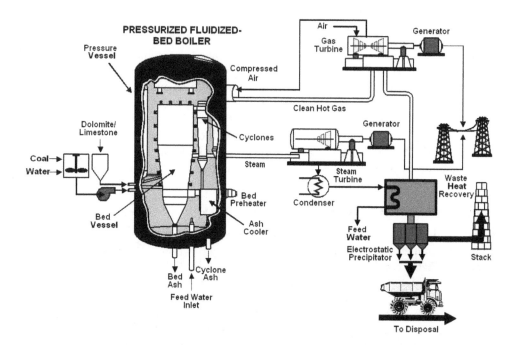

Figure 8.9 Large-scale PFBC technology at TIDD, Ohio, USA (*Source:* US DOE – Tidd PFBC Demonstration Project, June 1999).

(c) During the test period, NO_x emissions ranged from 0.15 to 0.33 lb/10^6 Btu, (typically around 0.20 lb/10^6) while CO emissions remained below 0.01 lb/10^6 Btu, and Particulate emissions were less than 0.02 lb/10^6 Btu.

(d) Combustion efficiency ranged, on average, from 99.3% at low bed levels to 99.5% at moderate to full bed height levels;

(e) Heat rate was 10,280 Btu/kWh based on HHV of the fuel and gross electrical output, or 33.2% efficiency (The low values were due to the use of a 1950s vintage steam generator without reheat, less than optimum performance of the prototype gas turbine, no attempt to optimize heat recovery, and the small scale of the demonstration unit);

(f) An Advanced Particulate Filter (APF) using a silicon carbide candle filter array on a flue gas slip stream achieved 99.99% filtration efficiency on a mass basis;

(g) Lack of erosion in the in-bed tube assembly and sustained operation with expected performance confirmed the PFBC boiler's readiness for commercial application;

(h) The ASEA Stal GT-35P gas turbine proved capable of operating commercially in a PFBC flue gas environment, but failed to satisfy performance requirements for reasons unrelated to the PFBC technology.

PFBC system can be used for cogeneration (steam and electricity) or combined cycle power generation. The combined cycle operation (gas turbine and steam turbine) improves the overall conversion efficiency by 5 to 8%. However, further research efforts are required to understand NO_x formation mechanisms, desulfurization reactions, and heat transfer.

Advanced PFBC

(1)   An early first generation PFBC system increases the gas turbine firing temperature by using natural gas in addition to the vitiated air from the PFB combustor. This mixture is burned in a topping combustor to provide higher inlet temperatures for greater combined cycle efficiency. The problem is with the use of natural gas, usually costly compared to coal, although becoming cheaper in recent years.

(2)   In more advanced second-generation systems, a pressurized carbonizer is incorporated to process the feed coal into fuel gas and char. The PFBC burns the char to produce steam and to heat combustion air for the gas turbine. The fuel gas from the carbonizer burns in a topping combustor linked to a gas turbine, heating the gases to the rated firing temperature of the combustion turbine. Steam is produced by the heat recovered from the gas turbine exhaust, and is used to drive a conventional steam turbine, resulting in a higher overall efficiency for the combined cycle power output. These systems are called APFBC, or advanced circulating pressurized fluidized-bed combustion combined cycle (ACPFBCC) systems. An APFBC system is entirely coal-fueled;

(3)   GFBCC: Gasification fluidized-bed combustion combined cycle systems have a pressurized circulating fluidized-bed (PCFB) partial gasifier feeding fuel *syngas* to the gas turbine topping combustor. The gas turbine exhaust supplies combustion air for the atmospheric circulating fluidized-bed combustor that burns the char from the PCFB partial gasifier

Advantages of PFBC

(a)   Improved Cycle efficiency (lower heat rate): The plant efficiency can be significantly improved by combining Rankine Cycle and Brayton Cycle. For the first generation PFBC combined cycles, efficiencies nearing 40% and heat rates of about 8500 Btu/kWh can be achieved. Second generation advanced combined cycles is projected to reach efficiencies more than 45% and heat rates as low as 7500 Btu/kWh;

(b)   Reduced emissions & Improved combustion: In addition to combined cycle operation and higher combustion rate, the increased pressure and corresponding air/gas density allow much lower fluidizing velocities (around 1 m/sec) which reduce the risk of erosion for immersed heat transfer tubes. The combined effect of lower velocity and deeper beds to hold the required heat transfer surface results in greatly increased in-bed residence time which reduces SOx emissions thereby improving combustion efficiency;

(c)   Reduced boiler size; and

(d)   Modularity.

PFBC system Components: The System comprises the following major components: Boiler and associated systems; Conventional steam turbine systems; Gas turbine; Hot Gas cleaning system; Fuel preparation and Feeding system; Ash Removal system. As the Gas turbine is driven by hot pressurized gases from the boiler and concurrently supply combustion air to boiler and generate electricity, the following characteristics are desirable: (a) it should provide volumetric-flow-characteristics which would permit nearly constant fluidizing velocity; excess air ratio and velocity into gas cleaning system (vital for cyclones), (b) it should balance the opposite requirements for a low air flow to boiler, a high air flow to gas turbine at low load; (c) withstand particulate loading in gases without significant damage; and (d) accept relatively low inlet gas temperature (around 8400°C) throughout the load range.

Gas cleaning systems have not yet been established in PFBC systems. Normally the high efficiency cyclones which are successfully tested, and Candle filters and ceramic tube filters which are still under test are employed.

Status of PFBC technology: The Tidd demonstration plant in the USA is already briefed. A demonstration plant, 130 MW$_e$ (co-generation) has been operating in Stockholm, Sweden since 1991 fulfilling all environmental conditions. Another demonstration plant, 80 MW$_e$ is operating in Escatron, Spain using 36% ash black lignite. A 70 MW$_e$ demo-plant operated at Wakamatsu from 1993 to 96.

Presently, a 350 MW$_e$ PFBC power plant is planned in Japan and another is on order to be operated at SPORN, USA. UK has gained enough experience with a 80 MW$_e$ PFBC plant in Grimethrope operating from 1980–1992 and is now offering commercial PFBC plants and developing second generation PFBC. ABB-Sweden is the leading manufacturer which has supplied the first three demonstration plants in the world and is now offering 300 MW$_e$ capacity plants. In India, BHEL-Hyderabad has been operating a 400 mm PFBC for nearly a decade providing useful data. IIT Madras has a 300 mm diameter research facility built with NSF (USA) grant. Indian government is considering a proposal by BHEL for setting up a 60 MW$_e$ PFBC plant (PS Jalkote).

8.4   PERFORMANCE AND COST COMPARISON

All the four primary coal generating technologies (subcritical, supercritical, ultra supercritical and CFB) discussed so far are *air-blown,* and compose essentially all the coal-based power generation units currently in operation and under installation. Table 8.3 summarizes representative operating performance and economics for these technologies (MIT Report 2007).

The data with CO_2 capture given in Table 8.3 will be discussed in a subsequent chapter. The performance data for the CFB unit in Table 8.3 is based on lignite. The lignite has a heating value of 17,400 kJ/kg and low sulfur; and the coal feed rate is higher than for the other technologies because of the lower heating value of the lignite.

Evaluation of technologies

To evaluate the technologies on a consistent basis, the design performance and operating parameters for the generating technologies (mentioned above) were based on the Carnegie Mellon Integrated Environmental Control Model, version 5.0 (IECM) (Rubin 2005), a modeling tool specific to coal-based power generation (Booras & Holt 2004). There are other modeling tools that could have been used. Each would have given rather different results because of numerous design and parameter choices, and engineering approximations included in each. IEC Model results are consistent with other models when operational differences are accounted for. The units all use

Table 8.3 Representative performance and economics of air-blown PC Generating technologies

	Subcritical PC		Supercritical PC		UltraSupercritical		Subcritical CFB[6]	
	w/o capture	w/ capture	w/o capture	w/ capture	w/o capture	w/ capture	w/o capture	w/ capture
Performance								
Heat rate[1], Btu/kW$_e$ − h	9950	13600	8870	11700	7880	10000	9810	13400
Generating η (HHV)	34.3%	25.1%	38.5%	29.3%	43.3%	34.1%	34.8%	25.5%
Coal feed, kg/h	208000	284000	185000	243000	164000	209000	297000	406000
CO$_2$ emitted, kg/h	466000	63600	415000	54500	369000	46800	517000	70700
CO$_2$ captured at 90%[2] kg/h	0	573000	0	491000	0	422000	0	36000
CO$_2$ emitted, g/kW$_e$ − h	931	127	830	109	738	94	1030	141
Costs								
Total Plant Cost, $/kW$_e^3$	1280	2230	1330	2140	1360	2090	1330	2270
Inv. Charge, ¢/kW$_e$ − h @ 15.1%[4]	2.60	4.52	2.70	4.34	2.76	4.24	2.70	4.60
Fuel, ¢/kW$_e$ − h @ $1.50/MMBtu	1.49	2.04	1.33	1.75	1.18	1.50	0.98	1.34
O&M, ¢/kW$_e$ − h	0.75	1.60	0.75	1.60	0.75	1.60	1.00	1.85
COE, ¢/kW$_e$ − h	**4.84**	**8.16**	**4.78**	**7.69**	**4.69**	**7.34**	**4.68**	**7.79**
Cost of CO$_2$ avoided[5] Vs same technolo/o capture, $/tonne	41.3		40.4		41.1		39.7	
Cost of CO$_2$ avoided[5] Vs supercritical w/o Capture, $/tone	48.2		40.4		34.8		42.8	

(*Source:*The Future of Coal, MIT Study 2007, Table 3.1)
The basis for the Table PC: 500 MWe net output; Illinois #6 coal (61.2% wt C, HHV = 25,350 kJ/kg), 85% capacity factor, (1) efficiency = 3414 Btu/kWe-h/(heat rate); (2) 90% removal used for all capture cases; (3) Based on design studies and estimates done between 2000 and 2004, a period of cost stability, updated to 2005$ using CPI inflation rate. 2007 cost would be higher because of recent rapid increases in engineering and construction costs, up 25 to 30% since 2004; (4) Annual carrying charge of 15.1% from EPRI-TAG methodology for a U.S. utility investing in U.S. capital markets; based on 55% debt @ 6.5%, 45% equity @ 11.5%, 38% tax rate, 2% inflation rate, 3 year construction period, 20 year book life, applied to total plant cost to calculate investment charge; (5) Does not include costs associated with transportation and injection/storage; (6) CFB burning lignite with HHV = 17,400 kJ/kg and costing $1/million Btu.

a standard Illinois #6 bituminous coal having a high-sulfur and a moderately high heating value, 3.25 wt% sulfur and 25,350 kJ/kg (HHV).

Levelized Cost of Electricity (COE): The levelized cost of electricity is the constant dollar electricity price that would be required over the life of the plant to cover all operating expenses, payment of debt and accrued interest on initial project expenses, and the payment of an acceptable return to investors. Levelized COE is comprised of three components: capital charge, operation and maintenance costs, and fuel costs. Capital cost is generally the largest component of COE. This study (MIT Report 2007) calculated the capital cost component of COE by applying a carrying charge factor of 15.1% to the total plant cost (TPC). Detailed description of the Models, assumptions, and analysis is given in Appendix 3B and 3C of MIT Report 2007, and the reader may refer to it.

Section B: Coal conversion technologies

Conversion technologies turn coal into a gas or liquid that can be used as fuel after required cleaning. One of the most advanced conversion technologies is the *combined-cycle coal gasification.*

8.5 GASIFICATION

Coal gasification means turning coal into more useful gaseous form (coal gas) containing a mixture of gases. It is a process to convert coal into fuel gas as well as liquid fuels which, after purification, can be used as clean fuels. Based on heat-content, coal gas is classified broadly into three types: coal gas with (i) low-heat-content ($<7\,MJ/m^3$) mainly consists of N_2 and CO_2 with components such as CO, H_2 and CH_4 (methane); (ii) high-heat-content ($37\,MJ/m^3$) mainly consists of methane, and (iii) medium-heat-content (7–$15\,MJ/m^3$) consists of CO and H_2 at the lower end, and more methane at higher end.

Crushed and dried coal is fed into a reactor (called gasifier) which reacts with steam and air or oxygen at high temperature, 800°-1900°C, and high pressures upto 10 MPa. When coal is burned with less than a stoichiometric quantity of air, with or without steam, the resulting gas is low-heat-content gas which can be used as fuel gas after purification. If oxygen is used instead of air, medium-heat-content gas that can be used as syngas results. Part of CO in the gas must be allowed to react with steam to get additional hydrogen – a reacton called 'shift conversion' which manipulates proper ratio of gaseous components depending on the requirements of different syngases for producing SNG (synthetic natural gas), liquid fuels, methanol and ammonia.

Originated in the late 1700s, gasification is now a fairly mature technology; vast experience available on coal gasification worldwide. Town gas was produced from coal as early as 1792; in 1913, the first process to produce methanol from syngas was installed (BASF). A high-temperature fluidized bed gasifier was patented in 1921 by Winkler, and synfuels production from coal was routine in Germany during World War II. Later, gasification was used to produce Fischer-Tropsch liquids and chemicals from coal and petroleum and in large-scale power generation. Today, gasification technology is widely used throughout the world. According to the Gasification Technologies Council, in 2007, some 144 gasification plants and 427 gasifiers were in operation worldwide, adding up to an equivalent thermal capacity of $56\,GW_{th}$, of which coal gasification accounted for about $31\,GW_{th}$ (Blesl & Bruchof 2010).

8.5.1 Chemistry of gasification

Gasification is basically a chemical process in which a carbon-based feedstock (here, coal) converts, in the presence of steam and oxygen at high temperature and moderate pressure in a gasifier to syngas, which is a mixture of carbon monoxide and hydrogen (Figure 8.10). Gasification occurs under reducing conditions.

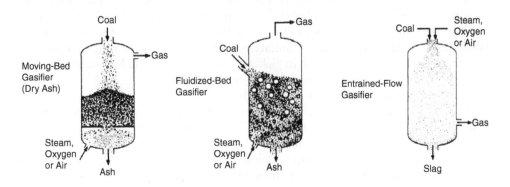

Figure 8.10 Three types of Gasification (*Source:* DOE/NETL site on Coal Power).

The chemistry of gasification is quite complex and involves many chemical reactions; some of the very important reactions are the following (Steigel *et al.* 2006):

Combustion with oxygen: $C + O_2 \rightarrow CO_2$ $\Delta Hr = -393.4 \, MJ/kmol$ (8.3)

Gasification with oxygen: $C + \frac{1}{2}O_2 \rightarrow CO$ $\Delta Hr = -111.4 \, MJ/kmol$ (8.4)

Gasification with steam: $C + H_2O \rightarrow H_2 + CO$ $\Delta Hr = 130.5 \, MJ/kmol$ (8.5)

Gasification with CO_2: $C + CO_2 \rightarrow 2CO$ $\Delta Hr = 170.7 \, MJ/kmol$ (8.6)

Water-Gas Shift reaction: $CO + H_2O \rightarrow H_2 + CO_2$ $\Delta Hr = -40.2 \, MJ/kmol$ (8.7)

Gasification with hydrogen: $C + 2H_2 \rightarrow CH_4$ $\Delta Hr = -74.7 \, MJ/kmol$ (8.8)

Reactions (Eqn 8.3 and Eqn 8.4) are exothermic oxidation reactions and provide most of the energy required by the endothermic gasification reactions (Eqn 8.5 and Eqn 8.6). The oxidation reactions occur very rapidly, completely consuming all of the oxygen present in the gasifier, so that most of the gasifier operates under reducing conditions. Reaction, Eqn 8.7, is the water-gas shift reaction, which in essence converts CO into H_2. The water-gas shift reaction alters the H_2/CO ratio in the final mixture but does not greatly impact the heating value of the syngas, because the heats of combustion of H_2 and CO on a molar basis are almost identical. Reaction Eqn 8.8, methane formation, is favored by high pressures and low temperatures and is, thus, mainly important in low temperature gasification systems. Methane formation is an exothermic reaction that does not consume oxygen and, therefore, increases the gasification efficiency and the final heating value of the syn gas. Methane is also formed via the following reaction:

$$CO + 3H_2 \rightarrow CH_4 + H_2O \qquad\qquad (8.9)$$

By and large, about 70% of the fuel heating value is associated with the CO and H_2 in the gas which can be even higher depending on the type of the gasifier used.

Many other chemical reactions also occur. In the initial stages of gasification, the rising temperature of the feedstock initiates devolatilization of the feedstock and the breaking of weaker chemical bonds to yield tars, oils, phenols, and hydrocarbon gases. These products generally react further to form H_2, CO, and CO_2. The fixed carbon that remains after devolatilization reacts with oxygen, steam, CO_2, and H_2.

Gasification may also be one of the best approaches to produce clean-burning hydrogen and other gases and liquid fuels. Hydrogen can be used in fuel cells to generate power; other coal gases can be used to fuel power-generating turbines, or for a wide range of commercial products, including diesel and other transport fuels (IEA 2006). To sum up, Coal gasification is an established technology that can utilize global coal reserves to produce clean electricity, fuels, and chemicals, and to provide a link to future hydrogen economy.

8.5.2   Generic types of gasifiers

The syngas produced from coal by a gasifier, composed mainly of CO and H_2, has a much lower heating value than natural gas. Gasifiers are broadly classified into three types: fixed-bed or movingbed, fluidized-bed, and entrained-flow (Figure 8.10) based on their designs and operational features: atmospheric or pressurized, dry coal or slurry feed, air or oxygen as oxidant, quench or heat recovery, and dry or liquid ash (slag) removal. The choice of the appropriate gasifier depends mostly on the fuel used and the desired gas usage among other factors.

In a moving-bed gasifier, large particles of coal move slowly down through the bed while reacting with gases moving upwards. The reactions taking place along the length of the gasifier are generally described in terms of 'zones'. In the drying zone at the top of the gasifier, the entering coal is heated and dried by syngas flowing in the opposite direction, while concurrently cooling before leaving the gasifier. The moisture content of the coal mostly controls the temperature of the discharge gas from the gasifier. Because of the countercurrent operation of this gasifier, hydrocarbon liquids can be found in the product gas which creates problems for downstream operations; however, techniques are available to capture the hydrocarbons and recycle them to the lower part of the gasifier. As the coal continues down the bed, it enters the carbonization zone where the coal is further heated and devolatilized by higher temperature gas. In the gasification zone, the devolatilized coal is converted to syngas by reacting with steam and CO_2. In the combustion zone, near the bottom of the reactor, oxygen reacts with the residual char to consume the left over carbon and generate the heat needed for the gasification zone. In the combustion zone, the gasifier can be made to operate either in *dry ash* or *slagging* mode. In the dry-ash version, the temperature is maintained below the ash slagging temperature by the endothermic reaction of the char with steam in the presence of excess steam. In addition, the ash below the combustion zone is cooled by the entering steam and oxidant. In the slagging version, much less steam is used so that the temperature of the ash in the combustion zone exceeds the ash fusion temperature of the coal forming molten slag. *The British Gas/Lurgi (BGL) process is an example of this type gasifier.* The key features of this gasifier are (i) low oxidant demand, (ii) production of hydrocarbon liquids – tars and oils, and (iii) high 'cold-gas' thermal efficiency, when the heating value of the hydrocarbon liquids is included. The reader may refer to Dolezal (1954, 1967) for an account on the technology of liquid slag removal and slag heat recovery.

Fluidized-bed gasifiers operate in a highly back-mixed mode thoroughly mixing the coal feed particles with those particles already undergoing gasification. Coal enters at the side of the reactor, while steam and oxidant enter near the bottom, thereby fluidizing the reacting bed. Char particles entrained in the raw gas leaving the top of the gasifier are recovered by a cyclone and recycled back to the gasifier. The ash particles removed below the bed give up heat to the incoming steam and oxidant. The gasifier operates under isothermal conditions at a temperature below the ash fusion temperature of the coal, thus avoiding formation of ashes and possible collapse of the bed. The low temperature operation of the gasifier allows utilizing relatively reactive feeds, such as low-rank coals and biomass, or lower quality feed stocks such as high-ash coals. In summary, this gasifier accepts a wide range of feed stocks; operate at uniform, moderate temperature; require moderate amounts of oxygen and steam, and involves extensive recycling of char particles. This is used in the *US Kellogg Rust Westinghouse (KRW)* and *Institute of Gas Technology (IGT) processes*, and the *German Winkler process*.

In entrained-flow gasifiers, more commonly used, fine coal particles react with steam and oxidant (normally, pure oxygen) at *high temperatures*, i.e., well above the fusion temperature of the ash and produce vitreous slag. The residence time of the coal in these gasifiers is very short, and high temperatures (more heat) are essential to achieve high carbon conversion. It means, more oxygen is consumed to combust more of the feedstock to generate the required heat. To minimize oxygen consumption, and thereby the cost, these gasifiers are usually supplied with higher quality feed stocks. These gasifiers can operate either in a down-flow or up-flow mode; the characteristics are: (i) ability to gasify all coals, although coals with lower ash content are favored, (ii) uniform temperature, (iii) very short feed residence time in the gasifier, (iv) very finely divided and homogeneous solid fuel, (v) relatively large oxidant requirement, (vi) large amount of sensible heat in the raw gas, (vii) high-temperature slagging operation, and (viii) entrainment of some ash/slag in the raw gas. Examples of this type are *GE Energy* and *ConocoPhillips* using wet slurry feed, and *Shell Siemens*, and *Mitsubishi* using dry feed. Except for the Mitsubishi gasifier, all others are oxygen-blown. Entrained flow gasifier technology is currently the most widely-used technology since it appeals to the chemical and petroleum refining industries for its ratio of H_2 to CO.

There is a 'transport gasifier' developed by KBR which is a hybrid between entrained flow and fluidized bed gasifiers. It can handle both lignite and bituminous coals, and operates at slightly higher temperatures than fluidized bed gasifiers. Both air and oxygen can be used as oxidants.

Gasifier is, thus, the heart of any gasification-based system which can process a wide variety of feedstocks into gaseous products at high temperature and usually at elevated pressure in the presence of steam and oxygen. The feedstock is not totally burned in the gasifier but partially oxidized, resulting in *product syngas containing mostly CO and H_2*.

Table 8.4 summarizes the characteristics and commercial examples of these three types of gasifiers.

These three types, though appropriate for a wide range of applications, some situations require a specialized gasification process (NETL: USDOE) such as, advanced

Table 8.4 Gasifier types and characteristics.

Flow Regime	Moving (or fixed) Bed	Fluidized Bed	Entrained Flow
Combustion analogy	Grate-fired combustors	Fluidized Bed combustors	Pulverized coal combustors
Fuel type	Solids	Solids	Solids or liquids
Fuel size	Big, 5–50 mm	0.5–5 mm	Fine, <500 microns
Residence time	High, 15–30 minutes	5–50 seconds	Small, 1–10 seconds
Oxidant	Air or oxy-blown	Air or oxygen-blown	only oxygen-blown
Gasoutlet temperature	400–500°C	700–900°C	900–1400°C
Ash handling	Slagging & non-slagging	Non-slagging	Slagging only
Commercial examples	Lurgi dry-ash (non-slagging); BGL (slagging)	GTI U-Gas, HT Winkler, KRW	GE Energy, Shell, Prenflo, ConocoPhillips, Noell
Comments	Moving beds are mechanically stirred, fixed beds are not; Gas and solid flows are countercurrent in moving bed gasifiers.	Bed temperature below ash fusion point to prevent agglomeration; Preferred for high-ash feed-stocks and waste fuels.	Not ideal for high-ash fuels due to energy penalty of ash-melting; Unsuitable for fuels that are hard to atomize or pulverize.

(*Source:* Drawn from Stiegel, G.J. 2005)

coal gasification (compact and low-cost with high carbon conversion and increased thermal efficiency), plasma gasification (suitable for feedstock such as agricultural or municipal or industrial wastes), and catalytic gasification (using a catalyst along with feedstock to control the products of the process).The nature of coal presents problems in gasification. For example, high-ash and high moisture coals produce an excessive quantity of molten ash inside the entrained flow gasifier creating operational problems and reducing efficiency. FB gasifier that operates at lower temperature is better suited for such coals. But the bed requires a regular draining of ash to maintain fluidization and prevent accumulation of agglomerating material; this step results in loss of solid carbon. Similarly, slurry-feed entrained gasifiers run into problems with low-grade, high moisture coals. The high porosity and oxygen content in these coals make it difficult to prepare slurry. Even the method of pre-drying combined with lock hoppering based feeding affects efficiency. The technology of coal-water slurry (CWS) feeding has been studied in detail during 1980s (Beer 1985, 1982).

Solids/Stamet pumps (continuous dry-feeding systems) could help to introduce high moisture coals directly to the gasifier; this method promises reduced cost, coal feed without lock hoppers and improvement of plant efficiency by about half-a-percentage point (Henderson 2008; Beer 2009).

8.5.3 Commercial gasifiers

Lurgi: Lurgi gasifiers use a moving bed design and operate below the ash melting point of the feed. The coal is only crushed because this process cannot handle coal fines. The coal is fed into the top of the gasifier via lockhoppers. Oxygen is injected at the bottom of the gasifier and reacts with the coal, pre-heated by the hot syngas rising through the coal bed. The ash drops off the bottom of the bed and is depressurized via a lockhopper.

Originally developed by Lurgi GmbH in the 1930s in Germany, more than 150 Lurgi gasifiers have been built with the largest unit capable of processing 1000 tpd of coal on a moisture and ash-free basis. Two well-known examples where Lurgi process is applied are the coal-to-gasoline refineries of Sasol in South Africa and the Dakota Gasification Synthetic Natural Gas plant in the USA. Both units feature a series of oxygen-blown gasifiers and use locally available low rank coal as the feedstock. In fact, Lurgi gasification process has gasified more coal than any other commercial gasification process.

The main drawback of the Lurgi process when applied to IGCC is the production of hydrocarbon liquids along with syngas. The liquids represent about 10% of the heating value of the feed and therefore must be utilized in order to achieve competitive efficiencies. Both at Sasol and Dakota Gasification, the size of the gasification operations (14,000 tpd at Dakota) makes it economical to recover the liquids and convert them into high value products such as naphtha, phenols and methanol. It is not yet certain that the economies of scale for a 500 to 600 MW IGCC (4000 to 5000 tpd of coal) would allow those by-products to be recovered competitively. The other drawback is limiting the operation to below the ash melting point of the coal. For lower reactivity coals such as bituminous coal, this will lower carbon conversion due to the lower gasification temperatures. It means, *low rank, high ash* coals provide a competitive advantage for the Lurgi process against its higher operating temperature competitors (Jeffry Phillips, EPRI).

GE Energy/Texaco: The GE Energy gasifier is most widely used in IGCC applications. The process, developed by Chevron-Texaco in the 1950s and presently owned by GE Energy, uses an entrained flow, refractory-lined gasifier which can operate at pressures in excess of 62 bar (900 psi). The coal water-slurry is injected at the top of the gasifier, and syngas and slag flow out from the bottom of the gasifier. Three options are available for heat recovery from the GE Energy process: Quench, Radiant, and Radiant and Convective. In the Quench option, both the syngas and slag are forced into a water bath where the slag solidifies and the syngas is cooled and saturated with water vapor. The slag is removed from the bottom of the quench section via a lockhopper while the saturated syngas is directed to gas clean-up equipment. In the radiant option, the syngas and slag enter a long, wide vessel which is lined with boiler tubes. The vessel is designed to cool the syngas below the melting point of the slag. At the end of the radiant vessel both the syngas and slag are quenched with water. In the radiant plus convective option instead of having a water quench for both the syngas and slag, only the slag drops into a water bath at the bottom of the radiant vessel. The syngas exits at the side of the vessel and enters a convective syngas cooler. Both fire tube boiler and water tube boiler designs have been used for the convective cooler.

Since 1950s, more than 100 commercial applications of the GE Energy gasification process have started. Some prominent examples: (a) the Cool Water project-built in the early 1980s; and Tampa Electric Company's Polk County Plant in the USA; (c) Coal and petroleum coke-to-chemicals units, such as, Kansas coke-to-ammonia plant in the USA, Ube Industries coke-to-ammonia plant in Japan, and the Eastman Chemical coal-to-chemicals facility in Tennessee, USA. These plants have operated with the very high availability factors expected in the chemicals industry for more than 20 years. The GE Energy gasifier can also be used to gasify petroleum refinery liquid by-products such as asphalt residues; and several IGCC projects based on these feedstocks have been built world-wide at refineries.

Shell: The Royal Dutch company, Shell, has developed two different gasification processes; the first, the Shell Gasification Process or SGP, to gasify liquid and gaseous feedstocks. It features a refractory-lined gasifier with a single feed injection point at the top of the gasifier. The gasification products pass through a syngas cooler before entering a wet scrubber. The second, the Shell Coal Gasification Process (SCGP), was developed specifically to gasify solid feeds. The SCGP gasifier features a water-cooled membrane wall similar to the membrane walls used in conventional coal boilers. There are four feed injectors oriented horizontally in the mid-section of gasifier vessel. Slag flows out of a slag tap at the bottom of the vessel where it falls into a water bath and syngas flows out the top of the vessel. As the syngas exits the gasifier it is quenched with cool, recycled syngas to a temperature well below the ash melting point of the coal. The quenched syngas is still quite warm (typically 900°C) and passes through a syngas cooler and a dry solids filter before a portion of the gas is split off for recycle to the quench zone. The coal is fed to the SCGP gasifier pneumatically using high pressure nitrogen as the transport medium. The coal must first be dried and finely ground in a roller mill where warm, inert gas flows through the mill to remove the coal's moisture. The dried coal is then pressurized via a system of lockhoppers. SCGP gasifiers operate at pressures up to about 40 bar.

Shell began development of the SGP process in the 1950s, and work on the SCGP process started as a joint project with Krupp Koppers in the mid-1970s. Both companies separated in 1981; and Krupp Koppers developed a dry-feed, membrane wall gasifier with the trade name PRENFLO. The only commercial application of the PRENFLO process has been the 280 MW Elcogas IGCC in Puertollano Spain. In 1999, Shell and Krupp Uhde joined again to continue work in coal gasification; however, only SCGP is offered commercially. The first commercial application of SCGP was the 250 MW Demkolec IGCC built in 1994 in Buggenum, The Netherlands. Nuon acquired the plant in the late 1990s, and is now operating as an independent power producer. Shell has also sold licenses for 12 SCGP gasifiers for use in coal-to-chemicals projects in China. The greatest advantage of Shell's process is its feed flexibility. The 240 tpd SCGP demonstration built at Shell's refinery in Deer Park, Texas in the 1980s was able to process a full range of feedstocks including lignite, sub-bituminous coal, bituminous coal and pet coke. The reason for SCGP's flexibility is the coal milling and drying process which eliminates the impact of moisture on the gasifier performance (however, the fuel for the drying process has a negative impact on thermal efficiency). The biggest issue is its higher capital cost, inherent in the more expensive nature of the gasifier design (boiler tubes are more expensive than refractory brick) and its dry feed system.

ConocoPhillips E-Gas: Originally developed by Dow Chemical, the E-Gas process features a unique two-stage gasifier design. The gasifier is refractory-lined and uses coal-water slurry feed. The first stage of the gasifier has two opposed, horizontally-oriented feed injectors. The syngas exits the top of the first stage and slag flows out of the bottom into a water bath. The syngas produced by the first stage enters the second stage at temperatures comparable to the exit temperatures of the other two entrained flow gasifiers, GE Energy and SCGP. Additional coal-water slurry is injected into this hot syngas in the second gasifier stage, but no additional oxygen is injected. Endothermic gasification reactions occur between the hot syngas and the second stage coal feed. This lowers the temperature of the syngas and increases the cold gas efficiency of the process. Upon exiting the top of the second stage of the gasifer, the syngas passes through a syngas cooler which features a firetube design. The cooled syngas then enters a rigid barrier filter where any unconverted char from the second stage is collected and recycled back to the first stage of the gasifier where the hotter temperatures ensure near complete carbon conversion.

Starting with a bench scale reactor in 1976, Dow developed the E-Gas process to a 550 tpd proto-type plant initially using lignite as the feedstock. The US government offered a price guarantee for syngas to be produced from a commercial scale E-Gas gasification plant built in 1984 (at Plaquemine), designed to process 1600 tpd (on dry basis) of sub-bituminous coal. Starting the plant operations in 1987 by a Dew subsidiary, Louisiana Gasification Technology Inc. (LGTI), the clean syngas generated was fed to two Westinghouse 501D gas turbines, already operating on natural gas at the Plaquemine complex. The total power output from the two turbines was 184 MW, and the operations went on upto 1995. ConocoPhillips is developing several new IGCC projects: the Mesaba IGCC and the Steelhead Energy project, with funding from USDOE and the State of Illinois respectively, will produce 600 MW of electricity and synthetic natural gas. There are several other manufacturers of gasifiers worldwide.

8.5.4 Syngas cleanup

The type of gasifier technology and the operating conditions influence the quantities of H_2O, CO_2, and CH_4 in the syngas and the minor and trace components. Under the reducing conditions in the gasifier, most of the sulfur in the fuel converts to hydrogen sulfide (H_2S), but around 3 to 10% converts to carbonyl sulfide (COS). Nitrogen in the fuel generally converts to gaseous nitrogen (N_2), and also form some ammonia (NH_3) and a little of hydrogen cyanide (HCN). Most of the chlorine in the fuel is converted to HCl leaving some chlorine in the particulate phase. Mercury and arsenic in trace amounts are released during gasification and partition among the different phases – fly ash, bottom ash, slag, and product gas. Typical gas compositions for commonly used commercial gasifiers are given in Table 8.5. These impurities need to be controlled either for burning syngas as a fuel or for converting to liquid fuels, or hydrogen or chemicals. The clean-up involves several steps: particulate removal, mercury removal, shifts conversion, COS hydrolysis, acid gas removal, sulfur recovery, tail gas treating, and SCOT process (Stiegel 2005).

Chemical solvents, such as mono-ethanolamine (MEA), di-ethanolamine (DEA), and methyl di-ethanolamine (MDEA), and physical solvents, such as methanol (Rectisol) and mixtures of dimethyl ethers of polyethylene glycol (Selexol), operating at ambient or lower temperatures are employed to clean the syngas. For processes using amines, primarily MEA, the commercial vendors are ABB/Lumus, Fluor, MHI (Japan), HTC Purenergy, Aker Clean Carbon (Norway), and Cansolv (Herzog 2009). The selection of the technology for gas cleanup is dependent on the purity requirements of downstream operations. To start with, cooling is necessary depending on the

Table 8.5 Composition of Raw Syngas from Coal Fired Gasifiers.

Gasifier technology	Sasol/Lurgi[1]	Texaco/ GE Energy[2a]	BGL[2b]	E-gas/Conoco Phillips	Shell/ Uhde[2c]
Bed type	Moving	Entrained	Moving	Entrained	Entrained
Coal feed	Dry	Slurry	Dry	Slurry	Dry
Coal type	Illinois #6	Illinois #6	Illinois #6	Illinois #6	Illinois #6
Oxidant	Oxygen	Oxygen	Oxygen	Oxygen	Oxygen
Pressure, MPa (psia)	0.101(14.7)	4.22 (612)	2.82 (409)	2.86 (415)	2.46 (357)
Ash form	Slag	Slag	Slag	Slag	Slag
Composition	In %	In %	In %	In %	In %
Hydrogen	52.2	30.3	26.4	33.5	26.7
CO	29.5	39.6	45.8	44.9	63.1
CO_2	5.6	10.8	2.9	16.0	1.5
CH_2	4.4	0.1	3.8	1.8	0.03
Other HC	0.3	–	0.2	–	–
H_2S	0.9	1.0	1.0	1.0	1.3
COS	0.04	0.02	0.1	0.1	0.1
$N_2 + Ar$	1.5	1.6	3.3	2.5	5.2
H_2O	5.1	16.5	16.3	–	2.0
$NH_3 + HCN$	0.5	0.1	0.2	0.2	0.02
HCl	–	0.02	0.03	0.03	0.03
H_2S/COS	20/1	42/1	11/1	10/1	9/1

Sources: (1) Rath: Status of Gasification Demonstration Plants, *Proc. 2nd Annul. Fuel Cells Contract Review Mtg.*, DOE/METC-9090/6112, p. 91; (2) *Coal Gasification Guidebook: Status, Applications, and Technologies*; Electric Power Research Institute, EPRI TR-102034, 1993. 2a: pp. 5–28; 2b: pp. 5–58; 2c: pp. 5–48 (Stiegel et al.)

temperature of the raw gas. Raw syngas from the high temperature gas cooling (HTGC) system needs to be cleaned to remove contaminants including fine particulates, sulfur, ammonia, chlorides, mercury, and other trace heavy metals to meet environmental emission regulations, as well as to protect the processes that follow.

Depending on the application, syngas is treated to obtain the needed hydrogen-to-carbon monoxide ratio to meet downstream process requirement. In applications where very low sulfur (<10 ppmv) syngas is required, converting COS to H_2S before sulfur removal may also be needed. Typical cleanup and conditioning processes include cyclone and filters for bulk particulates removal; wet scrubbing to remove fine particulates, ammonia and chlorides; solid absorbents for mercury and trace heavy metal removal; water gas shift (WGS) for H_2/CO adjustment; catalytic hydrolysis for converting COS to H_2S; and acid gas removal (AGR) for extracting sulfur-bearing gases and CO_2 removal.

For fine char and ash particulate removal, raw syngas leaving the HTGC system is normally quenched and scrubbed with water in a trayed column prior to recycle to the slurry-fed gasifiers. For dry feed gasification, cyclones and candle filters are used to recover most of the fine particulate for recycle to the gasifiers before final cleanup with water quenching and scrubbing. In addition, fine particulates, chlorides, ammonia, some H_2S, and other trace contaminants are also removed from the syngas during the scrubbing process. The scrubbed gas is then reheated for COS hydrolysis and/or a sour WGS if needed; or cooled in the low temperature gas cooling system by generating low pressure steam, preheating boiler feed water, and heat exchanging it against cooling water before further processing. Used up water from the scrubber column is directed to the sour water treatment system, where it is depressurized and decanted in a gravity settler to remove fine particulates. The solid-concentrated underflows from the settler bottom are filtered to recover the fine particulate as the filter cake, which is then either discarded or recycled to the gasifier depending on its carbon content. Water from the settler is recycled for gasification uses with the excess being sent to the wastewater treatment system for disposal. Catalysts have been developed which promote the reaction of HCN with water vapor to produce NH_3 and CO. Therefore, a HCN conversion catalyst is typically installed upstream of the NH_3 water wash column. HCl has the tendency to corrode in the GT hot section.

The next step is to deal with sulfur pollutants. Sulfur concentrations greater than 250 ppm cause high rapid corrosion of hot section parts, and also result in a relatively high sulfuric acid dew point in the exhaust gas. This would limit the recoverable heat quantity from the exhaust in the HRSG. Hence sulfur compounds usually present as H_2S are removed by converting it into elemental sulfur or sulfuric acid. Depending on the gasification temperature and moisture content, approximately 3–10% of the sulfur is converted to COS during gasification. To generate low sulfur syngas, the COS is converted to H_2S before sulfur removal via current commercial AGR processes.

In the AGR system, H_2S and CO_2 are removed from the syngas using either physical or chemical solvent absorption. For chemical synthesis applications which require syngas with less than 1 ppmv sulfur, physical solvent processes such as Rectisol and Selexol are normally used. For power generation applications, which allow higher sulfur levels (approximately 10 to 30 ppmv sulfur), chemical solvent processes such as MDEA and Sulfinol are normally used. The physical solvent absorption processes operate under cryogenic temperatures while the chemical solvent absorption processes operate slightly above ambient temperature. Physical solvent processes are two to four times more costly than MDEA-based chemical solvent processes.

Mercury vapor in the syngas can be removed by passing the syngas through an activated carbon bed, though mercury has no effect on gas turbine. This bed will also remove heavy metals such as arsenic which can cause problems in the gas turbine. Current commercial practice is to pass cooled syngas from LTGC through sulfided, activated carbon beds to remove over 90% of the mercury and a significant amount of other heavy metals. Due to the sulfur in the activated carbon, these beds are normally placed ahead of the AGR system to minimize the possibility of sulfur slipping back into and contaminating the cleaned syngas.

Although 'hot gas cleaning' which preserves the energy content of the flue gas is preferable, the process at commercial scale is not yet available. The reader may refer to proceedings of 6th International Symposium on Gas cleaning at High temperatures, Osaka, Japan, October 20–22, 2005, Kamiya, Makino and Kanaoka (eds), for more details and analysis on syngas cleaning technologies.

However, Syngas cleanup requires further improvements (Stiegel 2005):

(a)   Development of deep cleaning technologies required to meet future environmental regulations; bringing the costs of Rectisol performance at equal or lower than amine systems;

(b)   Improved definition of contaminants: particulates, H_2S, COS, CO_2, NH_3, volatile metals, carbonyls, and their mandatory levels;
(c)   Necessity to operate nearer to downstream process requirements;
(d)   To decrease the high cost of particulate and chemical removal by reducing the number of operations;
(e)   Cold Gas Cleanup (ambient): for e.g, opportunities to improve conventional technologies; removal of heat stable salts; to develop new approaches;
(f)   Warm Gas Cleanup (300°–700°F): for e.g., to develop technologies that operate above the dew point of the gas stream more efficiently; development of technologies for multi-contaminant removal (e.g., mercury, arsenic, selenium, ammonia, particulates, etc.);
(g)   Particulate filtration: development of more durable, reliable, and cost effective filters with useful life of three years; simpler, cheaper approaches to solids removal.

8.5.5   Partial gasification combined cycle:

Partial gasification combined cycle (PGCC) technology shows promise for power generation because PGCC can realize the benefits of gasification while achieving the reliability, flexibility and economics of a commercial power plant. PGCC utilizes proven combustion power technology to simplify the gasification process and overcome the technology hurdles associated with it.

In partial gasification (Bose et al. 2004; Robertson et al. 1989), in addition to syngas, a residual char is also produced for combustion, while in total gasification all the carbon in the feed coal is gasified. Partial gasification of coal in a Pressurized Fluidized Bed Gasifier produces syngas for a Topping combustor of the Gas turbine, and char for combustion in PFBC. The latter generates steam for the steam turbine and high pressure flue gas for the gas turbine in a Gas Turbine-Steam Combined Cycle (Figure 8.11) (Robertson et al. 1989).

Instead of PFB combustion boiler, a PC boiler can also be used. PGCC configuration utilizing PFB boiler technology is known as gasification fluid bed combined cycle (GFBCC), and the one utilizing PC boiler technology is known as gasification pulverized coal combined cycle (GPCCC).

The design shown in Figure 8.11 has the advantages of fluidized beds, such as, reduced sensitivity to fuel quality and sulfur capture by sorbents in the bed, and raising the gas turbine inlet temperature if needed by the use of natural gas. The combustion products of the char burning in the PFBC are cleaned to remove particulates and alkali at 870°C (1600°F), and are ducted to the gas turbine where the syngas from the partial gasifier is injected. The oxygen in the PFBC exhaust gas is sufficient enough to burn the syngas in the topping combustor of the gas turbine. The topping combustor has to be of special design capable of being cooled by the PFBC exhaust gas (which is at 870°C), instead of the usual compressor exit air at 411°C, without overheating, in addition to being a low NO_x combustor. Westinghouse's all metallic Multi Annular Swirl Burner (MASB) operating in Rich-Quench-Lean mode solves the cooling problem by creating thick layers of gas flow over the leading edges of overlapping concentric annular passages in the combustor, and gives NO_x emissions below 9 ppm at 15% O_2 for syngas as fuel (Beer & Garland 1997; Domeracki et al. 1995; Beer et al.

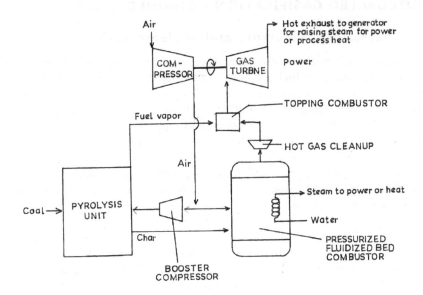

Figure 8.11 PFBC with Syngas Topping Combustor (Redrawn from Robertson *et al.* 1989).

1997, 2009). The plant's efficiency is estimated to be 48.2%. If the syngas and the air effluent of char combustion are cooled to 538°C (1000°F), commercially available porous metal filters can be used instead of ceramic filters for particulate cleanup, and no alkali getters are needed. This reduces plant cost and increases availability, although the efficiency may reduce to 46% (Beer & Garland 1997).

However, unlike other gasification technologies, PGCC does not remove the sulfur in the syngas; the H_2S in the syngas combusts to SO_2 in the CT. The sulfur is removed after the CT in either the PCFB boiler (GFBCC) or in a standard flue gas desulfurization (FGD) system downstream of the PC boiler (GPCCC). Experience with gas turbine manufacturers suitable for a wide range of fuels, it has been observed that the sulfur concentrations in the CT gases do not affect CT life and maintenance (Giglio *et al.*).

The main advantages of PGCC are its operational, fuel and design flexibility. Most other gasification technologies require 100% of the plant's fuel to be gasified; within PGCC, the amount of gasification and combustion of the fuel can be adjusted to optimize plant characteristics – plant output, reliability, efficiency and cost – to meet the needs of a specific project. This flexibility is a result of the limited integration among PGCC's three fundamental technologies, gasification, combustion turbine, and a steam plant. Foster Wheeler has developed and tested a pilot-scale design of PGCC, and prepared for a commercial demonstration project, 300 MWe GFBCC design utilizing the GE 6FA combustion turbine.

The 'gas turbine and steam turbine power output ratio' of about 15/85 for PGCC improves to 55/45 for IGCC as seen in the next section.

8.6   INTEGRATED GASIFICATION COMBINED CYCLE

8.6.1   Process, plant structure and implementation

Combined-cycle Coal gasification process consists of three stages: (i) *gasification* where coal is converted into a raw fuel gas, mostly composed of CO and H_2, by reacting with a gasification agent such as high temperature steam-oxygen mixture or oxygen or air in a reactor (Higman & van der Burgdt 2003; Holt 2001); (ii) *gas clean-up*, purification of the resulting gas after cooling to remove pollutants (DOE/NETL reports); and (iii) *power generation*, i.e., burning the cleaned gas in a *combustion-turbine* to produce power, and collecting the heat in the gas turbine exit gas in a heat recovery steam generator (HRSG) and sending to a *steam-turbine* generator to produce more electricity. Because of this dual combination of environment-friendly gasification technology and efficient Gas-steam combined cycle to generate power, this technology is known as 'integrated gas combined cycle (IGCC)' offering higher efficiency and lower emissions compared to PC technologies. The first two stages have already been covered. Around, two-thirds of the total electricity generated in the IGCC plant is produced by the gas turbine and one third by the steam turbine. The combined cycle efficiency improves through the reduced effect of the steam condenser's heat loss (Beer 2009).

Today, Natural Gas Combined Cycle (NGCC) is the most efficient power technology and expected to remain so in the future. Thermodynamically, the combined cycle utilizes the fuel's energy most efficiently; and NGCC, using a clean high-energy natural gas fuel with minimal energy losses to the cycle, sets the higher efficiency limit for all other power generation technologies (Giglio *et al.*, Foster Wheeler).

IGCC and 'Partial gasification combined cycle' fall in between NGCC and PC/CFB steam plants. The efficiencies of these cycles are much higher than steam plants since they utilize combined cycle technology. But, unlike NGCC plants, they carry the energy loss or price linked with gasifying solid fuel and capturing the pollutants resulting from solid fuels. They offer a substantial increase in efficiency (5–10 points or 10–25%) above conventional steam plants and also benefit from progress in combustion turbine (CT) technology. As CTs improve to more efficient and larger sizes, these gasification technologies also advance in size, efficiency, and economy of scale.

IGCC unit involves the integration of a number of processes – air separation, gasification, syngas cleanup (including AGR and sulfur recovery), HRSG, and power generation as shown schematically in Figure 8.12 (Figure author: Stan Zurek, Wikipedia). These processes broadly constitute two main subcomponents: a gasification island and a power island. The 'integrated' aspect comes from the steam and nitroge generation in the gasification island that supplies the power island. The IGCC power island consists of gas turbine, heat recovery steam generator (HRSG), steam turbine, and condenser along with other supporting systems as in the case of a NGCC plant. These elements generate power using the Brayton and Rankine cycles. The IGCC power island uses excess steam from the gasification island to power the steam turbine; the surplus nitrogen from the gasification island supplies the power island with nitrogen to reduce NO_x generation at the combustor. The gasification island consists of four main components, the coal preparation system, air separation unit (ASU), gasifier, and syngas cleaning setup, and produces syngas from coal to run the gas turbine. The coal preparation facilities are by and large similar to a PC plant. Coal is crushed to a very

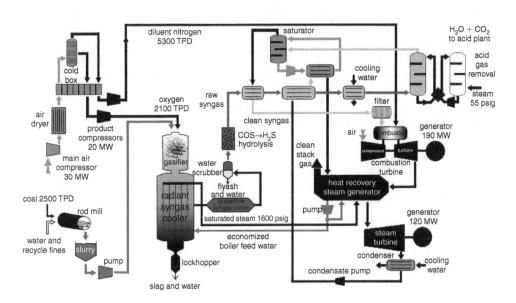

Figure 8.12 Diagram of IGCC Power plant, showing all the details (*Source:* Wikipedia, Author: Stan Zurek, retrieved March 2, 2013).

fine powder and injected into the gasifier. Depending on the gasifier, coal is either fed in dry form by nitrogen transport or as water-slurry. The gasifier converts the coal into syngas, composed of CO and H_2 which has a much lower heating value than natural gas. For this to occur, the gasifier uses oxygen of high purity supplied by cryogenic ASU which is costly as well as low efficient. The syngas cleanup represents the final component of the gasification island and has four major removal steps: particulate, sulfur, mercury, and carbon removal. These have been explained earlier. The carbon removal process involves a water shift reaction step that converts the CO in the syngas stream to CO_2 which is then separated and compressed (Lako, Paul 2010).

Though IGCC is currently expensive, the economic advantage of IGCC is in the removal of pollutants prior to combustion. Post combustion flue gas cleanup generally is more cost intensive because the pollutants are not in a concentrated, pressurized stream like syngas. Further, many advantages of IGCC over PC-fired technology such as the following make it a leading technology in the power generation sector (Stiegel *et al.*): (a) capable of handling almost any carbon-based feedstock; (b) sound environmental features, viz, low CO_2 emissions per unit of generated power, and ability to clean product gas to reach near-zero emissions of criteria pollutants, particulates, and mercury at much lower costs and higher efficiencies; (c) flexibility to utilize part of syngas to other applications apart from running a turbine to generate power; (d) extraction of coal energy up to 60% as against 30 to 35% with the existing PC-fired power plants; (e) high thermal efficiency on par with the best current PC plants and potential for further increase; (f) cost effective CO_2 recovery for sequestration; (g) pure hydrogen production, if desired; and (h) substantial reduction in water consumption, and more than 50% reduction in the solid by-products produced. IGCC uses less water since 60% of its power is derived from an air-based Brayton cycle reducing the heat load on the steam turbine condenser to only 40% of that of an equivalent rated PC-fired plant. Further, through the direct desulphurization of the gas, IGCC does not require a large FGD unit which consumes large amounts of water. Extra gains in reducing water use can be achieved when carbon capture is included into the plant (Barnes 2011). Shell estimates IGCC generation efficiency, based on their gasifier, of 46–47% (LHV, net) or 44–45% (HHV, net) for bituminous coals with an FB-class gas turbine (Van Holthoon, 2007, Barnes 2011). The highest reported efficiency for an IGCC is 41.8% (HHV, net) where Shell gasifier powering an F-class turbine fuelled with Pittsburgh coal, is used (Barnes 2011). Fuel cell Handbook 7th edition (2004) published by USDOE gives the 'thermal efficiency' ranges on HHV basis for PC, FBC and IGCC technologies: PC plant: 34–42%; FBC plant: 36–45%, and IGCC plant: 38–50%.

Currently, oxygen-blown, entrained- flow gasifiers which have the advantage of avoiding tar formation and related problems, are the preferred technology for IGCC plants, although other configurations are being evaluated. Further, due to an increased rate of gasification, it is easier to match the capacity of a single gasifier to that of a modern gas turbine. High pressure gasification reduces the cost of syngas clean-up because of the smaller size of equipment, and saves auxiliary power for the compression of syngas. High pressure and high temperature operation, however, has implications for the modes of coal feeding and ash removal; consequently, the coal property range that can be handled by a gasifier may be affected. As seen earlier, coal can be fed into a high pressure entrained flow gasifier both as dry feed and coal-water slurry.

There are many possible IGCC plant configurations because of varying gasifier designs; also, IGCC has a large number of process areas that can use different technologies.

MIT Report (2007) discusses a 500 MW_e oxygen-blown IGCC unit without CO_2 capture showing typical stream flows and conditions. A lower-pressure (4.2 MPa) GE radiant-cooling gasifier is used here, producing high-pressure steam for electricity generation. Nitrogen from the ASU is fed to the combustion turbine to produce increased power and reduce NO_x formation. Internal power consumption is about 90 MWe, and the net efficiency is 38.4%. The stack gas composition is also given. Solvent MDEA can achieve 99.4% sulfur removal (shown in the figure) from the syngas for 0.033 lb SO_2/million Btu, as low as or lower than for recently permitted PC units. Selexol can achieve 99.8% sulfur removal for an emission rate of 0.009lb SO_2/million Btu. Rectisol, which is more expensive, can achieve 99.91% sulfur removal for an emissions rate of 0.004 lb SO_2/million Btu (EPA 2006). NO_x emission control is strictly a combustion turbine issue and is achieved by nitrogen dilution prior to combustion to reduce combustion temperature. If SCR is added, it would result in NO_x reduction to significantly low levels.

Implementation of IGCC projects

There are several refinery-based IGCC plants in Europe that have demonstrated good availability (90–95%) after initial teething problems. Major coal-based IGCC projects operate in Europe and the USA. Two 'first generation' large-scale demonstration projects in the US had shown features of low emissions and integrated control of gasification process with a combined cycle: (1) Texaco Cool Water project of 125 MW_e

output using Texaco gasifier technology with bituminous coal feedstock (1000 tpd) operated during 1984–88; (2) Dow/LGTI project of 160 MW_e output using E-Gas technology with sub-bituminous coal feedstock (2200 tpd) operated during 1987–1995. These plants provided a lot of experience, though closed after the price guarantee periods were over.

The following are the improved IGCC systems ('second generation' plants) based on the learning experience from the earlier plants. These plants make use of different gasifier designs, gas cooling and gas clean setups and integration schemes between the plant units:

(a)   250 MW_e Buggenum plant (The Netherlands) of 253 MW_e output using Shell technology with bituminous coal feedstock; currently uses about 30% biomass as a supplemental feedstock; started in 1994 with a dry-feed oxygen-blown gasifier and 1060°C gas turbine; the developer, NUON, is paid an incentive fee by the government for using the biomass and now constructing a 1,300 MW IGCC plant. The Nuon Buggenum project is aimed at testing pre-combustion CO_2 capture in order to better select, design and optimize a capture system after gaining some operating experience. Both the watergas shift reactors and the CO_2 capture process will be optimized for their performance and efficiency; and different physical and chemical solvents will be tested.

(b)   300 MW_e capacity plant in Puertollano, Spain using Prenflo technology utilizes a mix of petcoke and coal (2500 tpd). Started in 1998 with a dry feed oxygen-blown gasifier and 1120°C gas turbine, this IGCC plant has captured its first tonne of CO_2 in late 2010 (Carbon capture journal Oct. 2010). Preliminary results from this pilot plant indicate operation as expected; expected to complete final testing in mid-2011.

(c)   Wabash River (USA) IGCC plant is a repowering facility built to replace an outdated conventional pulverized coal power plant. The repowering plant of 262 MW_e began operation in 1995 as a demonstration unit on Illinois bituminous coal, and since 1999 commercial operation has started using petroleum coke. The Plant is designed to use a variety of local coals with upto 5.9% sulfur content (dry basis) and a higher heating value of 13,500 Btu/lb (moisture and ash free). This plant uses E-GAS technology with feedstock (2544 tpd). The schematic of this plant is shown in Figure 8.13. Coal is first slurried with water and fed with 95%-pure oxygen to the first stage of the gasifier. The coal is partially combusted in this stage to maintain a temperature, around 2,500°F (1,371°C), and most of the coal reacts with steam to produce the raw syngas at this stage. Ash in the coal melts and flows out of the bottom of the gasifier vessel as slag. Additional coal slurry is added to the second gasification stage where it undergoes devolatization, pyrolysis, and partial gasification to cool the raw syngas and enhance its heating value. The raw syngas is then further cooled to produce steam for power generation. The steam is generated at a pressure of about 1,600 psia.

Particulates contained in the syngas are removed via candle filters and then recycled to the first stage of the gasifier to gasify any remaining residual carbon. A single gasifier processes 2,544 tpd of coal, although two gasifiers, each with the capability of 100% capacity, were installed. After the removal of particles, the syngas is further cooled and

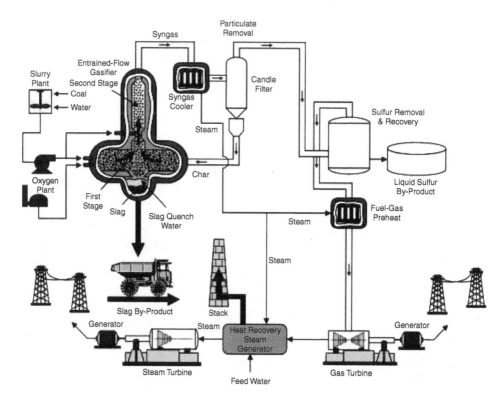

Figure 8.13 The Process flow diagram of The Wabash River Repowering Project (Downloaded from DOE/NETL: Gaspedia: Applications of Gasification – IGCC).

scrubbed to remove chlorides before passing through a catalyst bed which converts carbonyl sulfide to H_2S. High quality sulfur is recovered, at 99.99% purity, via an amine system. Both the recovered sulfur and the slag have commercial applications. Moisture is added to the cleaned syngas to control NO_x emissions and then fed to a GE192 MW gas turbine, where it is combusted to generate electricity. The advanced design of the turbine allows for combustion temperatures to reach 2,350°F (1,222°C), which is significantly higher than previous demonstrations. The heat from turbine exhaust is recovered in a HRSG to produce steam for further production of electricity; flue gas is then emitted to the atmosphere through a 222 ft stack. The plant is installed with a new air separation unit, gasifier, clean-up system, gas turbine and heat recovery steam generator, but the existing, 30-year old 100 MW steam turbine is retained. The coal boiler that was originally built to supply steam to the steam turbine was removed when the IGCC equipment started up. This energy-efficient technology removes more than 97% of the SO_2 and 82% of the NO_x from the stack emissions while producing 262 MW_e (net) of electricity. Over the demonstration period, this Project processed 1.5 million tons of coal, generating more than 4 million MWh of electricity. Thermal efficiencies of the plant were 39.7% for coal and 40.2% for petroleumcoke, HHV basis (DOE/NETL 2000); (d) 250 MW_e net Plant in Tampa, FL, USA, using GE technology

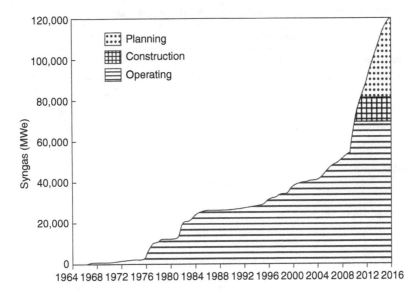

Figure 8.14 Gasification: Capacity and planned growth (Redrawn from Bhattacharya 2011).

with bituminous coal feedstock (2200 tpd); started in 1996 with slurry fed oxygen-blown gasifier and 1200°C gas turbine; (e) Nakoso demo Plant, 250 MW$_e$, in Japan using Mitsubishi technology; started in 2007 with a dry-feed air-blown coal gasifier and 1200°C gas turbine; employs not only F-type turbines but G-type as well; (f) 350 MW$_e$ plant in Czech Republic started in 1996 using Lurgi dry ash technology.

Present status: Although IGCC technology has started four decades ago, the growth has been impressive since 1990s as shown in Figure 8.14. As of December 2010, the worldwide figures are as follows: Syngas capacity: 70,800 MWth, 144 operating plants and 412 operating gasifiers; 11 plants and 16 gasifiers are under construction; 37 plants and 76 gasifiers are planned (Bhattacharya 2011).

8.6.2 Environmental performance of gasification

In Table 8.6 are listed the environmental contrasts in utilizing coal combustion and gasification technologies for power generation. By comparison, coal gasification is more environment-friendly. Modern power plants based on lignite coal have reached 43% efficiency; the medium-term potential of lignite-based steam power plants is up to 50% with pre-drying and 700°C technology, and >50% for hard coals. The IGCC plants can operate at much higher efficiency with less emission levels compared to conventional coal-fired plants as shown in Table 8.7. IGCC plants offering over 55% efficiencies are achievable in the near future (BINE Informationsdienst) with lower CO_2 addition compared to other advanced power plant technologies.

8.6.3 Economics

IGCC plants are presently not cost competitive with other advanced systems such as PC fired Supercritical Plants. Even so, there are other considerations such as efficient

Table 8.6 Environmental contrasts between Combustion and Gasification.

	Combustion	Gasification
Sulfur converted to...	SO_2	H_2S or COS
Sulfur capture	Flue gas scrubbers, boiler limestone injection	Absorbed in physical or chemical solvents
Sulfur disposal	Gypsum sold for wallboard	Sold as H_2SO_4 or elemental S
Nitrogen converted to...	NO_x	Traces of NH_3 in syngas (syngas combustion produces low levels of NO_x)
NO_x control	Required (e.g., low-NO_x burners, staged combustion, SCR/SNCR)	Currently not needed for IGCC (but tighter regulations could require SCR)
C is converted to...	CO_2	Mostly CO in syngas
CO_2 control	Post-combustion removal from diluted stream	Pre-combustion removal from concentrated stream
Water requirements	Much more steam cycle cooling water needed	Some water needed for slurry; steam cycle and process needs

(*Source:* NETL/Stiegel 2005)

Table 8.7 Emission levels from Steam power plants & IGCC plant (kg/MWh).

	Steam Power plant (Past)	Steam Power plant (State-of-the-art)	IGCC Power plant
SO_2	35	0.8	0.07
NO_x	2.6	0.8	0.35
Dust	30	0.1	0.001
CO_2	1250	1000	800
Efficiency	35%	45%	55%*

*Energy research sees 55% efficiency as feasible for IGCC, and is working towards the target; The data is compiled from the figures given in BINE Informationsdienst, Institut fur Energiever-fahrensetechnik und Chemieingenieurwesen der Technische Universitat (TU), Bergakademie Freiberg)

removal of CO_2 from the high pressure fuel gas, and controlling mercury emissions at significantly lower cost than in PC combustion, which may in the future favor IGCC. Results of an EPRI study (Holt 2004) on comparative average data of capital and operating expenses and on efficiency of subcritical steam and advanced power generation technologies, without CCS, are shown in Table 8.8.

Assumptions used to derive results in the Table: Book life = 20 years; Commercial operation date = 2010; Total Plant Cost (TPC) includes Engineering and Contingencies; Total Capital Requirement (TCR) includes Interest During Construction and Owners' Cost; Assumes EPRI's TAG financial parameters; All costs in 2003 dollars (Beer 2009). The values in the first three columns are slightly different from those given in MIT Study 2007.

Table 8.8 Costs for 500 MW Power Plants Using a Range of Technologies without CCS.

	PC Sub-critical	PC Super critical	CFB	IGCC (E-gas) w/spare	IGCC (E-gas) No spare	NGCC High CF	NGCC Low CF
Total Plant Cost, $/kW	1,230	1,290	1,290	1,350	1,250	440	440
Total Capital requirement, $/kW	1,430	1,490	1,490	1,610	1,490	475	475
Ave. Heat rate							
Btu/kWh (HHV)	9,310	8,690	9,800	8,630	8,630	7,200	7,200
η% (HHV)	36.7	39.2	34.8	39.5	39.5	47.4	6,840
η% (LHV)	38.6	41.3	36.7	41.6	41.6	50.0	47.4/50
Levelized Fuel Cost, $/MWtu (2003$)	1.50	1.50	1.00	1.50	1.50	5.00	5.00
Capital, $/MWh (Levelized)	25.0	26.1	26.1	28.1	26.0	8.4	16.9
O&M,$/MWh (Levelized)	7.5	7.5	10.1	8.9	8.3	2.9	3.6
Fuel, $/MWh (Levelized)	14.0	13.0	9.8	12.9	12.9	36.0	36.0
COE, $/MWh (Levelized)	**46.5**	**46.6**	**46.0**	**49.9**	**47.2**	**47.3**	**56.5**

(*Source:* Holt 2004; Beer 2009)

8.6.4 R&D areas for improvement

Compared to NGCC and pulverized coal, IGCC is a developing technology. Because of its potential to demonstrate higher efficiency, the rate of progress is expected to be very significant with continued research and development activity.

Research that offers significant improvements in fuel flexibility, increased availability and reliability, efficiency, economics and environmental sustainability has to be promoted (Barnes 2011):

(a) Fuel flexibility: The technical and economic performance of IGCC depends considerably on feedstock quality compared to pulverized coal plants (Dalton 2004). This aspect is especially critical given that future gasification plants could possibly process a wide range of low-cost feedstocks such as biomass, municipal and other solid wastes, or combinations of these, in addition to all types of coals;

(b) Development of efficient and cost-effective oxygen separation technologies: presently utilized cryogenic distillation process which is energy intensive and costly. Progress in developing Ion transport membrane (ITM) technology in which oxygen is transported at 540°C temperature through a ceramic membrane (perovskite or fluorite or mixed ceramic membranes) has been significant (Foy & McGovern 2005, at www.netl.doe.gov/publications/.../Tech%20Session%20Paper%2011) Successful development of ITM promises improvement of IGCC efficiency by one percentage point, and reduction of the capital cost by around 5%; i.e., \$75/kW (US National Research Council Panel on DOE's IGCC Program, Washington, DC 2006; Beer 2009).

(c)   Removing carbon dioxide from syngas: Research focus is currently on new types of pollutant-capturing sorbents that work at elevated temperatures and do not degrade under the harsh conditions of a gasification system. Also, new types of gas filters, and novel cleaning systems with low energy price are being examined;

(d)   Utilization of gasification by-products: These solid by-products can have commercial value; possible uses of the solid material produced when coal and other feedstocks (e.g., biomass, municipal waste) are utilized need to be investigated. Some plants produce adequately pure sulfur which can be marketed.

Significant developments and approaches having potential are emerging that can be applicable to future coal gasification processes (Barnes 2011):

(a)   Improved gasification cycles: Besides the established basic thermodynamic cycles related to coal gasification plant, a number of improved and non-conventional cycles and the use of alternative fluids to water/steam have been developed in recent times, which may contribute to future coal gasification plants with enhanced efficiency. These gains are at present *speculative*; in some cases require the development of technologies appreciably in advance of the current best practice. Moreover, some cycles (such as Matiant cycle) are susceptible to the unavoidable parasitic losses present in a highly integrated system such as a coal gasification plant (refer to Heitmeir *et al.* 2006; Houyou *et al.* 2001; Alexander 2007; Kalina 1982; Turboden 2011).

(b)   Improvements to gas turbine operation: Gas turbine technology is basic to the overall process efficiency of coal gasification plant. Continuous progress has enhanced inlet temperatures and pressures continuously, resulting in efficiency increase. Alongside, techniques have been developed to improve the efficiency of existing gas turbines, often under particular conditions local to the site of operation. Further, advances in emissions control technologies closely coupled to the turbine operation ensure conformity with existing as well as probable future regulations (refer to Hodrein 2008; USDOE NETL 2006; Guthee *et al.* 2008; Mee 1999; GE Aero Energy products 2011; Ikeguchi *et al.* 2010; Nakagaki *et al.* 2003; Smith 2009, Geosits & Schmoe 2005).

(c)   Cycles for Enhanced power generation: While the main types of gasifier keep advancing, a number of alternative approaches to coal gasification have emerged. These include systems with enhanced energy recovery, gasifiers operating in a CO_2-rich atmosphere, quench gasifier for low-grade coals, and innovative technologies and combinations of technologies; also includes materials improvements of refractory gasifier liners and metal component coatings to improved flame monitoring equipment (examples: Advanced IGCC/IGFC system; oxy-fuel incorporating the principles of flue gas recirculation to IGCC for CO_2 capture; 'Coal without carbon' next generation technologies; Integrated Gasification Steam Cycle, and MHD). Some of these units well-established at the pilot plant scale are showing promise for further scale-up. Even an earlier technology, such as MHD, is being revived after nearly five decades. Although it is unclear which of these approaches will finally emerge successful, most of these systems have the potential to offer significant advantages either in respect of lower capital costs, or in dealing with type of coals that may involve in niche applications, or in retrofits to the existing coal plants under stringent regulations on CO_2 emissions in future (Iki *et al.* 2009; Oki *et al.* 2011; Shirai *et al.* 2007; Calderon 2007; Griffiths *et al.* 2009; Pettus *et al.* 2009; Darby 2010; Alfoen 1942; Bazter *et al.* 2003; Hustad *et al.* 2009). The reader is advised to refer to the Barnes report (2011) (GS42-ccc187) for more technical details.

To accelerate wider deployment of IGCC, the 2011 IEA Report (Burnard & Bhattacharya 2011) has listed R&D needs, similar to what already mentioned.

The lignite-based IGCC Plants (Bhattacharya 2011) specially require the R&D focus on the following: (1) large-scale demonstration using all types of lignite coals, (2) need for separate combustors for fluidized bed or transport gasifiers, if carbon from bed discharge is to be converted, (3) demonstration of quench gasifier for lignite coals containing high levels of alkali, (4) study of ash-related problems in gasifier that are characteristic of the gasifier being used and the type of lignite coal, (5) description of solid wastes discharged from the gasifiers, and (6) development of reliable system for feeding lignite coals to pressurized systems.

8.6.5   Outlook for coal-based IGCC

Power generation capacity by coal gasification is insignificant compared to coal-fired power generation capacity, around 1760 GW_e. Even so, its potential for high efficiency and lower emission levels in the coming decades with further R&D explained above, its utilization will be significantly high. This is mentioned in the last chapter.

8.7   HYBRID GASIFICATION

One of the promising coal fired advanced cycles expected to be ready for demonstration in 2010–2015 is Hybrid Gasification Fuel Cell-Gas Turbine-Steam Combined Cycle (Figure 8.15). The Solid State Energy Conversion Alliance (SECA), USA, is entrusted with the development and demonstration of coal-fueled central-station electric power generation technology based upon the solid oxide fuel cell (SOFC). The fuel gas produced in an oxygen-blown gasifier at elevated pressure is cleaned at a temperature of 1273K and is sent to a solid oxide fuel cell (SOFC) on the anode side, while air from a compressor exhaust preheated in a recuperator enters on the side of the cathode. The hydrogen in the gas is used to generate electricity in the SOFC, and the CO burns in a combustion turbine that drives the compressor. Electric power is produced in another SOFC and a gas turbine, at a lower pressure, downstream of the high-pressure turbine with more power added by a bottoming steam cycle. The efficiency could reach 60% in this near zero emission schemes, together with a carbon capture of over 90% (excluding compression losses for pressurizing CO_2 for transport and storage). Demonstration of this novel plant is expected within the FutureGen program during the 2012–2020 (Beer 2009). Effective control of such fuel cell-turbine systems is a significant challenge.

Advanced Integrated Gasification Fuel Cell (IGFC) systems are projected to achieve $\geq 99\%$ CO_2 capture, near-zero emissions of criteria pollutants (less than

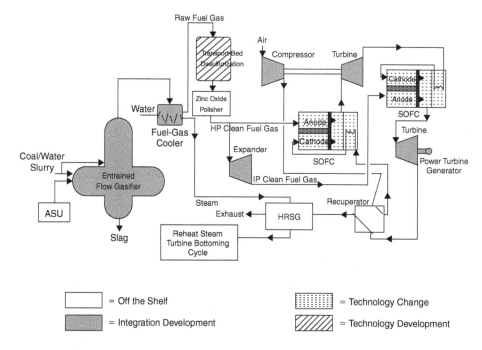

Figure 8.15 Schematic of Fuel Cell-Gas Turbine-Steam C.C (*Source:* Ruth 1998; Beer 2009, reproduced with the permission of Professor Janos Beer © Elsevier).

0.5 ppm NO_x emission) and considerable reduced raw water consumption. The coal gasifier enables the SOFC to use coal as fuel, and good synergy between the coal gasification and SOFC technologies is mainly expected when they are teamed in a carbon-capturing electric power system. For the SOFC application, there is no specific need to shift the CO to hydrogen and CO_2 before fuel cell entry.

These systems are capable of a coal-to-electricity efficiency exceeding 50% (HHV); and 60% HHV for advanced pressurized systems. The cost goal of $175/kW stacks and $700/kW power blocks (2007 US $ basis) will ensure that the cost of electricity to the user will not exceed today's typical rates. These advanced IGFC systems broadly address environmental, climate change, and water concerns associated with fossil fuel use while establishing concurrently a base for a secure energy future (NETL, USDOE 2011).

A recent study by Braun *et al.* (2012) examines the performance of a solid oxide fuel cell (SOFC) based integrated gasification power plant concept at the utility scale (>100 MW). The primary system concept evaluated was a pressurized ∼150 MW SOFC hybrid power system integrated with an entrained-flow, dry-fed, oxygen-blown, slagging coal gasifier and a combined cycle in the form of a gas turbine and an organic Rankine cycle power generator. The concepts analyzed include carbon capture via oxy-combustion followed by water removal and gas compression to pipeline-ready CO_2 sequestration conditions. The results of the study indicate that hybrid SOFC systems could achieve electric efficiencies approaching 66% (LHV) when operating using coal-derived clean syngas and without CO_2 capture.

8.8 UNDERGROUND COAL GASIFICATION

Underground coal gasification (UCG) converts coal in-situ into a gaseous product, syngas, through the same chemical reactions that occur in surface gasifiers. UCG is particularly useful in extracting energy from coal seams that are un-mineable deep or thin coals under many different geological settings, and could increase to a great extent the coal resource available for utilization. In the USA, it is estimated 300–400% increase in recoverable coal reserves is possible. At present, a large number of commercial projects are in various stages of development in the USA, Canada, South Africa, India, Australia, New Zealand, and China to produce power, liquid fuels, and synthetic natural gas (Elizabeth Burton et al. 2010). As of 2008, the number of UCG trials includes 200 in the former Soviet Union, 33 in the USA, and around 40 distributed among South Africa, China, Australia, Canada, New Zealand, India, Pakistan, and Europe (van der Reit 2008; Roddy & Gonzalez 2010).

In underground gasification, the oxidants, steam and oxygen, are injected into a coal seam through wells drilled from the surface. The coal seam is ignited and partially burned. The heat generated by the combustion gasifies additional coal to produce fuel-grade gases which are piped to the surface. Different temperatures, pressures, and gas compositions exist at different locations within a gasification channel, which is normally divided into three zones: oxidization, reduction, and dry distillation and pyrolysis (Perkins & Sahajwalla 2005; Yang et al. 2010; Self et al. 2012). As a result, the chemical reactions differ in the three zones. In the oxidization zone, multiphase chemical reactions occur involving the oxygen in the gasification agents and the carbon in the coal. The highest temperatures, 700–900°C occur in the oxidation zone; it may reach up to 1500°C also due to the large release of energy during the initial reactions (Yang et al. 2003; Elizabeth Burton et al. 2010; Sury et al. 2004). By using oxygen instead of atmospheric air, the heating value of the syngas gets increased (Perkins & Sahajwalla 2008), but producing pure oxygen requires additional energy. As the coal face burns and the immediate area is depleted, the injection of oxidants is controlled. Finally, the product gas consists of CO_2, H_2, CO, and CH_4 along with small quantities of contaminants including SO_x, NO_x, and H_2S (Sury et al. 2004). However, the composition of syngas is highly dependent on the gasification agent, air injection method, and coal composition (Stanczyk et al. 2011; Prabu & Jayanti 2012). Separate production wells are used to bring the product gas to surface. These gases are then cleaned and used to drive gas turbines to generate power (see Figure 8.16, Self et al. 2012).

Process: There are a variety of designs for underground coal gasification, designed to inject oxidant and possibly steam into the reaction zone, and also to provide a path for production gases to flow in a controlled manner to surface. The process is based on the natural permeability of the coal seam to transmit gases to and from the combustion zone; for high pressure break up of coal, an enhanced permeability created through reverse combustion, an in-seam channel, or hydro-fracturing may be used in varying degree (Gregg & Edgard 1978; Stephens et al. 1985a; Walker et al. 2001; Creedy & Garner 2004; Elizabeth Burton et al. 2010).

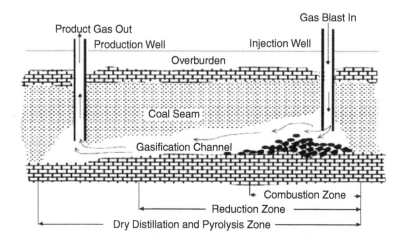

Figure 8.16 Schematic of *in-situ* UCG process (*Source:* Self *et al.* 2012).

The simplest design uses two vertical wells: one injection and one production. Sometimes it is necessary to establish communication between the two wells, and a common method to use reverse combustion to open internal pathways in the coal. Another alternative is to drill a lateral well connecting the two vertical wells (Portman Energy 2012).

The Soviet UCG technology, in which simple vertical wells, inclined wells, and long directional wells were used, was further developed by Ergo Exergy and tested at Linc's Chinchilla site in 1999–2003, in Majuba UCG plant (2007–present), in Cougar Energy's UCG plant in Australia (2010). The same technology is being applied by several others in New Zealand, Canada, the USA, India and other countries. During 1980s and 1990s, a method known as CRIP (controlled retraction and injection point) was developed by the Lawrence Livermore National Laboratory and demonstrated in the USA and Spain. This method uses a vertical production well and an extended lateral well drilled directionally in the coal. The lateral well is used for injection of oxidant and steam, and the injection point can be changed by retracting the injector (Portman Energy 2012).

Carbon Energy was the first to adopt a system, which uses a pair of lateral wells in parallel. This system allows a consistent separation distance between the injection and production wells while progressively mining the coal between the two wells. It provides access to the greatest quantity of coal per well set and also allows greater consistency in production gas quality.

Recently, Portman Energy developed a new technology wherein a method called SWIFT (Single Well Integrated Flow Tubing) uses a single vertical well for both Syngas recovery and oxidant delivery. The design has a single casing of tubing strings enclosed and filled with an inert gas to allow for leak monitoring, corrosion prevention and heat transfer. A series of horizontally drilled lateral oxidant delivery lines into the coal and a single or multiple syngas recovery pipeline(s) allow for a larger area of coal to be combusted at one time. The developers claim this method will increase the syngas production by up to ten times compared to earlier technologies. Further, the novelty in design approaches including the single well will significantly lower development costs, and as the facilities and wellheads are concentrated at a single point, the need for surface access roads, pipelines and other facilities will be reduced (Portman Energy 2012).

A wide variety of coals from lignite to bituminous is amenable to the UCG process and may be successfully gasified. In selecting appropriate locations for UCG, several factors that include surface conditions, hydrogeology, lithoglogy, coal quantity, and quality are considered. Other important criteria include depth of 100–600 m (330–2,000 ft), thickness exceeding 5 m (16 ft), ash content less than 60%, minimal discontinuities and isolation from aquifers (Beath 2006).

UCG has several advantages over conventional underground or strip mining and surface gasification:

(a)   Reduces operating costs and surface damage, and eliminates mine safety issues such as mine collapse and asphyxiation compared to conventional mining;
(b)   Greatly increases domestic resource availability;
(c)   Capital costs are substantially reduced as no surface gasification systems are needed;
(d)   No transport of coal at the surface resulting in reduced cost, emissions, and local footprint associated with coal shipping and stockpiling;
(e)   Most of the ash in the coal stays underground, thereby avoiding the need for excessive gas clean-up, and the environmental issues associated with fly ash waste stored at the surface;
(f)   No production of some criteria pollutants (e.g., SO_x, NO_x); other emissions such as mercury, particulates, sulfur species are greatly reduced in volume; and
(g)   Advantageous for geological carbon storage; the infrastructure can be used subsequently for geologic CO_2 sequestration operations; may be possible to store CO_2 in the reactor zone underground as well in adjacent strata (Elizabeth Burton 2010).

There are, however, several concerns to be addressed. These ignition processes produce a syngas stream which is similar to surface-produced syngas in composition. But CO_2 and hydrogen content may be higher due to several factors, including a higher than optimal rate of water flux into the UCG reactor and ash catalysis of water-gas shift. Because of the nature of in-situ conversion, UCG syngas is lower in sulfur, tar, particulates and mercury and very low ash content than conventional syngas. Other components are similar and can be controlled through conventional gas processing and cleanup methods.

Costs

The economics of UCG appear extremely promising. The capital costs of UCG plants appear to be considerably less than the equivalent plant fed by surface gasifiers because the gasifier's cost is saved. Similarly, operating expenses are also likely to be much lower because there are no coal mining and transportation activities; moreover, ash management facilities are significantly reduced. Even for configurations requiring a substantial environmental monitoring program and additional swing facilities, UCG plants retain many economic advantages. There is substantial literature on UCG and the reader may refer to them for technical and other details.

REFERENCES

Section A:

Allam, R.J. (2007): Improved oxygen production technologies Report 2007/14, Cheltenham, UK, IEAGHG R&D Program, p. 90, Oct. 2007.

Allam, R.J. (2009): Improved oxygen production technology, Proc. 9th Intl. Conf. on GHG control technologies, Washington, DC, 16–20 June 2008, Oxford, UK, Elsvier Ltd. pp. 461–470 (2009).

Anheden, M., Yan, J., & De Smedt, G. (2005): Denitrogenation (oxy-fuel concepts), Oil & Gas Science and Technology Revue de l'IFP, 60(3), 485–495.

Anderson, K., Birkstead, H., Maksinen, P., Johnsson, F., Stroemberg, L., & Lyngfelt (2002): *An 865 MW Lignite fired CO₂ free power plant-Technical feasibility study*, Proc. of the 6th Conf. on GHG Control Technologies, Kyoto, Japan 2002.

Armor, A.F., Viswanathan, R., & Dalton, S.M. (2003): 28th Int. Conf. on Coal Utilization & Fuel Systems, 2003, pp. 1426–38, US DOE, ASME.

Bazzuto, C., Scheffknecht, G., & Fouilloux, J.P. (2001): Clean power generation technologies utilizing solid fuels, *World Energy Council, Proc. of 18th Energy Congress*, The Industry Players Perspective: Discussion Session 1, 20–25 October, Buenos Aires, Argentina.

Beér, J.M. (1996): *Low NOₓ Burners for Boilers, Furnaces and Gas Turbines; Drive Towards the Lower Bounds of NOₓ Emissions, Invited Lect.*, Third Intl. Conf. on Combustion Technologies for a Clean Environment, Lisbon, Portugal, 1995; Combust. Sci.and Tech. **121**, 1–6, p. 169.

Beer, J.M. (2000): *Combustion Technology Developments in Power Generation in Response to Environmental Challenges:* Elsevier, Progress in Energy and Environmental Sciences, **26**, pp. 301–327.

Beer, J.M. (2009): High efficiency electric Power generation: The Environmental Role, Progress in Energy & Environmental Science, **33**, pp. 107–134, at http://web.mit.edu/mitei/docs/reports/beer-combustion.pdf

Blum, R., & Hald, J. (2002): ELSAM Skaerbaek, Denmark.

Bob and Wilcox (1992): Steam: its generation and use, Schultz, S.C. and J.B. Kitto (eds) pp. 24–10.

Booras, G., and N. Holt (2004): Pulverized Coal and IGCC Plant Cost and Performance Estimates, *Gasification Technologies Conf. 2004:* Washington, DC.

Büki, G. (1998): Magyar Energiatechnika, **6**, pp. 33–42.

Buhre, B.J.P, Elliott, L.K., Sheng, C.D., Gupta, R.P., & Wall, T.F. (2005): *Oxy-fuel combustion technology for coal fired power generation* Progress in Energy and Combustion Science, **31** (4), pp. 283–307.

Burnard, K., & Bhattacharya, S. (2011): Power generation from Coal: Ongoing developments and Outlook, Information paper, October 2011, IEA, Paris, France.

Cao, B.L., & Feng, Y.K. (1984): Technology of fluidized bed combustion boilers in China. In: *Fluidized Bed Combustion and Applied Technology,* R.G. Schwieger (ed.), First International Symposium, pp. 1–63.

Croiset, E., & Thambimuthu, K.V. (1999): *Coal combustion with flue gas recirculation for CO₂ recovery,* In: Greenhouse Gas Control Technologies, Riemer, P., Eliassonand, B. & Wokaun, A. (eds), Elsevier Science Ltd.

Croiset, E., Douglas, P. & Tan, Y. (2005): Coal oxy-fuel combustion: a review, In: *Proc. 30th Intl. Technical Conf. on Coal utilization and Fuel systems*, Clearwater, FL, April 2005.

Sakkested, B. (ed.), Gaithersburg, MD, Coal Technology Association, Vol. 1, 699–708, 2005.

COMTES700 – On Track towards the 50 plus Power Plant, Presentation at New Build Europe 2008, Dusseldorf, 4–5 March 2008.

Dalton, S. (2006): Ultra supercritical Progress in the US and in Coal fleet for tomorrow, 2nd Annual Conf. of the Ultra supercritical Thermal Power technology Collaboration Network, 27–28 Oct. 2006, Qingdao, China.

Davidson, R.M. & Santos, S. (2010): CCC/168, June 2010, IEAGHG Oxy-fuel combustion of Pulverized Coal, IEAGHG Program, 2010/07, August 2010.

Dillon D.J., Panesar R.S., Wall R.A., Allam R.J., White V., Gibbins J., & Haines M.R. (2004): *Oxy-Combustion processes for CO₂ capture from advanced supercritical PF and NGC-Cplant,* Proc. of the Seventh Conf. on Greenhouse Gas Control Technologies (CHGT-7), Vancouver, BC, Canada.

Eskin, N. & Hepbasli, A. (2006): Development and Applications of Clean Coal Fluidized Bed Technology, *Energy Sources, Part A,* **28**, 1085–1097, Copyright © Taylor & Francis Group, LLC.

Field, M.A., Gill, D.W., Morgan, B.B., & Hawksley, P.G.W. (1967): *Combustion of Pulverized Coal.* 1967, Leatherhead, England: BCURA Gierschner, G. (2008): Fluidized Bed Combustion Boiler (From 'Types of Boilers' at www.energyefficiencyasia.org/energyequipment/typesofboiler.html

Gaglia, B.N. & Hall, A. (1987): Comparison of Bubbling and Circulating Fluidized Bed Industrial Steam Generation, Proc. of the Intl. Conference on Fluidized Bed Combustion, May 3–7, 1987.

Giglio, R. & Wehrenberg, J. (2009): Electricity and CFB Power technology, Foster-Wheeler, January 2009 at TP_CFB_09_02.pdf

Henry, J.F., Fishburn, J.D., Perrin, I.J., Scarlin, B., Stamatelopoulos, G.N., & Vanstone, R. (2004): 29th Int. Conf. on Coal Utilization & Fuel Systems 2004, pp. 1028–42, US DOE, ASME.

Hesselmann, G. (2009): Challenges in Oxy-fuel: a boiler-maker perspective, First Oxy-fuel Combustion Conf., Cottbus, Germany, 8–11 September 2009.

Higman, C., and van der Burgt, M. (2003): *Gasification,* New York: Elsevier.

Hottel, H.C., & Sarofim, A.F. (1967): Radiative Heat Transfer, NY: McGraw-Hill.

IEAGHG (2005): Oxy-combustion processes for CO₂ capture from power plant, Report 2005/9, Cheltenham, UK, IEA GHG R&D Program, pp 212, July 2005.

Jantti, T., Lampenius, H., Ruskannen, M., & Parkkonen, R. (2009): Supercritical OUT CFB Projects – Lagisza 460 MWₑ and Novocherkasskaya 330 MWₑ, Presented at Russia Power 2009, Moscow.

Khare, S., Wall, T., Gupta, R., Elliott, L., & Buhre, B. (2005): *Retrofitting of air-fired pf plant to oxy-fuel: heat transfer impacts for the furnace and convective pass and associated oxygen production requirements,* 5th Asia-Pacific Conference on Combustion, Adalaide, July 2005.

Kjaer, S., Klauke, F., Vanstone, R., Zeijseink, A., Weissinger, G., Kristensen, P., McKenzie, E.C. (1978): Burning coal in fluidized bed, *Chemical Engineering,* 85(18):116–127.

Meier, J., Blum, R., & Wieghardt, K. (2001): Powergen Europe, 2001, Brussels.

MIT Report 2007: The Future of Coal – Options for a Carbon constrained World, Massachusetts Institute of Technology, August 1, 2007; at http://web.mit.edu/coal/

Nalbandian, H. (2009): Performance and Risks of Advanced Pulverized Coal Plants, UK Center for Applied Energy Research, *Energeia,* 20 (1), 6 pages.

NCC (2004): *Opportunities to Expedite the Construction of New Coal-Based Power Plants,* National Coal Council.

Nsakala, Liljedahl *et al.,* (2003): GHG Emissions Control in CFB boilers, US DOE/NETL Cooperative Agreement May 2003.

Oka, S. (2001a): Is the future of BFBC technology in Distributive Power generation?, Invited lecture at the 3rd SEEC Symposium: 'Fluidized Bed Technology in Energy Production, Chemical and Process Engineering and Ecology', September 2001, Sinaia, Romania; *Thermal Science*, 5(2), pp. 33–48.

Palkes, M. (2003): *Boiler Materials for Ultra Supercritical Coal Power Plants Conceptual Design ALSTOM Approach*, NETL-DOE, 2003, USC T-1.

Parsons, E.L., Shelton, W.W., & Lyons, J.L. (2002): *Advanced Fossil Power Systems Comparison Study*, NETL: Morgan town, WV.

Rice, R.L., Shang, J.Y., & Ayers, W.J. (1980): Fluidized bed combustion of North Dakota lignite, *Proc. of the Sixth Intl. Conf. on Fluidized Combustion*, 3, Technical Sessions, Atlanta, Georgia. April 9–11, pp. 211–219.

Riddiford, F., Wright, I., Espie, T., & Torqui, A., (2004): Monitoring geological storage: In Salah Gas CO_2 Storage Project, GHGT-7, Vancouver, B.C.

Rubin, E.S. (2005): *Integrated Environmental Control Model 5.0*. Carnegie Mellon University: Pittsburgh.

Santos, S. & Davison, J. (2006): Oxy-fuel combustion for power generation industry: review of the past 20 years of R&D activities and what are the future challenges ahead, In: 8th Intl. Conf. on GHG control technologies, Trondheim, Norway, 19–22 June 2006.

Santos, S. & Heines, M. (2006): Oxy-fuel Combustion application for coal-fired power plant: what is the current state of knowledge? Intl. Oxy-combustion Network for CO_2 capture report on Inaugural (1st) workshop, Cottbus, Germany, 29–30 November 2005, Report 2006/4, Cheltenham UK, IEAGHG R&D program, pp. 77–107, July 2006.

Santos, S., Heines, M., Davison, J., & Roberts, P. (2006): Challenges in the development of oxy-combustion technology for coal-fired plant, 31st Intl. Techl. Conf., on Coal utilization & fuel systems, Clearwater, FL. 21–26 May 2006.

Sarofim, A.F. (2007): Oxy-fuel combustion: Progress and remaining issues, 2nd meeting of the Oxy-fuel network, Windsor, CT, 25–26 January 2007, Report 2007/16, Cheltenham, UK, IEA GHG R&D program, pp. 92–128, Nov. 2007.

Schernikau, L. (2010): Coal-Fired Power Plants, *Economics of the International Coal Trade*, DOI 10.1007/978-90-481-9240-3_4, Copyright: Springer Science + Business Media B.V.

Schilling, H.D. (1993): VGB Kraftwerkstechnik, 73(8), pp. 564–76 (English Edn).

Shaddix, C.R. (2012): Coal combustion, gasification, and beyond: developing new technologies for a changing world, *Combustion and Flame*, 159, 3003–3006.

Smoot, L.D., & Smith, P.J. (1985): *Coal Combustion and Gasification*: The Plenum Chemical Engineering Series, ed. D. Luss, New York: Plenum Press.

Termuehlen and Empsperger (2003): *Clean and Efficient Coal-fired Power plants*, New York: ASME Press.

Toftegaard, M., Brix, J., Jensen, P., Glarborg, P. & Jensen, A. (2010): Oxy-fuel combustion of solid fuels, *Progress in Energy and Combustion Science*, 36(5), 581–625 (2010).

USDOE (1999): Tidd PFBC Demonstration Project, DOE/FE-0398, Project Performance Summary, CCTDP, June 1999.

Utt, J. & Giglio, R. (2012): The introduction of F-W's 660 MW_e supercritical CFBC technology for the Indian market, presented at *Power-Gen India & Central Asia Conference, New Delhi*, April 20, 2012.

Virr, M.J. (2000): The Development of a Modular System to Burn Farm Animal waste to generate Heat and Power, in: *Proc. of the 7th Annual Intl. Pittsburgh Coal Conf, University of Pittsburgh*, 11–14 September, 2000.

Weitzel, P. a.M.P (2004): *Cited by Wiswqanathan, et al.* Power, April 2004.

Williams, A., Pourkashanian, M., Jones, J., & Skorupska, N. (2000): *Combustion and Gasification of Coal*, Applied Energy Technology Series, NY: Taylor & Francis.

Wilson, M., & Monea M. (eds) (2004): *IEA GHG Weyburn CO_2 Monitoring & Storage Project Summary Report 2000–2004*, 273 p.

White, L.C. (1991): *Modern Power Station Practice – Volume G: Station Operation and Maintenance*, 3rd edition, British Electricity International, Pergamon Press, Oxford, UK.

Yamada *et al.* (2003): Coal Combustion Power Generation Technology, Bulletin of the Japan Institute of Energy, **82** (11), pp. 822–829.

ZEP (2008): Recommendations for RTD, support actions, and Intl. collaboration priorities within the next FP7 energy work program in support of deployment of CCS in Europe, ETP for zero-emission power plants, p. 27, 18th April 2008.

Zheng, L., Tan, Y. & Wall, T. (2005): Some thoughts and observations on Oxy-fuel technology developments, *Proc. 22nd Annual Coal Conf.*, Pittsburg, PA, 12–15 Sept. 2005, paper 307, p. 10.

Section B

Alexander, B.R. (2007): *Analysis and optimization of the Graz Cycle: a coal fired power generation scheme with near-zero carbon dioxide emissions*, Thesis (B.Sc.), Massachusetts Institute of Technology, Cambridge, MA, vp (Jun 2007).

Alfvén, H. (1942): Existence of electromagnetic-hydrodynamic waves. *Nature*; 150(3805); 405–406 (1942).

Baxter, E., Anderson, R.E., & Doyle, S.E. (2003): *Fabricate and test an advanced non-polluting turbine drive gas generator, Final report*; Contract DE-FC26-00 NT40804, Springfield, VA, National Technical Information Service, p. 47.

Barnes, I. (2011): *Next generation Coal gasification technology*, CCC/187, IEA Clean Coal Centre, September 2011. Beath, A. (2006): UCG Resource utilization efficiency, CSIRO Exploration & Mining.

Beér, J.M. (1985): *Coal-Water Fuel Combustion; Fundamentals and Application. A North American Overview*, Second European Conf. on Coal Liquid Mixtures, I. Chem. E. Symposium Ser. No. 95. London, UK, 1985.

Beér, J.M., & G. Vermes (1982): Gas Turbine Combustor for Coal-Water Slurries, *ASME Engineering Foundation Conf. on Tomorrow's Fuels*, at Santa Barbara, CA, November 7–12.

Beér, J.M., Dowdy, T.E., & Bachovchin, D.M.: US Patent No. 5636510 (1997).

Beér, J.M. & Garland, R.V. (1997): A Coal Fueled Combustion Turbine Cogeneration System with Topping Combustion, *Trans. ASME J. Engineering for Gas Turbines and Power*, **119** (1), pp. 84–92.

Burnard, K., & Bhattacharya, S. (2011): Power Generation from Coal: Ongoing Developments and Outlook, Information paper, october 2011, IEA, Paris, France.

Bhattacharya, S. (2011): Gasification Technologies – Status, Research & Development Needs, *Presented at INAE-ATSE Workshop on Energy efficiency*, New Delhi, April 11–12, 2011.

Bolland, O., Kvamsdal, H.M., & Boden, J.C. (2001): A thermodynamic comparison of the oxy-fuel power cycles water-cycle, Graz-cycle and Matiant-cycle. Paper presented at: *Intl. Conf. on power generation and sustainable development*, Liège, Belgium, 8–9 Oct 2001, p. 6.

Bose, A., Bonk, D., Fanand, Z., & Robertson, A. (2004): *Proc. 29th Int. Tech. Conf. on Coal Utilization & Fuel Systems*, US DOE, ASME, pp. 624–33.

Blesl, M. & Bruchof, D. (2010): *Syngas Production from Coal, IEA-ESTAP, Technology Brief-SO1*, May 2010, Simbolotti, G. & Tosoto, G. (project coordinators); at www.etsap.org

Braun, R.J., Kameswaran, S., Yamanis, J., & Sun, E. (2012): Highly Efficient IGFC Hybrid Power Systems Employing Bottoming Organic Rankine Cycles with optional Carbon Capture, *J. Eng. Gas Turbines Power*, **134** (2), 021801, 15 pages.

Burton, E., Friedmann, J., & Ravi Upadhye: Best practices in Underground Coal gasification, Lawrence Livermore National Laboratory, Livermore, CA, USDOE contract No. W-7405-Eng-48.

Calderon, R. (2007): Calderon gasification process. Presentation at: *Advanced gasification systems conference*, Houston, TX, 1 Nov 2007; Available at: www.syngasrefiner.com/agt/Pres/ReinaCalderon.pdf, 27 pp.

Carbon Capture Journal (2010): http://www.carboncapturejournal.com/displaynews/php?newsID=662, (Oct. 14, 2010).

Collodi, G., & Jones, R.M. (2003): The Sarlux IGCC Project, an Outline of the Construction and Commissioning Activities, in *1999 Gasification Technologies Conference*, 2003: San Francisco, CA.

Elizabeth Burton, Friedmann, J., & Ravi, U. (2010): Best Practices in UC gasification, *Lawrence Livermore National laboratory Report*, 2010.

Couch, G.R. (2009): *Underground coal gasification*. CCC/151, London, UK, IEA Clean Coal Centre, pp. 129 (Jul 2009).

Creedy, D.P., & Garner, K. (2004): Clean Energy from Underground Coal Gasification in China, *DTI Cleaner Coal Technology Transfer Program*, Report No. COAL R250 DTI/Pub URN 03/1611, February 2004.

Dalton, S. (2004): Cost Comparison IGCC and Advanced Coal, presented at *Round table on Deploying Advanced Clean Coal Plants*, July 2004.

Darby, A.K. (2010): Compact gasification development and test status, *Annual gasification technologies conference*, Washington, DC, USA, 31 Oct–3 Nov 2010; at: www.gasification.org/uploads/downloads/Conferences/2010/40Darby.pdf, Arlington, VA: Gasification Technologies Council, p. 13 (2010).

Dolezal, R. (1954): *Schmelzfeuerungen, Theorie, Bau und Betrieb*, VEB Verlag Technik, Berlin.

Dolezal, R. (1967): Large Boiler Furnaces, *Fuel and Energy Monograph Series*, (ed.) J.M. Beér, Elsevier: New York, 1967.

Domeracki, W.F., Dowdy, T.E., & Bachovchin, D.M. (1995): *Topping Combustor Status for Second Generation Pressurized Fluid Bed Cycle Application*, ASME Paper 95-GT-106 (1995).

DOE/NETL: Gasifipedia – Gasification in detail: at http://www.netl.doe.gov/technologies/coalpower/ gasification/gasifipedia/4-gasifiers/4-1_types.html.

DOE/NETL: Gasipedia – Applications of Gasification – IGCC, at http://www.netl.doe.gov/technologies/coalpower/gasification/gasipedia/6-apps/6-2-6-2_wabash.html.

DOE/NETL (2006): *The gas turbine handbook*, available at: www.netl.doe.gov/technologies/coalpower/turbines/refshelf/handbook/TableofContents.html, Morgantown, WV, National Energy Technology Laboratory.

DOE/NETL (2010): *Advanced gasification technologies*, Available at: www.netl.doe.gov/technologies/coalpower/gasification/index.html, vp (2010).

DOE/NETL (2011): Analysis of Integrated Gasification Fuel Cell Plant Configurations, DOE/NETL-2011/1482, February, 2011.

Edwards, S., & Chapman, R. (2005): IGCC technology: a promising and complex solution, *World Energy*; 8(3), 48–50.

EPA (2006): *Environmental Footprints and Costs of Coal-based Integrated Gasification Combined Cycle and Pulverized Coal Technologies*, in *Clean Air Markets*, EPA, Editor, 2006, Nexant, Inc.: San Francisco, pp. 4-1 to 4-14.

EPRI (2010): *Coal Fleet for Tomorrow – future coal generation options – Program 66*, Available at: http://mydocs.epri.com/docs/Portfolio/PDF/2010_P066.pdf, Palo Alto, CA, Electric Power Research Institute, p. 20 (2010).

Fernando, R. (2008): *Coal gasification*. CCC/140, London, UK, IEA Clean Coal Centre, p. 56, Oct 2008.

Foy, K. & McGovern, J. (2005): Comparison of Ion Transport membranes, *Proc. 4th Annual Conf. on Carbon Capture and sequestration*, DOE/NETL, May 2–5, 2005.

GE Aero Energy Products (2011): *LM6000 SPRINT™ gas turbine generator set*. Houston, TX, GE Aero Energy Products, p. 3.

Geosits, R.F., & Schmoe, L.A. (2005): IGCC – The Challenges of Integration, *GT2005 ASME Turbo Expo 2005*, June 2005.

Gregg, D.W., & Edgard, T.F. (1978): Underground Coal Gasification, *AIChE J.* 24, 753–781.

Griffiths, J. (2009): IGSC: using rocket science for increased power, 100% CO_2 capture, and no NO_x or Sox, *Modern Power Systems*; 29(2); 12–14, 16–18.

Güthe, F., Hellat, J., & Flohr, P. (2008): The reheat concept: the proven pathway to ultralow emissions and high efficiency and flexibility, *J. of Engineering for Gas Turbines and Power*; 131(2); 021503; 1–7.

Haupt, G. & J. Karg (1997): *The Role of IGCC in Advanced Power Generation* Power-Gen Asia'97 Singapore, Sep.1997; Siemens Power Generation 1997.

Heitmeir, F., Sanz, W., & Jericha, H. (2006): Section 1.3.1.1: Graz cycle – a zero emission power plant of highest efficiency, In: *The gas turbine handbook*; at www.netl.doe.gov/technologies/coalpower/turbines/refshelf/handbook/1.3.1.1.pdf, NETL, pp. 81–92.

Henderson, C. (2007): *Fossil fuel-fired power generation. Case studies of recently constructed coal and gas-fired power plants*. Paris, France, IEA, p. 176, 2007.

Henderson, C. (2008): *Future developments in IGCC*. CCC/143, London, UK, IEA Clean Coal Centre, p. 45, Dec 2008.

Herzog, H.J. (2009): Capture technologies for retrofits, *MIT Retrofit Symposium*, MIT, March 29, 2009.

Higman, C., & vanderBurgt, M. (2003): *Gasification*, New York: Elsevier.

Holt, N. (2001): Integrated Gasification Combined Cycle Power Plants, in *Encyclopedia of Physical Science and Technology*, Academic Press: New York.

Holt, N. (2004): Gasification Technology Conference, San Francisco Oct. 2004.

Holt, N. (2004): IGCC Technology Status, Economics and Needs, in *International Energy Agency Zero Emission Technologies Technology Workshop*. 2004a: Gold Coast, Queensland, Australia.

Hodrien, C. (2008): Advanced gas turbine power cycles, Presented at *New and unusual power generation processes*, Cardiff, UK, 18 Jun 2008, available at www.britishflame.org.uk/calendar/ New2008/CH.pdf, p. 36.

Houyou, S., Mathieu, P. & Nihart, R. (2001): Techno-economic comparison of different options of very low CO_2 emission technologies, In: *Greenhouse gas control technologies: Proc. of the fifth Intl. Conf. on greenhouse gas control technologies*, Cairns, Qld, Australia, 13–16 Aug 2000; Collingwood, Vic, Australia, CSIRO Publishing, pp. 1003–1008.

Huberg, R. (2009): Modeling of IGFC system – CO_2 removal from the gas streams, using membrane reactors, Master's thesis, University of Iceland.

Hustad, C-W., Coleman, D.L., & Mikus, T. (2009): *Technology overview for integration of an MHD topping cycle with the CES oxyfuel combustor*. Available at: www.co2.no/download.asp?

Ikeguchi, T., Kawasaki, T., Saito, H., Iizuka, M., & Suzuki, H. (2010): Development of electricity and energy technologies for low-carbon society, *Hitachi Review*, 59(3), pp. 53–61.

Iki, N., Tsutsumi, A., Matsuzawa, Y., & Furutani, F. (2009): Parametric study of advanced IGCC, *Proc. of ASME Turbo Expo 2009: Power for Land, Sea and Air: GT2009*, Orlando, FL, 8–12 Jun 2009; Fairfield, NJ, American Society of Mechanical Engineers, Paper GT2009-59984, p. 9.

Kalina, A.I. (1982): *Generation of energy by means of a working fluid, and regeneration of a working Fluid*, US Patent 4346561, Alexandria, United States Patent and Trademark Office, vp (31 Aug 1982).

Kleiner, K. (2008): Coal-to-gas: part of a low-emissions future? Available at: www.nature.com/climate/2008/0803/full/climate.2008.18.html, *Nature Reports Climate Change*; (3); vp (Mar 2008).

Lako, P. (2010): Coal-fired Power, IEA-ETSAP-Technology Brief EO1, April 2010, Simbolotti, G., & Tosato, G. (project coordinators), at www.etsap.org.

Makino, H., Mimaki, T., & Abe, T. (2007): *Proposal of the high efficiency system with CO_2 capture and the task on an integrated coal gasification combined cycle power generation*, CRIEPI report M07003, Tokyo, Japan, Central Research Institute of Electric Power Industry, vp (2007) (In Japanese).

Mee, T.R. (1999): Inlet fogging augments power production, *Power Engineering*, 103(2), pp. 26–30.

MIT Study (2007): *The Future of Coal – Options for a Carbon constrained World*, Massachusetts Institute of Technology, Cambridge, MA, 2007.

Minchener, A.J. (2005): Coal gasification for advanced power generation. *Fuel*; 84(17), 2222–2235.

Nakagaki, T., Yamada, M., Hirata, H., & Ohashi, Y. (2003): A study on principal component for chemically recuperated gas turbine with natural gas steam reforming (1st report, system characteristics of CRGT and design of reformer), *Trans. of the Japan Soc. of Mech. Engineers Series B*, 69(687), 2545–2552.

NCC (2012): Harnessing Coal's carbon content to Advance the Economy, Environment and Energy Security, June 22, 2012; Study chair: Richard Bajura, National Coal Council, Washington, DC.

Oki, Y., Inumaru, J., Hara, S., Kobayashi, M., Watanabe, H., Umemoto, S., & Makino, H. (2011): Development of oxy-fuel IGCC system with CO_2 recirculation for CO_2 capture, *Energy Procedia*, 4, 1066–1073.

Ono, T. (2003): NPRC Negishi IGCC Startup and Operation, in *Gasification Technologies 2003*, San Francisco, CA.

Perkins, G. & Sahajwalla, V. (2008): Steady-state model for estimating gas production from underground coal gasification, *Energy Fuel*, 22, 3902–3914.

Perkins, G. & Sahajwallaa, V. (2005): A mathematical model for the chemical reaction of a semi-infinite block of coal in underground coal gasification, *Energy Fuel*, 19, 1679–1692.

Pettus, A., & Tatsutani, M. (eds) (2009): *Coal without carbon: an investment plan for federal action*, available at www.catf.us, Boston, MA, USA, Clean Air Task Force, p. 97 (Sep 2009).

Portman Energy (2012): UCG – the 3rd way, 7th UCG Association Conf, London.

Prabu, V., & Jayanti, S. (2012): Integration of underground coal gasification with a solid oxide fuel cell system for clean coal utilization, *Int. Journal of Hydrogen Energy*, 37, 1677–1688.

Reid, W.T. (1971): External Corrosion and Deposits, Boilers and Gas Turbines, *Fuel and Energy Science Monograph Series*, (Ed.) J.M. Beér, New York: Elsevier 1971.

Robertson, A., R. Garland, R. Newby, A. Rehmat and L. Rebow (1989): *Second Generation Pressurized Fluidized Bed Combustion Plant*, Foster Wheeler Dev. Corp., Report to the US DOE DE-AC-21-86MC21023.

Roddy, D. & Gonzalez, G. (2010): Underground coal gasification (UCG) with carbon capture and storage (CCS), In: Hester, R.E., Harrison, R.M. (eds) *Issues in Environmental Science and Technology*, 29, pp. 102–125; Royal Society of Chemistry, Cambridge (2010).

Ruth, L.A. (1998): US DOE Vision21 Workshop, FETC Pittsburgh, PA, Dec.1998.

Sallans, P. (2010): Choosing the Best coals in the Best location for UCG, *Advanced Coal technologies Conference*, University of Wyoming, Larmie.

Self, S., Reddy, B.V. & Rosen, M.A. (2012): Review of underground coal gasification technologies and carbon capture, *Intl. J of Energy and Environmental Engineering*, 3, 16, Springer Open Journal, doi: 10.1186/2251-6832-3-16.

Shirai, H,. Hara, S., Koda, E., Watanabe, H., Yosiba, F., Inumaru, J., Nunokawa, M., Smith, I.M. (2009): *Gas turbine technology for syngas/hydrogen in coal-based IGCC*. CCC/155, London, UK, IEA Clean Coal Centre, pp. 51 (Oct 2009).

Stańczyk, K., Howaniec, N., Smoliński, A., Świadrowski, J., Kapusta, K., Wiatowski, M., Grabowski, J., & Rogut, J. (2011): Gasification of lignite and hard coal with air and oxygen enriched air in a pilot scale ex-situ reactor for underground gasification, *Fuel*, **90**, 1953–1962.

Stephens, D.R., Hill, R.W., & Borg, I.Y. (1985): Underground Coal Gasification Review, Lawrence Livermore National Laboratory, Livermore, CA, UCRL-92068.

Stiegel, G.J. (2005): Gasification Technologies – Clean, Secure and Affordable Energy Systems, *IGCC and CCTs Conference*, Tampa, FL, June 28, 2005.

Stiegel, G.J., Ramezan, M., & Mcllvried, H.G.: Integrated Gasification Combined Cycle, at 1.2IGCC.pdf

Sury, M., *et al.* (2004): Review of Environmental issues of UCG, report no. COAL R272, DTI/Pub.URN 04/1880, November 2004.

Turboden (2011): *The organic Rankine cycle*; at www.turboden.eu, Brescia, Italy.

UCG Association (2011): More information on: www.ucgp.com, Woking, UK.

Van Holthoon, E. (2007): Shell gasification processes, Presentation at: *Annual gasification technologies Conf.*, San Francisco, CA, 15–17 Oct 2007; Available at: http://www.gasification.org/uploads/downloads/Conferences/2007/43EVANH.pdf, Arlington, VA, Gasification Technologies Council, vp (2007).

Van Nierop, P. & Sharma, P. (2010) Plasma gasification – integrated facility solutions for multiple waste streams, *Annual gasification technologies Conference*, Washington, DC, 31 Oct–3 Nov 2010; Available at: www.gasification.org/uploads /downloads/Conferences/ 2010/ 14VANNIEROP.pdf; Arlington, VA, Gasification Technologies Council, 26 p.

Van der Riet, M. (2008): Underground coal gasification, *In: Proceedings of the SAIEE Generation Conference*, Eskom College, Midrand (19 Feb 2008).

Walker, L.K., *et al.* (2001): An IGCC Project at Chinchilla, Australia, based on UCG, *Gasification Technologies Conference*, San Francisco, USA.

World Coal Association (2011): *Underground coal gasification*; Available at:www.worldcoal. org/coal/uses-of-coal/underground-coal-gasification/ vp (2011)

Wikipedia (2013): IGCC Power plant, figure 8.2: author – Stan Zurek, at http://en.wikipedia. org/wiki/integrated_gasification_combined_cycle.

Yang, L.H., Pang, X.L., Liu, S.Q., & Chen, F. (2010): Temperature and gas pressure features in the temperature-control blasting underground coal gasification, *Energy Sources*: Part A **32**, 1737–1746.

Yang, L., Liang, J., & Yu, L. (2003): Clean Coal technology – study on the pilot project experiment of underground coal gasification, *Energy*, **28**, 1445–1460.

Ziock, H., Guthrie, G., Lackner, K., Ruby, J., & Nawaz, M. (2001): *Zero emission coal, a new approach and why it is needed*. Available at: http://library.lanl.gov/cgi-bin/getfile? 00796497.pdf, NM: Los Alamos National Laboratory, LA-UR-01-5865, p. 19 (2001).

Chapter 9

Carbon capture and storage

9.1 INTRODUCTION

It is evident from numerous studies that fossil fuels, especially coal will remain dominant in global energy mix for powering the global economy for many more decades, probably till the end of the century (for e.g., USDOE 1999; MIT Study 2007; Morrison 2008; Herzog 2009). Hence the efforts to achieve the needed emissions reductions become crucial to mitigate the impact of global warming due to coal usage. The broad approach suggested is: improve the efficiency levels of coal-fired power plants to start with and apply technologies for carbon dioxide capture and storage (CCS). According to MIT Coal Study (2007), 'CCS is the critical enabling technology that would reduce carbon dioxide$_2$ emissions significantly while also allowing coal to meet the world's pressing energy needs'. In IEA evaluations, CCS was shown as a cost-effective means which could play an increasing role, incentivized by stable CO_2 price (Morrison 2008).

Among several sources of CO_2 emissions, power generation plants account for more than one-third of the emissions worldwide, and coal-fired plants represent the largest set of CO_2 sources. Estimates of the worldwide total new construction of coal-fired plants by 2030 are around 1,400 gigawatts (WEO 2006), and such a major expansion globally would dramatically increase greenhouse gas emissions. A new 1,000 MW coal power plant using the latest conventional pulverized coal technology produces about 6 million tons of CO_2 annually (Socolow 2005). In the absence of CO_2 emission controls, the proposed new capacity of about 1400 GW would generate as much as 7.6 billion metric tons of CO_2 each year. This means, around 30% of increase over the current annual global emissions of 25 billion metric tons of CO_2 from the fossil fuels consumption (IEO 2006). Worldwide emissions from these new plants between now and 2030 would be equal to between 50% of all fossil fuel emissions during the past 250 years (Socolow 2005; Berlin & Sussman 2007).

Several industrial processes produce highly concentrated streams of CO_2 as a byproduct and are a good capture target. Ammonia manufacturing, fermentation, and hydrogen production in oil refining, and gas-producing wells are a few examples suitable for carbon capture. Fuel-conversion processes also offer opportunities for CO_2 capture. For example, producing oil from the oil sands in Canada is currently very carbon intensive and adding CCS facility to the production process can reduce the carbon intensity. Producing hydrogen fuels from carbon-rich feed-stocks, such as

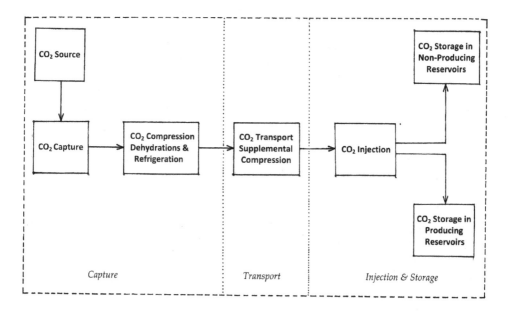

Figure 9.1   Basic CCS Project Schematic (Redrawn from Mike McCormick, C2ES 2012).

natural gas, coal, and biomass are other instances for CO_2 capture. In these cases, the CO_2 by-product would be highly concentrated (in many cases, >99% CO_2) and the incremental costs of carbon capture would be relatively low compared to capture from a power plant (usually just requiring compression).

CCS refers to the capture of CO_2 from power plants and large industrial sources, its compression into a fluid and transportation to suitable locations, and injection into deep underground geological formations for long-term storage (Figure 9.1). Each step of CCS – *capturing, compressing, transporting, injection and storing* – is very important and involves several issues.

Presently, CCS is not yet commercially demonstrated in the primary application, viz., large-scale coal or natural gas fired power plants for which it is conceived.

9.2   TRANSPORT AND STORAGE

To facilitate easier and cheaper transport and storage, captured CO_2 is first compressed to a dense 'supercritical' liquid. The resulting high pressures, typically 11–14 MPa, are also required to inject CO_2 deep underground for geological sequestration (Benson & Cole 2008). As part of the capture system, compression of CO_2 is done within the plant premises. The captured carbon dioxide must be stored in a way that prevents its release into the atmosphere at least over a few centuries. Such storage is technically straightforward, although no one is sure how to guarantee against accidental releases in the distant future. Generally, geological and oceanic sites are chosen for storage of CO_2. Since many point sources of captured CO_2 would not be close to geological or oceanic storage facilities, transportation would be required in such cases.

9.2.1 Transportation

Safe and reliable transportation of CO_2 from the site of capture to a storage site is an important stage in the CCS process. Transportation of gases including CO_2 is a already occuring around the world. The main form of transportation is through pipelines although shipping would be a possibility. Trains and trucks are thought to be too small-scale for projects of this size.

The total transportation distance that would be covered by the 75 LSIPS currently under development and in operation is around 9000 km. More than 80% of these projects are planning to use onshore pipelines, particularly in the USA and Canada, where enormous experience in CO_2 transportation already exists. The construction of so called 'trunk lines' connecting one or two LSIPs with a proven storage formation could enable subsequent smaller projects to follow more easily. Projects in European countries, The Netherlands, Norway, and UK, are considering offshore pipelines. These countries are looking to transport their CO_2 via pipeline or ship to various offshore storage locations in the North Sea. The only offshore pipeline for CO_2 currently in use is part of the Snøhvit project in Norway (GCCSI 2012). Transportation of CO_2 by pipeline occurs in the Netherlands, with about 85 km of pipeline supplying 300 kt per annum of gaseous CO_2 to greenhouses, as well as other pipelines in Hungary, Croatia, and Turkey for the purpose of EOR (Buit *et al.* 2011). The IEA estimates that Europe, China and the USA could need a transportation capacity in the order of some $GtCO_2$ per year by 2030 for CO_2 deployment. Pipelines use could be in the order of 10 to 12 thousand km in the next ten years (to transport 300 $MtCO_2$ from 100 CCS projects), 70 to 120 thousand km by 2030, and 200 to 360 thousand km by 2050, with investment of about US$0.5–1 trillion. Factors such as tax benefits, incentives for CCS projects, and data on potential storage sites may accelerate investment in pipelines. The CO_2 can also be transported by ship either in semi-refrigerated tanks ($-50°C$, 7 bars) or in compressed natural gas (CNG) carriers. CO_2 transportation by ship is considered a viable option only for small CCS projects (Simbolotti 2010).

Pipelines would require new regulatory measures to ensure that proper materials are used (for e.g., CO_2 combined with water is highly corrosive to some pipeline materials); and measures for monitoring for leaks, protection of health and providing overall safety are adequate. Now, these are all technically possible, and pipelines in general currently operate in a mature market.

In order to better facilitate the development of this new CO_2 transportation infrastructure, there are a few areas that require further studies: (a) development of international standards and design codes to further promote safe and efficient operation of CO_2 transport infrastructure; (b) Creation of novel financial and commercial systems for CO_2 networks and activity centers to allow several partners and their priority access within a network; (c) Financing for assets that will initially appear 'oversized' but anticipating future addition of CO_2 volumes to the network; and (d) validation of detailed thermodynamic modeling of CO_2 streams containing impurities (GCSSI 2012).

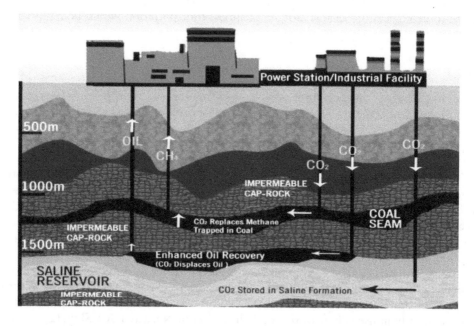

Figure 9.2 Injecting CO$_2$ into geological formations; available at www.wri.org/project/carbon-dioxide-capture-storage/ccs-basics (*Source*: WRI, www.wri.org, reproduced with permission from WRI).

9.2.2 Storage

The *storing* is done by injecting CO$_2$ into the geological formations, i.e., depleted oil and gas reserves, deep saline aquifers and unminable coal beds where long-term sequestration of CO$_2$ is possible (Figure 9.2).

CO$_2$ injection underground is a standard technology in the oil industry since 1970s, and used worldwide as part of routine oil-field operations (Bennion & Bachu 2008) for enhanced oil recovery (EOR). This results in additional oil revival, and concurrently generates revenue to compensate the costs of capture and storage (Lake 1989). CO$_2$ flooding is an effective and well-established oil recovery method that can use existing injection infrastructure and the experience of the oil industry to extend the lifetime of many reservoirs. As the CO$_2$ is injected at pressure into sedimentary rocks of these formations, generally at depths greater than 800m, it seeps into the pore spaces of about 10% to 30% of the total volume (Blunt 2010) in the surrounding rock. Its escape to the surface is blocked by a caprock or overlaying impermeable layer. Further, CO$_2$ remains a liquid at that depth and displaces liquids, such as oil (or water) that are present in the pores of the rock. Most companies, for economic reasons, recycle the injected CO$_2$ which results in the CO$_2$ being effectively stored permanently within the oil reservoir.

There are approximately 70 CO$_2$ injection projects worldwide devoted to EOR, most situated in West Texas where over 1,000 km of pipelines have delivered CO$_2$ to many oilfields in the region for over 30 years. Here, mostly, natural sources of CO$_2$

from underground reservoirs are used. CO_2 for EOR is now being expanded as CCUS (carbon capture, utilization and storage) initiative to tap huge amounts of 'stranded' oil underground, and to enlarge coal-to-liquid fuels production activity in the USA (NCC 2012). The other application of CO_2 injection is for recovering natural gas. These projects are not meant for CO_2 storage, and capacity-wise, they are inadequate to store CO_2 gas emitted by power plants operating globally. Since many oil and gas fields exist in the USA and offshore UK, storage in depleted hydrocarbon reservoirs is an attractive option for these countries. India and China, although have significant coal-fired power generation, do not have such widespread oil and gas resources; hence, saline aquifer storage may be the only solution for large-scale storage in these countries.

The deep saline aquifers are believed to be everywhere at depths generally below one kilometer and are estimated to underlie at least one-half of the area of populated continents (Williams 2005). They have the greatest capacity to store CO_2 and would play a useful role in any large-scale CCS program. Large-scale injection and geological storage of CO_2 has been safely operated in saline reservoirs for more than 15 years and in oil and gas reservoirs for decades. There is also the possibility of injecting CO_2 into the deep ocean. CO_2, however, reacts with water to produce carbonic acid, which turns the water more acidic. Many aquatic organisms are highly sensitive to changes in acidity, making oceanic storage more problematic than geological storage from an environmental point (IPCC 2005). In some types of formations, the CO_2 may dissolve in water and react with minerals in the host rock to form carbonates, becoming permanently entrained (Figure 9.3). In general, geological storage is preferred over oceanic storage, as it is a more mature technology with fewer ecological implications. It currently represents the only option to substantially address the GHG emissions from fossil fuel-fired power plants and large industrial facilities (McCormick 2012). Each project, however, must be examined for its viability, in particular with regard to limiting leakage rates to acceptable levels.

9.2.3 Estimated capacity

According to the IPCC Special report (2005), from a technical standpoint, there is a lower limit of about $1,700\, GtCO_2$, with very uncertain upper limits (possibly over $10,000\, GtCO_2$). From an economic point of view (considered realistic), however, the report declares that there is almost certainly a minimum of $200\, GtCO_2$ capacity and likely $2,000\, GtCO_2$. According to the IEA Greenhouse R&D Program (IEA 2008), oil and gas reservoirs have an estimated CO_2 storage capacity of about 920 Gt – with a considerable margin of uncertainty. For comparison, the estimated total global emissions of CO_2 from fossil fuel use are around 30 Gt per year, while deep saline aquifers could store between 400 to 10,000 Gt in total (Gale 2003).

The reservoirs with the largest potential capacity are the deep saline formations. At present, capacity estimates for CO_2 storage in deep saline formations are highly uncertain because the data available to do rigorous capacity calculations is meagre. Data are typically obtained through drilling wells. Unlike drilling in oil and gas reservoirs, drilling in deep saline formations do not generate revenue; so the number of wells drilled into these formations is limited. That is why much more data exists for

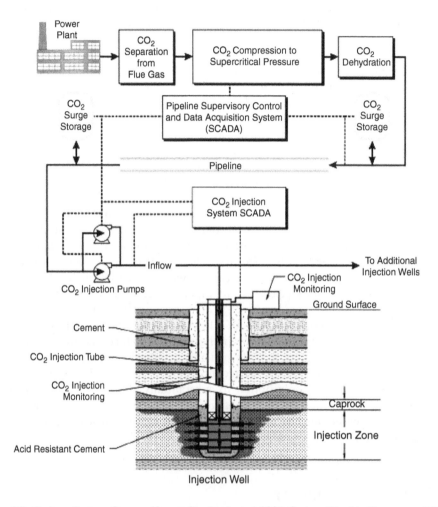

Figure 9.3 Carbon Capture Storage (*Source*: Dooley, J. *et al.* 2006, Carbon Dioxide Capture and Geologic Storage: A key Component of a Global Energy Technology Startegy to Address Climate
Change, Joint Global Change Research Institute, PNNL, College Park, MD, p. 67; Reproduced
with the permission of Dooley, J.).

storage estimates in oil and gas reservoirs. However, owing to a current lack of field
data needed to confirm the methodology used to calculate storage capacities, there is
still uncertainty in these numbers.

Calculating storage capacities in coal seams is very difficult, in part because the
feasibility of large-scale storage in coal seams has not been demonstrated. The global
storage potential in unminable coal seams is estimated to range between 100 and
200 Gt (Simbolotti 2010). Any capacity estimates for these formations must be taken
as very highly uncertain.

The US DOE has completed a *Carbon Sequestration Atlas of the United States
and Canada* (at http://www.netl.doe.gov/technologies/carbon_seq/refshelf/atlas/).

It provides capacity estimates for oil and gas reservoirs as 82 $GtCO_2$, while for saline formations, it gives a range of 920–3,400 $GtCO_2$. The high end of this range is greater than the worldwide capacity reported by the IPCC. Although the IPCC was conservative in its estimates, the notable point is that IPCC estimates highlight the uncertainty in making these estimates (Herzog 2009).

Six fully integrated large-scale CCS projects are currently in commercial operation globally which are discussed in the later pages. All these projects are injecting about 1MPa (million tonnes per annum) of CO_2. These capture projects relate to industrial processes, and hence the cost of capture is relatively small (see for e.g., GHG R&D Program IA 2008, Simbolotti 2010, IEA /CSLF Report 2010). The technologies and operational aspects of injecting and storing CO_2 in geologic formations are established processes. Experiences at these sites are no different from each other. These projects have not only provided opportunities to learn but confirmed that CO_2 can be safely stored in geological formations (Herzog 2009). MIT Coal Study (MIT Report 2007) also reports that 'geological carbon sequestration is likely to be safe, effective, and competitive with many other options on an economic basis'.

CO_2 injection specifically for geologic sequestration, however, involves different technical issues and potentially much larger volumes of CO_2 and larger scale projects than in the past. Each geologic storage site is unique and must be screened and extensively characterized, which takes years of time incurring heavy budget, before a decision can be made to start a commercial project. Geologic storage is also a long-term financial burden associated with a CCS project. More importantly, it poses a most important challenge: public acceptance.

The industry, academia, and government have to direct their research efforts to successfully demonstrate that large quantities of CO_2 can be stored without leaks over long periods and under a range of geologic conditions. The large scale sequestration projects now started are expected to provide supportive evidence that leakage from CO_2 storage formations will be unlikely (Berlin & Sussman 2007).

So far, there is not as much of experience with capturing and sequestering CO_2 generated in fossil fuel-fired power plants; a number of promising projects which are at different stages of development, planning and execution may provide with time (Global CCSI 2012). For example, two power generating projects, namely, *Boundary Dam Integrated CCS Demo Project* in Canada with post-combustion capture, 1 Mt/year, to be used for EOR, and *Kemper County IGCC Project* in the USA, with pre-combustion capture, 3.5 Mt/year, also to be used for EOR are expected to begin operation in 2014 (G CCSI 2012). Widespread implementation of CCS at coal-fired power plants would greatly expand the scale of CO_2 sequestration because of the massive amounts of CO_2 that would be captured and then stored on a permanent basis. Several other examples are discussed in Chapter 9. The storage mechanism, injection design and pressure response from aquefers, risk assessment, and so on, are discussed in many publications, and recently by Martin Blunt (2010) and Manchao He *et al.* (2011).

IPCC Special Report on Carbon Dioxide Capture and Storage (IPCC 2005) and a study by Joint Global Change Research Institute (Dooley *et al.* 2006) have reported that there is considerable worldwide geological storage capability for CO_2. Figure 9.4 demonstrates the initial assessments of theoretical global CO_2 storage capacity as studied by Dooley and his group (2006).

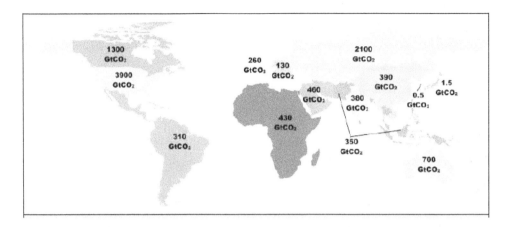

Figure 9.4 Initial assessments of theoretical Global CO_2 Storage capacity (Source: Dooley, J., Davidson, C., Dahowski, R., Wise, M., Gupta, N., Kim, S., & Malone, E., 2006, Carbon Dioxide Capture and Geologic Storage: A key Component of a Global Energy Technology Startegy to Address Climate Change, Joint Global Change Research Institute, Pacific Northwest National Laboratory, College Park, MD, p. 67; Reproduced with the permission of Dooley, J.).

9.2.4   Storage times

Storage times of injected CO_2 mainly depend on two factors: (i) the tendency of a geological formation to leak that includes 'during' seismic events, and (ii) its chemical behavior. Carbon dioxide can dissolve over hundreds or thousands of years into the surrounding water, which then tends to sink due to its heaviness, rather than rise. In addition, chemical reactions over millions of years can permanently trap the CO_2 as carbonates, provided the rocks have suitable minerals. CO_2 may also displace methane in some coal beds and adsorb to the coal. Storage times for geological formations are generally measured in several hundreds, up to tens of thousands of years (Lackner 2003). Oceanic storage times depend on depth of injection. After about 100 years, about 20% of injected CO_2 would be expected to leak if the injection depth were 800m, while at a depth of 3,000 m, only 1% CO_2 would leak (IPCC 2005).

Impacts of Leaks: Since liquid CO_2 is lighter, it tries to move upwards; hence, suitable geological formations must have a 'cap' rock to act as a barrier to its movement. If the cap rock is not sufficiently wide, CO_2 could leak out around the edges. In this case, mechanisms would be required to prevent such leakage. The viability of any such project would have to be established on an individual basis. If CCS technology is going to be foremost in reducing CO_2 emissions, presumably some hundreds of Gt of carbon will need to be stored over a period of 100 years. Leak rates of even 1% per year would mean an emission of over 1 Gt of carbon per year, fairly a large amount. Terrible leaks arising from seismic events or other failures of geological storage or broken pipelines would not only release large amounts but create possibly lethal CO_2 concentrations in the locality of such an incident. These emergencies have to be carefully discussed in deciding where to store CO_2; then only, deep ocean sequestration becomes a viable option (Williams 2006).

Cumulative leakage rates of 1%–10% over 100 years or 5%–40% over 500 years would maintain storage as a feasible option for reducing emissions (IPCC 2005). However, if coal remains a significant fraction of the world's energy mix and is mostly combusted (as the current projections reveal), a few thousand Gt of carbon would have to be stored that could result in potential leak. Hence, if dependence on storage is greater, the leakage rates will have to be drastically reduced so that future generations need not be troubled with efforts to control.

9.2.5 Monitoring and verification

To ensure a safe and effective storage, a regulatory supervision, careful management of site selection and acquisition of monitoring data, and verification of CO_2 emission reductions are imperative. Proper monitoring and verification is perhaps the most important method of obtaining public approval for geologic storage of CO_2. The main types of measurements that constitute effective monitoring and verification of CCS, in general, are: (1) measurement of CO_2 concentrations in the workplace (separation facility and wellfield) to ensure the safety of workers and the public; (2) measurement of emissions from the capture system and surface facilities to verify emission reductions; (3) measurement of CO_2 injection rates, which are used to determine how much CO_2 has been injected into the underground formation; and if EOR is taking place along with CO_2 storage, monitoring of any CO_2 produced with the oil to calculate the net storage; (4) measurement of the condition of the well using well logs and wellhead pressure measurements; and (5) measurement of the location of the column of CO_2 as it fills up the storage formation; this type of measurement can also be used as an early warning system in the event of CO_2 leakage from the storage reservoir (Bensen, S.M: CCS in underground geologic formations, Workshop at Pew center, available at http://www.c2es.org/docUploads/10–50_Benson.pdf).

Also, the methods available for studying the natural cycling of CO_2 between the atmosphere and the Earth's surface can be used as they have sufficient sensitivity to detect CO_2 leaks in the event they reach the surface. Deploying surface flux monitoring may be helpful in creating public acceptance for CO_2 storage. For the first four measurements, the measurement technology used in several applications, including electrical generation plants, the oil and gas industry, natural gas storage, and so on can be adopted. The fifth, monitoring plume migration, is challenging because the sensitivity and resolution of existing measurement techniques need to be evaluated and improved. Certain types of plumes, namely, narrow vertical plumes of rising CO_2, for instance, may be difficult to detect (Myer 2003). Seismic imaging, developed for oil and gas exploration, is the primary method today for monitoring migration of CO_2 plumes in geologic storage projects. The differences in the images taken on a periodic basis can be used to detect the location of CO_2. Electromagnetic and gravitational measurements are available with lower sensitivity and resolution, but may be used in combination with seismic techniques to fine-tune the interpretation of the data or in the interim between seismic measurements. Seismic imaging has been successfully applied for monitoring the location of the CO_2 plume at the Sleipner West project in the North Sea. Even so, research and demonstration in a wide variety of geologic settings are needed because the accuracy and sensitivity of monitoring techniques differs in different geologic environments. Pilot projects, ranging from small to large scale, will facilitate to demonstrate the accuracy, reliability and sensitivity of existing techniques. In addition, developing new cost effective techniques and methods for providing early warning that a storage project is failing will benefit tremondously.

9.3   CAPTURE TECHNOLOGIES

Separation or capture of CO_2 from industrial processes has long been practiced, like removal of CO_2 impurities in natural gas treatment and the production of hydrogen, ammonia and other industrial chemicals. In most cases, the captured CO_2 stream is simply vented to the atmosphere; and in a few cases used in the manufacture of other chemicals (IPCC 2005). CO_2 also has been captured from a portion of the flue gases produced at coal or natural gas-fired power plants, and is sold as a commodity to industries such as food processing. Overall, only a small amount of CO_2 is utilized globally to produce industrial products and most of it is emitted to the atmosphere.

Three basic approaches to capturing carbon dioxide from fossil fuel burning power plants are available: (1) post-combustion (also referred as flue gas separation), (2) oxy-fuel combustion, and (3) pre-combustion, shown in Figure 9.5.

Common to all three approaches is the process of capturing the CO_2 from the other major constituents in the flue gas or syngas into a form that can be geologically stored or beneficially used/converted; and the basic differences are how the CO_2 is concentrated.

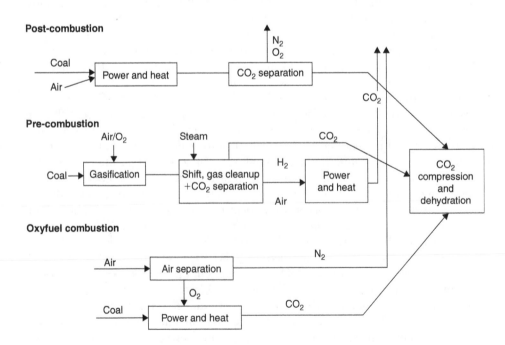

Figure 9.5 Technical options for CO_2 capture from coal-fired power plants (Redrawn from Global CCS Institute 2012a).

Each process has its own advantages, disadvantages, applicability to various coal-based generation technologies, and opportunities for EOR or coproduction.

Each of these 'capturing' processes carries both an energy and economic price, contributing significantly to the total costs of a complete CCS system. The IPCC has estimated that the increase in energy required to capture CO_2 is between 10% and 40% depending on the technology – the NGCC requiring the least and PC requiring the most (IPCC 2005).

The fraction of output power used in the capture as a function of base power plant efficiency is shown in Figure 9.6 (Morrison 2008). Higher the efficiency of the power plant, lower the output power utilized for CO_2 capture.

9.3.1 Post-combustion capture (flue gas separation)

The post-combustion capture refers to capture of carbon dioxide in the flue gas from a conventional coal-based power plant, PC or CFBC unit. The CO_2 content in the flue gas is around 10 to 15% at 1 atmospheric pressure which is turned into relatively pure CO_2. For a small/medium fraction of CO_2 in the flue gas, the best capture method with highest TRL is chemical absorption in amines (organic solvents), particularly monoethanol amine, MEA (Rao & Rubin 2002). This technology is commercially used in a few industrial units. The amine separation system is arranged at the downstream of other pollution control technologies (Figure 9.7).

In chemical absorption process, a gas is absorbed in a liquid solvent by formation of a chemically bonded compound. In a power plant, the flue gas is bubbled through the solvent in a packed absorber column, where the solvent preferentially removes the CO_2 from the flue gas. The CO_2-contained solvent is then pumped to a second vessel, called a regenerator, where heat releases a stream of CO_2 gas, and the solvent is allowed

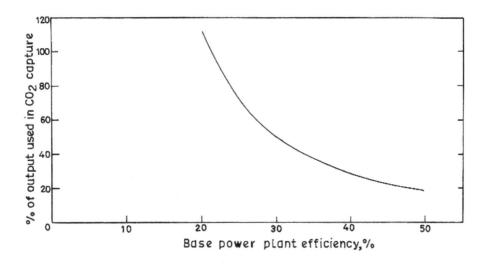

Figure 9.6 Percent of plant power used in CO_2 capture [Redrawn from RWE npower, Morrison 2008].

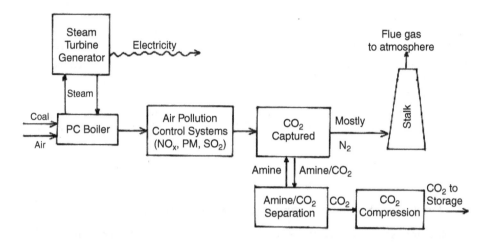

Figure 9.7 Schematic of a pulverized coal–fired (PC) power plant with post-combustion CO_2 capture using an amine system (Redrawn from Rubin 2008).

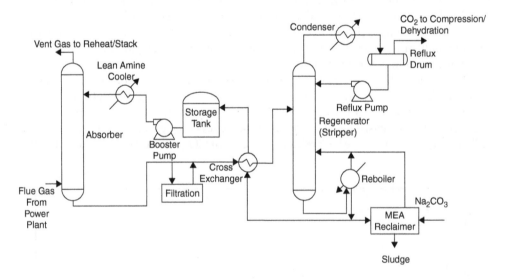

Figure 9.8 Schematic of a typical chemical absorption process (amine separation process) for flue gas from power plant (*Source:* Herzog & Golomb, MIT 2004; Reproduced with the permission of Howard Herzog).

to recycle. The concentrated CO_2 gas stream is compressed into a supercritical fluid for transport to the sequestration site. The basic chemical reaction for this process:

$$C_2H_4OHNH_2 + H_2O + CO_2 \leftrightarrow C_2H_4OHNH_3^+ + HCO_3^- \qquad (9.1)$$

A schematic of a typical chemical absorption process for power plant flue gas is shown in Figure 9.8.

MEA was originally developed as a general, non-selective solvent to remove acid gases, CO_2 and H_2S, from natural-gas streams. The process was modified incorporating inhibitors that reduce solvent degradation and equipment corrosion when applied to CO_2 capture from flue gas. To deal with degradation and corrosion the solvent strength is kept relatively low, resulting in relatively huge equipment and solvent regeneration costs. Methyldiethanolamine (MDEA), Designer amines, Catacarb and Benfield are the other chemical absorbents tried. One challenge is that solvents with lower energy requirements generally exhibit slower absorption rates.

The same post-combustion process is used for CO_2 capture at a natural gas-fired boiler or combined cycle (NGCC) power plant. Even though the flue gas CO_2 concentration is more dilute than in coal plants, high removal efficiencies can still be achieved with amine-based capture systems. The absence of impurities in natural gas also results in a clean flue gas stream, so that no additional cleanup is needed for effective CO_2 capture. The absorption process, though expensive, is profitable due to the market value of the captured CO_2, used in various industrial processes – the urea production, foam blowing, carbonated beverages, and dry ice making. This chemical absorption process has been in use for decades to recover byproduct CO_2 or directly manufacture CO_2 from fossil fuel combustion but never demonstrated at a coal-fired power plant. The advantage of the technique lies in its applicability to low-CO_2 partial pressures and high recovery rates of up to 98% and product purity >99 vol% (Jones 2007).

There is, however, a problem: breaking the chemical bond between the CO_2 and the chemical solvent is energy intensive. As a result, the thermal efficiency of the plant is derated. Besides, contaminants found in flue gases (SO_2, NO_x hydrocarbons, and particulates) have to be removed prior to capture which otherwise slow down the process of solvent absorption, and possibly allow formation of non-regenerable, heat-stable salts depending on their concentrations by combining with the amine. More details on the design, performance and operation of amine-based capture technologies are available in the literature (Rao & Rubin 2002; IPCC 2005; USDOE/NETL 2007).

Subcritical and supercritical PC units with CO_2 capture

A subcritical PC unit *with CO_2 capture* that produces 500 MW$_e$ power requires a 37% increase in plant size and in coal feed rate compared to a subcritical PC without CO_2 capture (MIT Report 2007). In a typical design, the key material flows considered are: feed air – 335,000 kg/hr, & coal feed – 284,000 kg/hr; Boiler/Super heater steam conditions – 16.5 MPa/38°C; Flue Gas clean-up: removal – particulates 99.9%, SO_2 99+%, NO_x as permitted; CO_2 capture 90% removal, CO_2 after compression, 573,000 kg/hr at 150 atms; and Stack gas – 3,210,000 kg/hr at 63°C, 0.10 MPa.

The generating efficiency of the plant with capture is reduced from 34.3% to 25.1% (Table 8.3); and the factors that cause reduction in efficiency are the following: (a) the thermal energy required to recover CO_2 from the amine solution reduces the efficiency by 5%, (b) the energy required to compress CO_2 from 0.1 MPa to about 15 MPa (to a supercritical fluid) reduces the efficiency by 3.5%, and (c) all other energy requirements add to less than one percentage point (MIT Report 2007).

In an ultra-supercritical PC unit with CO_2 capture (Figure 9.9) that produces a net power output of 500 MW$_e$ requires a 27% increase in unit size and in coal feed rate (44,000 kg/h more coal) compared to similar USC PC unit without CO_2 capture;

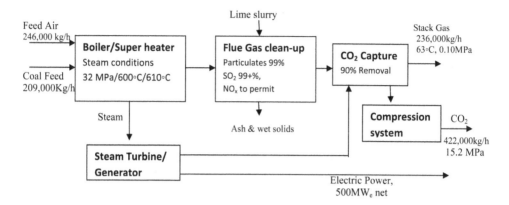

Figure 9.9 Ultra Supercritical 500 MW$_e$ Pulverized coal plant with CO_2 Capture (Redrawn from MIT Report 2007).

and the generating efficiency is reduced from 43.3% to 34.1% with CO_2 capture. The overall efficiency reduction is 9.2 percentage points in both cases, but the ultra-supercritical unit with no capture starts at a sufficiently high efficiency that with CO_2 capture, its efficiency is essentially the same as that of the subcritical unit without CO_2 capture (MIT Report 2007).

Improved Processes: Although chemical absorption process using liquid solvents are currently the most advanced options for post-combustion capture, the process imposes an energy penalty of 30% to 60% for coal-fired plants (Herzog, Drake, & Adams 1997; Turkenburg & Hendriks 1999; David & Herzog 2000). In order to reduce the capital and energy cost, and the size of the absorption and regenerator columns, *new processes* are being developed. The attempts to reduce the overall costs and/or energy penalty for CO_2 capture include using improved aqueous amines, non-aqueous solvents, adsorption using dry sorbents, membranes, etc, (NCC 2012). The current status of post-combustion CO_2 capture technologies is analysed by Bhown and Freeman of EPRI (2011, 2012).

9.3.2   Pre-combustion capture

In pre-combustion capture, the fossil fuel is partially oxidized under high pressure, which is accomplished in a plant by reacting coal with steam and oxygen at high temperature and pressure (gasification). The resulting gaseous fuel consisting mainly of carbon monoxide and hydrogen can be burned in a combined cycle power plant to generate electricity. After particulate impurities are removed from the syngas, a two stage shift reactor converts the carbon monoxide to CO_2 (15–60% by volume) via a reaction with steam resulting in a mixture of CO_2 and hydrogen. The CO_2 from the gas can be removed through acid gas removal (AGR) process using solvents such as Selexol. The separation of CO_2 produces a hydrogen-rich gas that is burned in a gas turbine to produce electricity.

Capturing of CO_2 using AGR process is practised commercially at full-scale in gas processing, and chemicals plants where CO_2 is separated as part of the standard industrial process. This process, however, is slightly different to pre-combustion for power generation. Capturing CO_2 prior to combustion has two advantages: CO_2 is not getting diluted by the combustion air, and the CO_2 stream is normally at elevated pressure. Therefore, more efficient separation methods can be applied, e.g. using pressure-swing-absorption in *physical solvents* such as Rectisol and Selexol. The use of physical solvents has several merits such as (a) low utility consumption, (b) Rectisol uses inexpensive and easily available methanol, (c) Selexol has a higher capacity to absorb gases compared to amines, and can remove H_2S and organic sulfur compounds, and (d) both provide simultaneous dehydration of the gas stream; and important issues such as (a) high costs of Rectisol refrigeration, (b) co-absorption of hydrocarbons in Selexol often requiring recycle compression, and reducing product revenue, and (c) the need of refrigeration for the lean Selexol solution (Jones 2007). These problems need to be addressed.

IGCC plants with CO₂ capture

Pre-combustion CO_2 capture is applicable to IGCC power plants, plants that gasify coal to produce liquid fuels, and coproduction plants capable of generating significant amounts of electricity and liquid fuels. In IGCC power plants, coal is gasified and CO_2 removed at high pressure. Figure 9.10 shows the schematic representation of IGCC without and with CO_2 capture (Katzer 2007). The gasifier is the biggest variable in the IGCC Plant in terms of type of feed (water-slurry feed or dry feed), operating pressure, and the amount of heat removed from it. For electricity generation without CO_2 capture, radiant and convective cooling sections that produce high-pressure steam for power generation lead to efficiencies of 40% or more.

To achieve CO_2 capture with IGCC, two shift reactors are added where CO in syngas react with steam over a catalyst to produce CO_2 and hydrogen. As the gas stream is at high pressure having a high CO_2 concentration, a weak CO_2-binding physical solvent, such as Selexol, can be used to separate out the CO_2.

The gas clean-up process is, in fact, doubled in size in this system, to first remove sulfur and then CO_2. As the capture is performed at high concentration and pressure, it is cheaper than for CO_2-capture from flue gas capture. Moreover, higher pressure in the gasifier improves the energy efficiency of both separation and compression of CO_2. The gas stream now being chiefly hydrogen, the gas turbine *requires modifications* for efficient operation. All the technologies of an IGCC plant with carbon capture, such as the air separation unit, gasification, gas cooling, shift reaction, sulfur control, and CO_2 capture are commercially developed and available; they are yet to be integrated together and demonstrated at the scale of operation.

The material flows for a 500 MW_e IGCC unit designed for CO_2 capture are discussed in MIT Study Report 2007. For CO_2 capture, a Texaco/GE full-quench gasifier is currently considered the optimum configuration. The overall generating efficiency is 31.2% which is a 7.2 percentage point less from the IGCC system without capture. Adding CO_2 capture requires a 23% increase in the coal feed rate. This compares with coal feed rate increases of 27% for ultra-supercritical PC and 37% for subcritical PC when amine-based CO_2 capture is used which are fairly higher. CO_2 compression

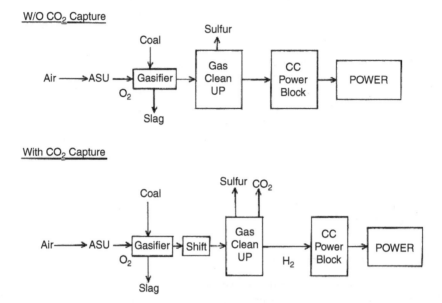

Figure 9.10 IGCC without and with CO₂ capture (Redrawn from Katzer, MIT, 2007).

and water gas shift are affected significantly with capture. CO_2 compression is about two-thirds that for the PC plants because the recovery of CO_2 is at an elevated pressure. As the energy required for CO_2 recovery is lower than that for the PC plant, less energy-intensive separation processes are adequate. The total efficiency reduction for IGCC is 7.2% as compared with 9.2% for the PC plants. This smaller delta between the no-capture and the capture cases is one of the attractive features of IGCC for application to CO_2 capture. More details regarding design, performance and operation of pre-combustion systems can be found in the literature (e.g. Metz _et al._ IPCC 2005; Chen & Rubin 2009).

Potential of IGCC with CCS: The organic combination of CCS and IGCC is central for clean use of coal (Harry 2007). IGCC with CCS is the _zero-CO₂ emission coal based_ power generation technology, and is expected to be the new approach of power generation in the future in response to global climate changes (Li Xianyong _et al._ 2009). With carbon capture, the cost of electricity from an IGCC plant would increase about 30%, whereas for a natural gas CC, the increase is 33%, and for a pulverized coal plant, 68%. This high potential for less expensive carbon capture makes IGCC an attractive choice.

Next generation IGCC plants with CO_2 capture will be simplified systems compared to earlier IGCC and are likely to have higher thermal efficiency and low costs. The main feature is that instead of using oxygen and nitrogen to gasify coal, oxygen and CO_2 are expected to be used. The foremost advantage is the possibility to improve the performance of cold gas efficiency and to reduce the unburned carbon (char). With a 1300°C class gas turbine it is possible to achieve 42% net thermal efficiency with CO_2 capture, as against 30% in a conventional IGCC system. Further, it is possible

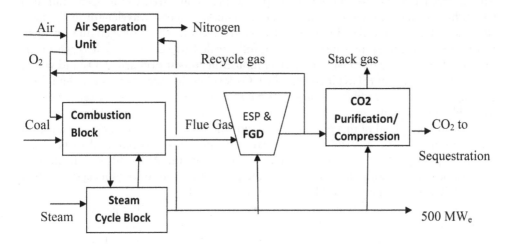

Figure 9.11   Oxy-fuel combustion with CO_2 capture (Redrawn from Katzer 2007).

to increase thermal efficiency to 45% in the new IGCC plant with a 1500°C class gas turbine. The CO_2 extracted from gas turbine exhaust gas will be utilized in this system. Using a closed gas turbine system capable of capturing the CO_2 by direct compression and liquefaction obviates the need for a separation and capture system (Guo Yun *et al.* 2010).

9.3.3   Oxy-fuel combustion capture

In this process, fuel is burned using pure oxygen instead of air as the primary oxidant (Buhre *et al.* 2005; Dillon *et al.* 2004; Jordal *et al.* 2004; and Andersson *et al.* 2003). This process is developed as an alternative to post-combustion capture for conventional PC power plants. In the system, shown schematically in Figure 9.11, nitrogen is removed from input air so that a stream of roughly 95% oxygen is available for firing which would reduce fuel consumption, but result in a very high flame temperature. The mixture is diluted by adding CO_2-rich flue gas from the boiler exit, so that both the flame temperature and furnace exit gas temperature can be brought back to air combustion levels (as happens in air-fired pulverized coal boiler). Most of the oxy-fuel combustion research to-date has focused on this aspect, viz. diluting oxygen to produce similar radiant and convective heat transfer profiles as exist in conventional air-fired coal boilers.

New designs have been proposed for oxyfueled boilers, to reduce or eliminate external recycle through other means such as slagging combustors or controlled staging of non-stoichiometric burners. The recycled flue gas can also be used to carry fuel into the boiler and ensure adequate convective heat transfer to all boiler areas. For similar conditions of heat transfer in the combustion chamber, about three pounds of flue gas has to be recirculated for every pound of flue gas produced, resulting in oxygen volume concentration of about 30%, compared to 21% for air fired combustion.

This difference is due to CO_2's higher specific heat compared to nitrogen, and also CO_2's high radiative emissivity at the flame's emission mean wavelength in the near IR region (Hottell and Sarofin 1967). To avoid excess levels of oxygen and nitrogen in the flue gas, the system has to be carefully sealed to prevent any leakage of air into the flue gas. The exhaust from oxy-combustion consist primarily CO_2 and water. Flue gas recirculation (FGR) increases the CO_2 level in the flue gas to more than 90% making *the flue gas ready for sequestration without energy intensive gas separation* as in air-blown PC plant.

The major appeal of this process is it avoids the need for a costly post-combustion CO_2 capture system. Instead, an air separation unit (ASU) is required to generate relatively pure (95%–99%) oxygen needed for combustion. ASU adds significantly to the cost because roughly three times more oxygen is required for oxyfuel systems than for an IGCC plant of comparable size (Rubin *et al.* 2012).

Buhre and others (2005) have observed the CO_2 in the flue gas of a PC coal boiler could reach higher than 95% if operated in lean combustion condition and in very minimal air in-leakage. Tan and others (2005) have reported that CO_2 levels in the flue gas from various industrial oxy-fuel pilot plants studied during 1980–2000 have achieved higher than 90% and reached as high as 95%. The corrosion of the compressor and pipeline, however, requires some post-combustion gas clean up.

MIT Report 2007 discusses a 500 MW_e oxy-fuel generating unit with CO_2 capture utilizing supercritical boiler and steam cycle. The coal feed rate is higher than that for supercritical PC without capture because of the power consumption of the air separation unit, but lower than that for a supercritical PC with amine-based CO_2 capture (see Table 8.3). In this design, wet FGD is used prior to recycle to remove 95% of the SO_x to avoid boiler corrosion problems and high SO_x concentration in the downstream compression/separation equipment. The non-condensables are removed from the compressed flue gas via a two-stage flash. This design produces a highly-pure CO_2 stream, similar to that from the PC capture cases, but requires additional cost to achieve this purity level. If this extra purification were not required for transport and geologic sequestration of carbon dioxide, oxy-fuel combustion could gain up to 1% in efficiency, and the COE could be reduced by up to 0.4 ¢/kW$_e$h. The generating efficiency is 30.6% (HHV), which is about 1% higher than supercritical PC with amine-based CO_2 capture (post-combustion).

Studies of the design aspects suggest that the process can be further simplified with SO_x and NO_x removal occurring in the downstream compression and separation stage at reduced cost (Allam 2006). Another main issue is about the heat balance between steam generation in the furnace by radiation, and superheating the steam mainly in the convection section of the boiler. As the mass flow rate of re-circulated flue gas decreases, the furnace temperature rises and more steam is generated by radiation in the furnace section. Under the same conditions the heat transfer in the convective section decreases because of the reduced gas mass flow rate, causing the superheat temperature to drop (Beer 2009).

The oxy-fuel PC unit has a gain over the air-driven PC unit due to improved boiler efficiency and reduced emissions control energy requirements, but the energy consumption of the ASU that results in a 6.4% reduction, outweighs this efficiency improvement. The overall efficiency reduction is 8.3% from supercritical PC makes the technology economically unviable.

Development of efficient new technologies such as membranes, chemical looping combustion, discussed later, to replace the conventional oxygen production systems can reduce costs (Zeng *et al.* 2003; Marion *et al.* 2004; Anheden *et al.* 2005; Allam 2007, 2009).

Oxy-fuel PC shows potential of lower COE and lower CO_2 avoided-cost than other PC capture technologies (MIT Report 2007). Several studies (DOE Oxy-combustion, 2007) including the Interagency report on CCS (2010) also noted that new oxy-combustion for CO_2 capture may result in a lower LCOE than new pre-combustion or new facilities with post-combustion CO_2 capture. Laboratory and pilot-scale studies have demonstrated that it has the advantage of reliability, availability and familiarity to industry, although there is a need to develop a 'clean coal' image for oxy-fuel combustion (Sarofim 2007). Several publications and reviews on oxy-fuel combustion are available, for example, Zheng, L. (ed.) (2011), Schheffknecht *et al.* (2011), Wall *et al.* (2011), Toftegaard *et al.* (2010), Davidson & Santos, IEAGHG (2010), Levendis and co-workers (2011, 2012) and earlier papers by Croiset *et al.* (2005), IEAGHG (2005), MIT Report (2007), ZEP (2008), Tan *et al.* (2005), Santos & Haines (2006), Santos & Davison (2006), and Santos *et al.* (2006).

Advantages of oxy-fuel over air-blown PC plants: Oxy-fuel PC combustion, by and large, has other advantages over air-blown PC plants: (a) mass and volume of the flue gas are reduced by around 75% enabling reduced size of the flue gas treatment apparatus, (b) less heat is lost in the flue gas because of less volume, (c) flue gas being primarily CO_2, ready for sequestration, (d) concentration of pollutants in the flue gas is higher, making separation easier, (e) most of the flue gases are condensable thereby making compression/separation possible, (f) heat of condensation can be captured and reused rather than lost in the flue gas, (g) NO_x pollutant is reduced to greatest extent because of nitrogen removal from the inlet air stream (Levendis and coworkers 2011). There are, of course, issues to be resolved, such as (i) treatment of the recycle stream – whether SO_2 be removed before recycle and whether the recycle stream be dried, and (ii) the purity requirements of the CO_2 stream for transport and sequestration.

9.3.4 Energy price in capture technologies

There is considerable energy penalty in the capture technologies compared to other environmental control equipment used at modern power plants. As a result, the efficiency is decreased and the cost is increased. For PC plants with capture, this means that proportionally more solid waste is produced and more chemicals – ammonia and limestone, are needed (per unit of electrical output) to control NO_x and SO_2 emissions. Water needs also increase significantly due to the extra cooling water needed for current amine capture systems. Because of the efficiency loss, a capture system that removes 90% of the CO_2 from the plant flue gas winds up reducing the net (avoided) emissions per kilowatt-hour by a smaller amount, typically 85 to 88% (Metz *et al.* 2005). Table 9.1 provides data on efficiency and energy penalty of different coal-based power technologies with and w/o carbon capture. There is a range of reported efficiencies and associated energy penalties around the values shown in the Table for each plant type (Metz *et al.* 2005; MIT 2007; CarnegieMellon Univ, 2009; Finkenroth 2011; Rubin *et al.* 2012).

Table 9.1 Representative values of power plant efficiency and CCS energy penalty.

Power plant & capture system type	Net plant η (%), HHV, w/o CCS	Net plant η (%) HHV, with CCS	CCS energy penalty: Addl. enrgy input (%) per net kWh output[a]	CCS energy penalty: Reduction in net kWh output (%) for a fixed energy input
Existing subcritical PC, post-combustion capture	33	23	43%	30%
New SC PC, post-combustion capture	40	31	29%	23%
New SCPC, Oxy-combustion capture	40	32	25%	20%
NewIGCC (bituminous), pre-combustion capture	40	33	21%	18%
New NGCC, post-combustion capture	50	43	16%	14%

(*Source:* Rubin *et al.* 2012, reproduced with permission of author)

Since higher efficiency plants have smaller energy penalty and associated impacts, replacing or repowering an old, inefficient plant with a new, more efficient unit with CO_2 capture can still provide a net efficiency gain that decreases all plant emissions and resource use. The net impact of the CO_2 capture energy penalty must, therefore, be assessed in the context of a particular situation or strategy for reducing CO_2 emissions. Any novel steps that raise the efficiency of power generation can reduce the impacts and cost of carbon capture.

The overall energy requirement for PC and IGCC plants is due to two factors: (a) electricity needed to operate system components like fans, pumps and carbon dioxide compressors which is approximately 40% of the total energy penalty, and (b) thermal energy required (or losses) for solvent regeneration in PC plants and for water-gas shift reaction in IGCC plants which is approximately 60% of the total energy penalty (Rubin *et al.* 2012). Thermal energy requirements are clearly the largest source of net energy losses and hence the priority area for research to reduce those losses. For oxy-combustion systems, the energy requirement for oxygen production is the biggest contributor to the energy penalty.

9.3.5 Partial CO_2 capture

The level of capture is controlled by adjusting the extent of CO to CO_2 conversion, as the acid gas removal (AGR) will capture a relative portion of the CO_2 in the stream. Some conversion naturally occurs in the gasifier, so there is some amount of CO_2 in the syngas exiting the gasifier which can be removed in the AGR. This process, known as 'skimming, can result in capture levels up to 25%. Installing a single-stage shift and removing the resulting carbon dioxide can achieve 50–80% capture' (Phillips 2007). Levels of capture beyond 80% require a two-stage shift. As mentioned already, the extent of capture achievable with different numbers of shifts is highly dependent on the type of gasifier, its operating parameters, plant specifics, and type of solvent used.

Table 9.2 Benefits of Partial Capture compared to Full Capture for IGCC Plants.

Technological Distinctions	Associated Performance and Economic Benefits
(1) Reduced number, size of equipment	Reduced capital cost
(2) Reduced auxiliary load	Improved plant output
(3) Reduced consumables and water use	–
(4) Reduced steam consumption	Improved electrical output or heat integration
(5) Reduced or avoided turbine derating	Improved plant output and efficiency

While the number of shift reactors can be arranged to achieve distinct capture levels, adjustments in the amount of catalyst or steam used in each shift may be utilized to shift a specific amount of CO to CO_2, allowing intermediate capture levels to be achieved. Bypassing a portion of the syngas around the shift reactors may also be used to control the capture level, but there are issues to be resolved with this option. The technological differences between partial and full capture and their associated benefits are summarized in Table 9.2 (Hildebrand & Herzog 2008).

There are important capital cost implications. Lower capture levels will have lower associated capital costs because equipment requirements are less. The amount of investment in the AGR will also depend on the level of capture, as at lower levels expansion into two fully-integrated stages may be unnecessary. Some equipment, like the flash tanks for solvent regeneration, carbon dioxide compressors, and peripheral components such as pumps and blowers, can be smaller or single, saving capital cost. Similar to the pulverized coal case, partial capture also saves auxiliary loads associated with capture, reduces water demand, and saves considerably on solvent and catalyst (Hildebrand & Herzog 2008).

As explained earlier, full capture has a significant impact on the syngas turbine; the shift of CO to CO_2 results in up to a 15% decrease in heating value of the syngas (Bohm *et al.* 2006). The syngas now being primarily hydrogen, the firing temperature is increased. This makes it necessary to derate the turbine to preserve its life and reduce NO_x formation which is usually achieved by diluting the gas with compressed nitrogen. This also serves to increase mass flow rate through the turbine, thus obtaining better output than without dilution. However, turbine output is still up to 10% lower than with unshifted syngas (Kubek *et al.* 2007). Partial capture can reduce or prevent this derating up to a certain capture level, and some of the CO heating value can be maintained. Furthermore, partial capture will reduce the amount of steam needed in the shift, which can be used for generation in the steam turbine or for heat integration.

9.4 PERFORMANCE AND COSTS

Technology performance and cost studies with and w/o CO_2 capture have been carried out by a number of groups (e.g., MIT, IEA, IPCC, US EIA, US DOE and US NREL, EPRI, Carnegie-Mellon University). These studies each use differing methodologies and assumptions regarding key economic and technology criteria. The different assumptions and methods used make it difficult arriving at significant comparisons of

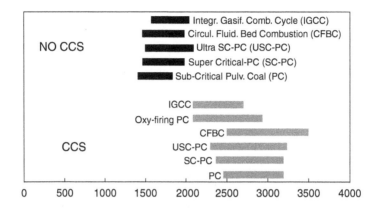

Figure 9.12 Coal Fired Power Plant Investment Cost ($2006/kW) with and w/out CCS (*Source:* International Energy Agency 2008).

technologies described in different reports, and comparisons of the findings of different reports for specific technologies. Moreover, no existing reports present absolute costs for CO_2 capture in which one can have a high degree of confidence. Meaningful estimates of absolute capture costs will not be feasible without the experience of early mover commercial-scale projects. But, it is feasible to understand relative costs for alternative technologies based on information in the literature if this information is analyzed in a 'self-consistent manner'. The economic analysis on 'self-consist basis' prepared by Robert Williams (2006) is discussed in 2012 NCC Report. NOAK costs estimates are used in this analysis to explore the potential for NOAK plants. Depending on the TRL of the technology, further development, demonstrations, and/or early commercial projects must be accomplished to achieve NOAK costs. While the economic analysis enables exploring the future role of different capture technologies, it must be noted that none of these technologies have achieved full commercial maturity for the costs (used in the economic analysis) to be certain. For extensive details of this analysis starting with the metrics used to describe relative economics and so on, the reader may refer to NCC report 2012.

Other studies related to (a) investment costs for different generating technologies with and w/o capture (IEA), (b) performance and COE for different technologies with capture (MIT), (c) performance and costs of *new plants* with capture (Rubin *et al.*), and (d) the cost and performance for a *range of CO_2 capture levels* for new SCPC and IGCC power plants (DOE/NETL) are outlined here. For more details of these studies, please see the original papers cited.

(a) Simbolotti (2010) discusses the *investment cost* for different types of coal power plants with and w/o CCS, given in Figure 9.12 (IEA 2008). The additional investment cost for CCS ranges from $600/kW to $1700/kW, an increase of 50% to 100% as compared to the plant with no CCS. Because of the high cost and loss of efficiency by 8–12% in coal power plants, the CCS in power generation is economically feasible only for large, highly-efficient plants.

Table 9.3 Costs and performance of generating technologies.

Parameter	Subcritical		Supercritical		Oxy-fuel	IGCC	
	w/o cap.	w/cap.	w/o cap.	w/cap	w capture	w/o cap.	w/cap.
PERFORMANCE:							
Heat rate	9950	13600	8870	11700	11200	8890	10900
Efficiency (HHV)	34.3%	25.1%	38.5%	29.3%	30.6%	38.4%	31.2%
CO_2 emitted	931	127	830	109	104	824	101
COSTS:							
Total Plant Cost	$1580	$2760	$1650	$2650	$2350	$1770	$2340
Cost of Electricity:							
(a) Inv. Charge	3.20	5.60	3.35	5.37	4.77	3.59	4.75
(b) Fuel	1.49	2.04	1.33	1.75	1.67	1.33	1.64
(c) O&M	0.75	1.60	0.75	1.60	1.45	0.90	1.05
Total COE	**5.45**	**9.24**	**5.43**	**8.72**	**7.89**	**5.82**	**7.44**
Cost of CO_2 avoided Vs same technology w/o capture, $/tone	47.1		45.7		34.0	22.3	

(Units: Heat Rate: Btu/kW$_e$h; CO_2 emitted: g/kW$_e$h; Total Plant cost: $/kW$_e$; Inv. Charge: ¢/kW$_e$h @15.1%; Fuel: ¢/kW$_e$h @$1.50/mmBtu; O&M: ¢/kW$_e$h; COE: ¢/kW$_e$h)
[Basis: 500 MWe plant; Illinois #6 coal, 85% capacity factor, COE at bus bar. Based on design studies between 2000 and 2004, a period of cost stability, indexed to 2007 $ using construction cost index (*Source*: Katzer 2007)]

The incremental investment cost for CCS demonstration in existing coal power plants is currently estimated between $0.5 and $1 billion, 50% of which for the CCS equipment. In its most ambitious CO_2 mitigation scenarios the IEA estimates that 3400 plants with CCS will be needed by 2050 to meet the emissions reduction targets (IEA, 2008b and 2010b). As a result, the related incremental CCS cost is likely to be in the order of 40% of the cost without CCS and would be between $2.5 and $3 trillion ($1.3trillion for capture, $0.5–1 trillion for transport and $0.5 trillion for storage).

(b) Katzer has discussed the performance and costs of generating technologies w/o and with CO_2 capture (Katzer 2007). The performance parameters, total plant cost, COE, O&M costs, cost of CO_2 avoided, are given in Table 9.3.

Please note that the parameter values for 'subcritical' and 'supercritical' given in Table 8.3 (Chapter 8) and here are the same, except for Total Plant cost and Inv. Charges. As a result, the COE are slightly higher in Table 9.3. These are indicative costs to allow comparison among technologies indexed to 2007 $ construction costs. Without capture, PC has the lowest COE, and about 8% higher for IGCC.

The COE with CO_2 capture is the lowest for IGCC. The cost of capture and compression for supercritical PC is about 3.3 ¢/kW$_e$-h; that for IGCC is about half of that or about 1.6 ¢/kW$_e$-h. The cost of CO_2-avoided for PC is approximately $46/tonne of CO_2; $22/tonne for IGCC; $34/tonne for oxy-fuel which is in between these two. These numbers include the cost of capture and compression to a supercritical fluid, but *do not* include the cost of CO_2 transport and injection. The lower COE for IGCC would appear to make it the technology of choice for CO_2 capture in power generation. But, oxy-fuel has significant development potential. Further, if lower rank coals are

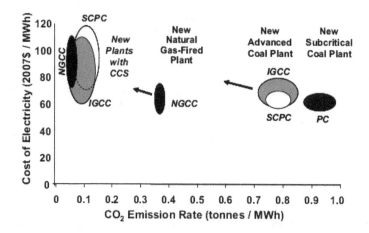

Figure 9.13   Cost of electricity generation (constant 2007 US$/MWh) as a function of the CO_2 emission rate (tonnes CO_2/MWh) for new power plants burning bituminous coal or natural gas. PC – subcritical pulverized coal; SCPC: supercritical pulverized coal; IGCC: integrated gasification combined cycle; NGCC: natural gas combined cycle (*Source*: Rubin *et al.* 2012, Reproduced with the permission of Professor E.S. Rubin, Copyright © Elsevier).

considered and move up with the rank, the cost difference between IGCC and PC narrows. As such, potentially significant reductions in the capture/recovery cost for PC could make it cost-effective with IGCC with capture, in certain applications. As seen earlier, issues for IGCC still exist, about operability and availability associated with it in the power generation. As of now, it appears that none of these technologies can be rejected and all of them require a close observation.

(c) Rubin and others have analysed the costs of generating electricity for *new power plants with and without CCS*, reported in recent studies based on current status of commercial post-combustion and pre-combustion capture processes (Rubin *et al.* 2012). The total cost of electricity generation (COE, in $/MWh) is shown in Figure 9.13 as a function of the CO_2 emission rate (tonnes CO_2/MWh) for power plants burning bituminous coal or natural gas.

The COE includes the costs of CO_2 transport and storage, but most of the cost, 80–90%, is for capture including compression. All plants capture and sequester 90% of the CO_2 in deep geologic formations. The broad range of costs shown for each plant type is a result of difference in assumptions made about the design, operation and financing (i.e., technical, operational, financial and economic) of the power plant to which the capture technology is applied. For example, factors such as higher plant efficiency, larger plant size, higher fuel quality, lower fuel cost, higher annual hours of operation, longer operating life and lower-cost of capital all reduce both the cost of electricity and the unit cost of CO_2 capture. The assumptions about the CO_2 capture system design and operation further contribute to variations in the overall cost. As mentioned above, these assumptions vary across studies, and since no single set of assumptions applies to all situations or all sites in the world, there is no universal estimate for the cost of CO_2 capture. Cost ranges would be even broader if additional factors such as a range of coal types or a larger range of boiler efficiencies were considered [Rubin (2010), Metz *et al.* (2005), USDOE (2007), MIT (2007), Carnegie Mellon University (2009), Holt (2007), Rubin *et al.* (2007)].

Thus, overlapping ovals do not imply that one technology can be more (or less) costly than another under the same set of assumptions (for example, while different studies report overlapping cost values for different SCPC and IGCC plants without CCS, individual studies show IGCC plants to be systematically more costly than a similarly-sized SCPC plant when all other assumptions are held constant). CCS is estimated to increase the cost of generating electricity by about 60%–80% at new coal combustion plants and by about 30%–50% at new coal gasification plants on a relative basis. On an absolute basis, this turns out to roughly $40–70/MWh for supercritical coal plants and $30–50/MWh for IGCC plants using bituminous coal. The cost per tonne of CO_2-avoided for a plant with capture relative to one without also can be calculated from the figure. This cost is equivalent to the 'carbon price' or CO_2 emissions tax, and with this, the CCS plant is more economical than the plant w/o capture. For new supercritical coal plants this cost is currently about $60–80/tonne CO_2. For IGCC plants with and without CCS, the avoidance cost is smaller, about $30–50/tonne CO_2. Since the cost of CO_2 avoided depends on the choice of 'reference plant' with no CCS, it is also useful to compare an IGCC plant with CCS to an SCPC reference plant, since without capture SCPC is less expensive than IGCC for the same design basis. In this case the cost of CO_2 avoided increases to roughly $40–60/tonne CO_2. In all cases, if the CO_2 can be sold for EOR with subsequent geological storage, the costs are lower (due to revenue generation in EOR). For plant using low-rank coals (subbituminous or lignite coal) the avoidance cost may be slightly higher than values based on the figure (Rubin *et al.* 2007; USDOE 2009).

(d) DOE NETL (2011a) has undertaken a detailed study to establish the cost and performance for a *range of CO_2 capture levels* for new SCPC and IGCC power plants. The SCPC cases utilize Fluor's Econamine FG Plus process for post-combustion CO_2 capture via temperature swing absorption with a 30 wt% MEA solution as the chemical solvent. The IGCC cases employ a slurry-fed, entrained-flow GEE gasifier and leverage UOP's two-stage Selexol process for bulk CO_2 capture and selective H_2S removal via pressure swing absorption. The Selexol process uses dimethyl ether of polyethylene glycol as a physical solvent for pre-combustion capture. The CO_2 capture sensitivity analysis includes an evaluation of the cost, performance, and environmental profile of these facilities while operating at ISO conditions and using Illinois #6 medium-sulfur bituminous coal. The study includes compression of the captured CO_2 to pipeline pressure of 15.2 MPa (2,215 psia), transporting 80 kilometers by pipeline, storage in a saline formation at a depth of 1,239 meters, and monitoring for 80 years. The site will have proper access to rail transportation for delivering coal. All local coal handling facilities are included in the design and cost estimation aspect of this study. The start-up date is taken as 2015 and all cases are considered 'first-of-a-kind'. The methodology included performing steady-state simulations of the technology using the ASPEN Plus modeling program. Each configuration was custom-made to achieve a specific level of carbon capture. The study draws the following conclusions: (a) At all levels of carbon capture considered, an IGCC plant will operate more efficiently than a SC PC plant. This higher efficiency will lower fuel costs and reduce coal handling costs for the same level of power output; (b) At lower levels of CO_2 capture, the

Table 9.4 Emission levels.

Technology (case)	Particulates (lb/mmBtu)	SO₂ (lb/mmBtu)	NOₓ (lb/mmBtu)	Mercury (% removed)
PC Plant:				
(a) Typical	0.02	0.22	0.11	
(b) Best Commercial	0.015 (99.5%)	0.04 (99+%)	0.03 (90+%)	90
(c) Design with CO₂ capture	0.01 (99.5%)	0.0006 (99.99%)	0.03 (95+%)	75–85
IGCC Plant:				
(a) Best commercial	0.001	0.015 (99.8%)	0.01	95
(b) Design with CO₂ capture	0.001	0.005 (99.9%)	0.01	>95

(*Source:* Katzer 2007)

COE for a SCPC plant is less than an IGCC plant, but starts rising as capture level increases. At capture levels greater than around 90%, both have nearly equal costs of electricity. The same trend applies to plant capital costs also (expressed in $/kW); (c) CO_2 avoided costs, using similar non-capture plants as reference, are substantially lower for IGCC than for SCPC, for all levels of capture. This is indicative of somewhat lower cost impact associated with adding CO_2 capture to an IGCC plant compared to SCPC. However, the avoided CO_2 costs for IGCC capture plants, using the SC PC non-capture reference plant, are higher due to the higher COE of IGCC plants w/o capture; (d) CO_2 avoided costs for SC PC plants are high for low levels of CCS, but start reducing as the capture level increases reaching a minimum for around 95% CCS, and then start rising slightly. That means, if a SC PC plant invests additional capital to install CO_2 controls, that equipment could be used most cost-effectively by capturing as much carbon dioxide as possible and up to about 95%; (e) CO_2 avoided costs for IGCC (relative to an IGCC w/o capture plant) are comparatively less unstable over a wide range of capture. Although they generally decrease as the level of CCS increases (similar to SCPC), the spread between the high and low values is small relative to SCPC. Since an IGCC system is more complex, it allows itself to optimization better than a SC PC plant. To hold a precise level of CO_2 capture, there are several variables to manipulate simultaneously (such as the number of water gas shift reactors, the amount of shift steam required, solvent circulation rate, and the syngas bypass rate) to minimize the cost and performance impact. The SCPC plant which is generally less expensive in most cases than IGCC is a simple system and hence exhibits less potential for optimization. The Report has given many details which the reader may find useful.

Emissions Performance: Coal's main problems are with emissions. In Table 9.4 are given the commercially demonstrated and projected emissions performance of PC and IGCC plants (Power Clean 2004; USEPA 2005). ESP or bag houses are used on all PC units, and PM emissions are typically very low. Further reductions can be obtained with improved ESP or wet ESP but it adds to the cost. FGD is not applied on the total generation capacity and thus typical emissions are quite high. The best IGCC commercial performance gives the levels of emissions reductions that have been demonstrated in full-scale commercially operating units, and can be further improved.

In fact with CO_2 capture, emissions levels are expected to be reduced even further (Holt 2007). The best commercial performance levels with IGCC are 3 to 10-times lower. IGCC with CO_2 capture should have even lower emissions (Katzer 2007).

9.5 POLICY FRAMEWORK TO DEPLOY CCS SYSTEMS

The pathway to reduce the costs of CO_2 capture from coal utilization is through the pursuit of R&D and, more significantly, as a result of the cost reducing experience of demonstration and early mover capture projects (learning by doing). To ensure a proper R&D plan that leads to demonstration of capture projects, a strong legal and policy framework is essential that include the following elements:

1 Accelerating the innovation and development of CCS technologies by funding large-scale demonstration projects that inform economic, technical, and regulatory decision making. Careful monitoring at large-scale demonstration sites can help to gain a better understanding of what happens to large volumes of CO_2 over long periods of time. Such experience will decrease costs and make CCS more competitive with other mitigation technologies. The funding recipients are obliged to expand knowledge-sharing activities to share lessons learnt faster;
2 Requiring all new coal power plants to meet 'emissions standards' that limit CO_2 emissions to levels achievable with CCS systems; this would include enhanced R&D for capture technologies to improve performance and reduce costs, large-scale experience with sequestration for a range of geologic formations, a national inventory of potential storage reservoirs, and a new regulatory framework for evaluating sequestration sites and allocating liability for long-term CO_2 storage;
3 An incentive for power plants developers using CCS: Currently, there is no cost for CO_2 emissions into the atmosphere. Subsidies that offset the cost differential between conventional plants and those with CCS, and market incentives and complimentary coal policies are needed to attract private investment in CCS and use of these technologies, e.g., GHG cap-and-trade program;
4 A legal and regulatory framework to protect human and environment health and safety; Issues like property rights and long-term liability have to be properly addressed to implement CCS as effectively as possible;
5 Engagement with the public to ensure acceptance of these practices because of little public awareness about CCS.

9.6 PRESENT STATUS OF CAPTURE TECHNOLOGY

Research and development activities on CCS have been undertaken by several groups because of its vital role in global warming mitigation. The progress in technology development and current status are reported regularly by major organizations and research groups (USDOE/NETL 2011; IEAGHG 2011; MIT 2011; GCCSI 2011; Rubin *et al.* 2012; and GCCSI 2012).

The concept of 'technology readiness level (TRL)' is used to decide the 'commercial'/'demonstration'/'pilot-scale'/'laboratory' level status of a project. i.e., the maturity of a technology or system is described by the scale of TRLs. The US Electric Power Research Institute (EPRI) and DOE has recently adopted TRLs, originally developed by NASA, to describe the status of new post-combustion carbon capture technologies. At the lower end of the scale (TRL 1), a technology consists of observed basic principles only, while at the top end (TRL 9), the technology has matured into a system successfully used in its actual operating atmosphere. The description of each TRL level, as defined by the DOE (2009), is the following (NCC 2012):

1   Scientific research begins translation to applied R&D – Lowest level of technology readiness: Examples might include paper studies of a technology's basic properties;
2   Invention begins – Once basic principles are observed, practical applications can be invented;
3   Active R&D is initiated: Examples include components that are not yet integrated or representative;
4   Basic technological components are integrated;
5   The basic technological components are integrated with reasonably realistic supporting elements so it can be tested in a simulated environment;
6   Model/prototype is tested in relevant environment;
7   Prototype near or at planned operational system – Represents a major step up from TRL 6, requiring demonstration of an actual system prototype in an operational environment;
8   Technology is proven to work – Actual technology completed and qualified through test and demonstration; and
9   Actual application of technology is in its final form – Technology proven through successful operations.

The Global CCS Institute identified 75 large-scale integrated CCS projects (LSIPs) globally as at September 2012 (Global CCSI 2012). LSIPs are the projects involving the capture, transport and storage of CO_2 at a scale of: (a) at least 800,000 tonnes of CO_2 annually for a coal-based power plant; or (b) at least 400,000 tonnes of CO_2 annually for other emission-intensive industrial facilities (including gas-based power generation). This definition of LSIPs will be regularly reviewed and adapted as CCS matures; as clear CCS legislation, regulation, and standards emerge; and as discussions progress on project boundaries, lifecycle analysis, and acceptable use of CO_2. More than half of all newly-identified large-scale integrated projects are located in China. 16 of these are currently operating or in construction, with a combined capture capacity of around 36 Mtpa of CO_2. A further 59 LSIPs are in the planning stages of development, with an additional potential capture capacity of more than 110 Mtpa. All newly-identified projects are studying EOR options, for an additional source of revenue. The distribution of LSIPs in different regions of the globe is given in Table 9.5. The peak in large-scale projects online expected to occur in 2015–16 has shifted over the past two years and is now projected to start from 2018–20. In addition, there are many projects of a smaller scale (or which focus on only part of the CCS chain) that are important for research and development, for demonstrating individual elements of CCS and to develop local capacity (Global CCSI 2012).

Table 9.5 Large scale integrated CO_2 capture plants as at September 2012.

CO_2 Capture	North America	Europe	Asia	Australia & New Zealand	Middle East/ N. Africa	Total
Power generation						
Pre-combustion	7	3	3	–	1	14
Post-combustion	4	11	2	1	1	19
Oxyfuel combustion	1	3	2	–	–	6
Not decided	–	–	2	1	–	3
Others						
Natural gas processing	6	2	1	1	1	11
Iron & Steel produc.	–	1	–	–	1	2
Cement production	–	–	–	–	–	0
Other industries	14	1	3	2	–	20
TOTAL						75

(Source: Global CCSI 2012)

There are five large scale fully integrated CCS projects (in which CO_2 is captured and sequestered in deep geological formations) operating in different locations globally. But *none* of them are from power generation.

(a) Sleipner project: Injection of CO_2 into sandstones of the Utsira Formation in the North Sea, about 250km off the coast of Norway began in 1996 as part of Statoil's offshore Sleipner Project. This gas production facility in the North Sea is the longest-running commercial CCS and now stored over 13 $MtCO_2$. The CO_2 content extracted with natural gas from the gas field is about 9% which must be reduced to 2.5% in order to meet commercial specs. CO_2 is separated by amine scrubbing (MDEA) and injected (1 $MtPaCO_2$) through horizontal wells in a deep saline sandstone formation located 800 to 1000 m below the sea floor in the near Utsira formation. A total of 20 Mt of CO_2 is expected to be stored over the lifetime of the project, and the Formation is estimated to have a capacity of about 600 billion tones of CO_2. The injected CO_2 has been monitored via time-lapse 3D seismic techniques over the years to get a better understanding of the behavior of CO_2 in the reservoir. Though the seismic monitoring strongly suggests that the CO_2 is safely stored and not leaking from the reservoir, the resolution of the data is inadequate to confirm that 100% of the CO_2 has remained in the reservoir. This was the first commercial CCS demo-project, started to avoid a carbon tax equivalent to $55/tCO_2$. The operation costs including CO_2 compression and monitoring have been estimated at about $16/tCO_2$ injected.

(b) In-Salah Project: Operating since 2004, the CO_2 (10% in volume) from an onshore natural gas field in the Sahara Desert in Algeria is separated from the gas by an amine-based process and injected (around 1 MtPa) into a 2000-meter deep saline acquifer in the Krechba Formation, close to the natural gas field. The Formation is expected to receive 17 million tones of CO_2 over the life of the project. Based on the CCS investment and operation costs (including extensive monitoring and seismic analysis), the total CCS cost, about $6/tCO_2$ injected, is significantly lower than similar offshore operations.

(c) Weyburn-Midale project: In this onshore project managed by EnCana oil company, 1.7 to 2.8 MtPa of CO_2 is captured from a coal gasification plant located in

North Dakota and is transported by a 320km pipeline to Saskatchewan, Canada and injected into depleting oil fields where it is used for EOR since 2001. The project was the first CCS project with systematic scientific studies and monitoring of the CO_2 behavior underground. The incremental oil production has been estimated to be in the order of 155 million barrels. Further EOR operations in the Midale oil field have started in 2005, and the project currently injects 6500 tonnes CO_2 per day, of which 3000 recycled. The Canadian authorities provided an initial financial incentive of some \$20/bbl. Some 30 to 40 million tCO_2 will be stored over the project life time. An international research consortium under IEA GHG R&D Program monitors the CO_2 behavior and assesses storage techniques and risks (Simbolotti 2010). The Weyburn project has now stored in excess of 18 Mt CO_2 (Williams 2006). The results for 2000–2004 show that CO_2 storage in the oil reservoirs is viable and safe. This project will produce a best-practice manual for carbon injection and storage.

(d) Snøhvit project: This off-shore project operating since April 2008 in the Barents Sea, Norway is the first liquefied natural gas plant that captures CO_2 for injection and storage. The natural gas and CO_2 from the Snøhvit field are piped for 150km to the LNG onshore plant of Hammerfest where the CO_2 (0.7MPa) is separated and piped back to the offshore site to be injected in the 2600-meters deep Tubasen saline formation below the gas field.

(e) Rangely project: This project, operating since 1986, has been using CO_2 for enhanced oil recovery. Gas is separated and reinjected with CO_2 from the LaBarge field in Wyomimg. From the start, nearly 23 to 25 million tonnes of CO_2 have been stored. It is found from computer modeling that nearly all of it is dissolved in the formation water as aqueous CO_2 and bicarbonates.

Recently, another major project, Century Plant, a gas processing unit with pre-combustion capture of CO_2 [5 MtPa (+3.5 MtPa in construction)] used for EOR has started operating from 2010 in the United States (Global CCSI 2012).

Progress in the three types of capture processes

(A) Post-combustion CO_2 capture systems: Amine-based systems have been used commercially to meet CO_2 product specifications in several industries, from natural gas production to food processing (Rochelle 2009). Several coal-fired and gas-fired power plants with CO_2 capture, where *a portion of the flue gas* stream is fitted with a CO_2 capture system are listed in Table 9.6. Sleipner project and Snohvit project discussed already are not included in the Table.

Although several suppliers of commercial amine-based systems exist, only ABB Lummus (now CB&I Lummus) has flue gas CO_2 capture units operating at coal-fired power plants; and Fluor Daniel and MHI have commercial installations at gas-fired plants. Both Fluor and MHI now also offer commercial guarantees for post-combustion capture at coal-fired power plants.

Capture systems have operated commercially on a fraction of power plant flue gases; and *no capture units* have so far been applied to the *full flue gas stream* of a modern coal- or gas-fired power plant. Hence, full-scale *demonstration units* (correspond to TRL 7 and 8) under conditions of using regular equipments need to be installed to establish operability and reliability of operation. This could help gaining approval of this technology by power utilities, the financing institutions and the developers and

Table 9.6 Fully integrated commercial post-combustion capture at power plants and Iindustrial processes.

Project & location	Fuel type	Start up year	Plant capacity	Capture system type	CO_2 captured (10^6 tonnes/yr)
Projects in USA:					
(a) IMC Global Inc. Soda Ash Plant (Trona, CA)	Coal & petroleum coke-fired	1978	43 MW	Amine (Lummus)	0.29
(b) AES Shady Point Power Plant (Panama city, OK)	Coal-fired	1991	9 MW	Amine (Lummus)	0.06
(c) Bellingham Cogen facility (Bellinham, MA)	Natural gas-fired	1991	17 MW	Amine (Fluor)	0.11
(d) Warrior Run Power Plant (Cumberland, MD)	Coal-fired	2000	8 MW	Amine (Lummus)	0.05
Projects outside the USA:					
(a) Soda Ash Botswana Sua Pan plant	Coal-fired	1991	17 MW	Amine (Lummus)	0.11
(b) Sumitomo Chemicals Plant, Japan	Coal & gas-fired	1994	8 MW	Amine (Fluor)	0.05
(c) PetronasGas Processing Plant, Malaysia	Natural gas-fired	1999	10 MW	Amine (MHI)	0.07
(d) BP Gas processing Plant, Algeria	Natural gas separation	2004	N/A	Amine (Multiple)	1.0
(e) Mitsubishi Chemical Kurosaki Plant, Japan	Natural gas-fired	2005	18 MW	Amine (MHI)	0.12
(f) Huaneng Co-generation Power Plant (Beijing, China)	Coal-fired	2008	0.5 MW	Amine (Huaneng)	0.003

(Reproduced with the permission of Prof. Rubin; Source: Rubin et al. 2012)

operators of the plants. Several such projects were announced in Europe and the USA (Kuuskraa 2007; ETP 2010; MIT 2011); due to very high costs involved for full-scale projects, some of the announced ones were cancelled or delayed (Source Watch 2010).

The major postcombustion capture *demonstration* projects planned globally, as of September 2011, are listed in Table 9.7 (EPRI 2011; Rubin *et al.* 2012). Most of these systems, to begin in 2014/2015, would be installed at existing coal-fired plants, with the captured CO_2 transported through pipeline to a geological storage site, mostly for EOR to generate revenue for offsetting part of the project costs.

These projects are at the very early stages of detailed design, to project the extent of increase in capture efficiency or cost reductions relative to current commercial systems. Rubin *et al.* (2012) also listed a number of *pilot-scale* (correspond to TRL 6 and TRL 7) post-combustion CO_2 capture projects that are currently operating, or in the design or construction stage, or have recently been completed. These projects are testing/developing improved amine-based solvents or testing ammonia-based solvents (Ciferno *et al.* 2005; Figueroa *et al.* 2008) or calcium-based sorbents (Blamey *et al.* 2010). The pilot projects also include testing of capture processes based on concentrated piperazine, amino acid salts, solid sorbents and membrane-based systems.

A large number of new processes and materials required for post-combustion capture are currently at the laboratory level of development (Clean Air task Force 2010). These can be broadly grouped as (i) liquid solvents (absorbents) that capture

Table 9.7 Planned demonstration projects at Power plants with post-combustion capture.

Name & location	Fuel type	Startup year	Plant capacity (MW)	Capture System (vendor)	Annual CO_2 captured (10^6 t/yr)
Projects in the USA:					
(a) Tenaska Trailblazer Energy Center(Sweetwater, Texas)	Coal-fired	2014	600	Amine (Fluor)	4.3
(b) NRG Energy WA Parish Plant (Houston, Texas)	Coal-fired	2015	240	Amine (Fluor)	1.5
Projects outside the USA:					
(a) SaskPower Boundary Dam Polygon (Estevan, Canada)	Coal-fired	2014	115	Amine (Cansolv)	1.0
(b) Vattenfall Janschwalde (Janschwalde, Germany)	Coal-fired	2015	125	Amine (undecided)	N/A
(c) PGE Bechatow Power Station, Poland	Coal-fired	2015	360	Amine (Alstom, Dow Chemical)	1.8
(d) Porto Tolle (Rovigo, Italy)	Coal-fired	2015	200 MW[a]	Amine (undecided)	1.0
(e) SSE Peterhead Power Station, Peterhead, UK	Coal-fired	2015	385	N/A	1.0

[a] Estimated from other reported data
(*Source:* Rubin *et al.* 2012, reproduced with the permission of Prof. Rubin)

CO_2 via chemical or physical mechanisms (DOE/NETL 2010, Freeman *et al.* 2010, Chapel *et al.* 1999, Wappel *et al.* 2009, Knuutila *et al.* 2009, Cullinene *et al.* 2004, Oexmann *et al.* 2008, GCCSI 2011), (ii) solid adsorbents thatcapture CO_2 via physical or chemical mechanisms (Figueroa *et al.* 2008, EPRI 2009, DOE/ NETL 2008, Grey *et al.* 2008, Plaza *et al.* 2007, Sjostrom *et al.* 2010, Radosz *et al.* 2008), and (3) membranes that selectively separate CO_2 from other gaseous species (Figueroa *et al.* 2008, EPRI 2009, Kotowicz *et al.* 2010, Favre *et al.* 2009, Zhao *et al.* 2010, Clean Air task Force & Consortium for Science). Several potential approaches to reduce the cost and/or improve the efficiency of CO_2 capture relative to current commercial systems are pursued within each group. For example, using 'organic/inorganic membrane' (e.g., polytetra-fluoroethylene micro-porous membrane) which separates the flue gas from the solvent. Actually, the membrane allows for greater contacting area within a given volume and it is the solvent that selectively absorbs CO_2. The use of a gas membrane offers advantages such as (a) high packing density, (b) high flexibility with respect to flow rates and solvent selection, (c) no foaming, channeling, entrainment and flooding which are regular problems in packed absorption towers, (d) ready transportation (e.g. offshore), and (e) significant savings in weight. Another example is 'Designing a once-through scrubbing process (without regeneration step)' in which one could scrub CO_2 from flue gas with seawater and then return the whole mixture to the ocean for storage. In the seawater scrubbing, the large volumes of water that are required create large pressure drops in the pipes and absorber (Herzog & Golomb, MIT 2004). Pilot studies undertaken to improve the absorption process show that the use of new solvent technologies (such as membranes) and better integration of capture technologies can lower energy penalties to about 20% for conventional coal (Herzog, Drake, & Adams 1997; David and Herzog 2000). It is expected that near-term technical improvements (i.e., 2012 technology level) could reduce the costs considerably (David and Herzog 2000). However, it is too early to reliably quantify the potential benefits or the possibility of success in taking to a commercial process. In addition, computer-based modeling studies of novel capture technology concepts whose fundamental principles are usually well understood, but lack the experimental data needed to test or verify the worth of the idea are also undertaken. Three such novel but untested approaches to carbon capture – novel sorbents, hybrid systems and novel regeneration methods – are under study (Eisamann 2010, ARPA 2010, 2010a, Pennline *et al.* 2010). The US EPRI has reviewed over a hundred post-combustion capture active projects and ranked them on the TRL scale (Bhown & Freeman 2009, 2011). The EPRI study finds that most of the new processes under progress employ absorption methods (i.e., solvents), and very few new processes and concepts utilize membranes or solid sorbents (adsorption) for CO_2 capture. Obviously, bigger issues are confronting these approaches which need to be investigated.

(B) Pre-combustion CO_2 capture systems: No commercial-level power plants with precombustion CO_2 capture are currently operating; two projects in the US and one in China are under construction. Mississippi Power Company is building the most developed commercial scale IGCC plant with capture, designed to capture around 3.5 MPa and to demonstrate the technical and commercial viability of full-scale IGCC with CCS. Scheduled to start in 2014, the project will generate 524 MW of electricity and around 65% of its emissions will be captured using a Selexol AGR unit. Another one, the Texas Clean Energy Project (TCEP), a 400 MW (gross) IGCC poly-generation plant with CCS, is being developed by the Summit Power Group; the produced syngas will be used for power generation as well as for the production of granulated urea for commercial sale. The project will capture 90% of the CO_2 from the production of urea using Rectisol. Using poly-generation will create additional revenue compensating part of the project costs. A 250 MW GreenGen IGCC/CCS project in China is at an advanced stage of development. After the completion of the first stage, the project will be enlarged to 650 MW_e by adding another 400 MW_e. The exact duration of the subsequent R&D operational program for the plant (rather than subsequent commercial power generation) has not yet been finalized (Global CCSI 2012). The Great Plains synfuels plant operated by the Dakota Gasification Company which uses coal gasification to produce synthetic natural gas is expected to capture approximately 3 MPa of CO_2 using the methanol-based Rectisol process. CO_2, released to the atmosphere earlier, is presently compressed and transported via a 205-mile pipeline to a Canadian oil field, where it is used for EOR (Rubin *et al.* 2012). At the demonstration level, several of the announced IGCC power plants with CO_2 capture have been delayed or cancelled except one under construction in China. Some of the large-scale projects planned in the USA and other countries and likely to be in place in the next few years, with costs shared between the public and private sectors are listed in Table 9.8. Most of them due to begin in 2014 or later include fuels production plants and IGCC power plants. The amounts of CO_2 captured vary widely, from 50 to 90% of the carbon in the feedstock, across the projects, and in most cases the captured CO_2 would be sequestered in a depleted oil reservoir for EOR.

Most of the projects will serve to demonstrate key technical issues of IGCC technology, viz, the reliability of gasifier operations and the large-scale use of hydrogen to power the gas turbine following CO_2 capture.

Table 9.8 Precombustion CO_2 capture Demonstration projects planned.

Project & Location	Plant & Fuel type	Startup year	Plant capacity	CO_2 capture system	Annual CO_2 captured (10^6 tonnes)
Projects in the USA:					
(a) Baard Energy Clean Fuels (Wellsville, Ohio)	Coal + biomass-to-liquids	2013	53,000 barrels/day	Rectisol	N/A
(b) DKRWEnergy (Medicine Bow, WY)	Coal-to-liquids	2014	20,000 barrels/day	Selexol	N/A
(c) SummitPower (Pennwell, Texas)	Coal IGCC & Polygen (urea)	2014	400 MW$_e$	Rectisol	3.0 (EOR)*
(d) Kemper county IGCC (KemperCounty, Miss.)	Lignite IGCC	2014	584 MW	N/A	~3 (EOR)*
(e) Wallula (Washington)	Coal IGCC	2014	600–700 MW	N/A	N/A
(f) TaylorvilleEnergyCenter (Taylorville, Ill.)	Coal to SNG + IGCC	2014[a]	602 MW	N/A	3.0
(g) HydrogenEnergy (Kern county, CA)	Petcoke IGCC	2016	390 MW	N/A	2.0 (EOR)*
Projects outside the USA:					
(a) GreenGen (Tianji Binhai, China)	Coal IGCC & polygeneration	2011, stage1, no CCS	250/400 MW	N/A	Saline*
(b) Estron Grange IGCC, (Teesside, UK)	Coal IGCC	2012	800 MW	N/A	5.0
(c) Don Valley IGCC (Stainforth, UK)	Coal IGCC	2014	900 MW	Seloxol	4.5 (EOR)*
(d) Genesee IGCC (Edmonton, Canada)	Coal IGCC	2015	270 MW	N/A	1.2
(e) RWE Goldenbergwerk (Hurth, Germany)	Lignite IGCC	2015[b]	260 MW	N/A	2.3
(f) Kedzierzyn Zero Emission Power and Chemicals (Poland)	Coal-Biomass IGCC & polygen	2015 500 kton/yr methanol	309 MW+	N/A	2.4
(g) Nuon Magnum (Netherlands)	Multi-fuel IGCC	2015[c]	1200 MW$_g$	N/A	N/A (EOR & EGR)*
(h) FuturGas (Kingston, Australia)	Lignite-to-liquids	2016	10,000 barrels/day	N/A	1.6

MW$_g$: megawatts gross generated; [a]This project is on hold pending future state funding; [b]Depends on outcome of the Carbon Storage Law; [c]Depends on performance of the Buggenum pilot plant (*Source*: Rubin et al. 2012; *MIT 2012).

Also at pilot-scale are two projects in Europe, mentioned in the previous chapter. At the laboratory level, research in pre-combustion capture is focused mainly on improving the capture efficiency in order to lower the size and cost of equipment. Areas of research for post-combustion capture are also relevant here, viz, liquid solvents that separate CO_2 from a gas stream by absorption; solid sorbents that separate CO_2 by adsorption onto a solid surface; and membranes that separate CO_2 by selective permeation through thin layers of solid materials. Research on physical solvents is aimed at improving the CO_2 carrying capacity and reducing the heat of absorption, while research on new or improved solvents seeks to develop solvents that allow CO_2 to be captured at higher pressures and temperatures. For example, ionic liquids which are liquid salts at room temperature have high CO_2 absorption potential and do not evaporate at temperatures as high as 250°C. Research work on solid sorbents is primarily focused on identifying the most promising sorbent materials and conducting lab-level experiments. Lehigh University, RTI International, TDA Research, the University of North Dakota and the URS Group are among the organizations currently working on development of solid sorbents (USDOE/NETL 2010a). Till now, the membrane technology has not been used for pre-combustion CO_2 capture in IGCC plants or industrial processes that require a high CO_2 recovery rate with high CO_2 purity. It would be interesting to apply to IGCC since the mixture of CO_2 and H_2 following the shift reactor is already at high pressure, unlike post-combustion situation which requires extra energy to create a pressure differential across the membrane.

Another area of research currently in progress is the development of sorbents and membranes that can be used within a WGS (water-gas shift) reactor to allow both the shift reaction and CO_2 capture to occur simultaneously (Van Selow et al. 2009). That is, in a sorbent-enhanced water-gas shift, the WGS catalyst is mixed with a CO_2 capture sorbent in a single reactor vessel so that sorbent removes CO_2 as soon as it is formed, allowing increased conversion of CO to CO_2. By this approach, CO_2 capture is achieved along with an efficient WGS reaction, and can lower the overall capital cost of the system (van Dijk et al. 2009). In addition to WGS reactor, studies are being made in a variety of other IGCC plant components that also affect CO_2 capture costs, such as the air separation unit, gasifier, and gas turbines are in progress. Novel concept development – advanced plant designs employing new plant integration, and advanced technologies such as solid oxide fuel cells – are other areas of active research (Chen & Rubin 2009, Klara & Plunkett 2010).

(C) Oxy-fuel combustion CO_2 Capture: There is substantial industrial experience with oxygen and oxygen-enriched combustion systems, but none of them separate or capture CO_2 from the resulting flue gas streams; hence, no direct industrial experience with oxy-combustion for large-scale CO_2 capture is available. So also, there are *no full-scale demonstration* plants. Commercial-scale demonstrations are needed to better quantify opportunities and challenges associated with oxy-combustion. However, the promise behind oxy-combustion for the United States Gulf Coast region was recognized by the Global CCS Institute when it reported that 'oxy-combustion combustion has the lowest breakpoint' when compared to post-combustion, pre-combustion, and NGCC with CO_2 capture (NCC 2012). A few plants that are recently planned in the USA and Europe projected to start in 2015, are listed in Table 9.9.

In DOE's FutureGen 2.0 program, a 170 MW commercial-scale oxycombustion power plant has completed Phase 1 and has moved into Phase 2 (FEED) in July 2012. These plants would be using energy and cost-intensive conventional ASU (cryogenic process) for oxygen supply. Integration of ASU with a large size coal-fired boiler with large amount of flue gas recirculation, essential to control furnace temperature under different operating conditions, would be a major issue in these plants. If these projects are constructed and operated successfully, the respective technologies will be considered to be at a TRL of 8 (GCCSI 2012b). There are, however, a few pilot-scale projects under operation, mostly in Europe. In September 2008, Vattenfall began operating a 30 MW_{th} oxy-combustion pilot plant at its Schwarze Pumpe site (along side their lignite-fired 2×800 MW plant) in Germany. This pilot-scale experiment with an

Table 9.9 Large-scale planned Demonstration plants of oxy-combustion capture of CO_2.

Project name & Location	Plant & Fuel type	Startup year	Capacity (MW)	Annual CO_2 captured (10^6 tonnes)
Projects in the USA:				
FutureGen 2.0 (Meredosia, Illinois)	Coal-fired	2015	200	1.3
Projects outside the USA:				
(a) Boundary Dam (Estevan, Canada)	Coal-fired	2015	100	1.0
(b) Datang Daqing (Heilongjiang, China)	Coal-fired	2015	350	~1.0
(c) OXYCFB3000 (Cubillosdel Sil, Spain)	Coal-fired	2015	300	N/A
(d) OxyCCS Demonstration (North Yorkshire, UK)	Coal-fired	2016	426 MW$_g$	~2.0

(*Source*: Rubin et al. 2012; reproduced with the permission of Prof. Rubin)

investment of US$96 million captured 75,000 tonnes of CO_2 a year and transported 400km by road tanker and injected into the depleted Altmark gas field. As of June 2010, the plant has been in operation for over 6500 hours since its start in mid-2008 (MIT, updated March 2013). According to Vattenfall, for full-scale operations, the cost of oxy-combustion capture will be 40 euros/tonne CO_2 or less (Herzog 2009). The oxyfuel pilot plant operated by Total in France is of comparable size to the Vattenfall unit but operates on a gas-fired boiler. The most recent project in Spain, 20 MW$_{th}$ with coal-fired boiler, began operation in 2011 (Lupin 2011). Hamilton, Ontario is developing a 24 MW$_e$ oxy-fuel electricity generation project (Bobcock & Wilcox 2006). Other projects are: 40 MW$_{th}$ capacity unit operating on coal-fired boiler in Renfrew, Scotland, started in 2009; and 30 MW$_e$ project operating on coal-fired boiler (demonstration-scale retrofit power plant) in Biloela,Central Queensland, Australia, just commissioned in December 2012. In terms of development, a 170 MWe coal-fired boiler is being retrofit to operate with high flame temperature oxy-combustion and the DOE NETL's Integrated Pollutant Removal (IPR^{TM}) system for CO_2 capture. The CO_2 from this project will be used for EOR (NCC 2012). Pressurized oxy-combustion is still under development as a system, but many of the key unit operations and major components have been tested at gasification plants at large-scale (Crew 2011; Weiss 2011).

Laboratory-level research and development in oxy-fuel combustion with CO_2 capture are required to improve efficiency and reduce costs. R&D activity initiated in Canada, Australia, Europe and the USA and more recently China and South Korea include: (i) studies of fundamental mechanisms that affect the performance and design of oxygen-fired boiler systems, such as oxy-combustion flame characteristics, burner design and fuel injection systems; (ii) development of advanced boiler materials to withstand high temperatures associated with oxygen combustion; (iii) small-scale experiments coupled with computational fluid dynamic (CFD) modeling of oxy-combustion processes in a number of areas (USDOE/NETL 2011); (iv) development of advanced flue gas purification systems to find lower-cost methods to remove contaminants such as SO_x, NO_x and trace elements and mercury, and (v) to remove such pollutants during the CO_2 compression process (White et al. 2010). The details of the R&D activities are available in literature (e.g., Shaddix 2012; DOE/NETL 2011; Allam 2010 & IEAGHG 2009a).

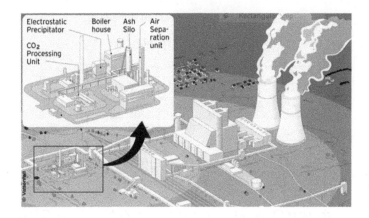

Schwarze Pumpe 30 MW$_{th}$ oxy-fuel combustion Thermal Pilot Plant (*Source*: Vattenfall website, www.vattenfall.com/en/ccs/pilot-plant.htm).

Callide Oxy-fuel Project entered demonstration phase in December 2012 (*Source*: CSEnergy website; at www.csenergy.com.au/content-(43)-callide-oxyfuel-project.htm).

The maximum focus, however, is on the development of improved, low-cost processes to obtain large quantities of oxygen, the major cost item in the current oxyfuel system, as in other technologies. The two main approaches under study are the 'Ion transport membrane' (ITM) and 'oxygen transport membrane' (OTM) that promise lower energy penalty and cost. Details of these studies can be found in literature (Amstrong *et al.* 2004; DOE/NETL 2011; Hashim *et al.* 2010). Another area of investigation is using solid sorbents such as perovskite and manganese oxide to absorb oxygen from air (DOE/NETL 2011).

Other developments, such as advanced plant designs that incorporate new plant integration concepts and novel oxygen combustion methods, such as chemical looping combustion (CLC), show the potential. In this process, shown in Figure 9.14, oxygen

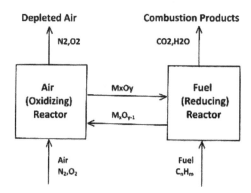

Figure 9.14 Schematic of Chemical Looping Combustion (Redrawn from DOE/NETL 2011).

carrying-sorbent, a metal oxide, is used to capture oxygen from air in one reactor (with a nominal flue gas of depleted air) and then to subsequently donate that captured oxygen to fuel molecules in a second reactor. If the fuel reactor is operated lean, the resulting exhaust contains only CO_2 and water vapor, as in other oxy-combustion schemes, allowing ready CO_2 capture.

A variety of metal oxide carrier systems are possible, though most of R&D has focused on iron, nickel, and copper oxides. Chemical looping has the potential to make carbon capture appreciably cheaper because CLC can decrease the costs and efficiency penalty associated with conventional air-separation technology for oxygen production. The development is at an early stage, and several key challenges in the areas of heat and mass transfer, ash separation, materials handling and oxygen carrier selection, coke formation, corrosion etc make CLC a rich area of research now (DOE/NETL 2012). A chemical looping combustor of 120-kW capacity, the largest unit presently, is being tested in Austria (Kolbitsch *et al.* 2009). The US DOE is funding projects that include two chemical looping studies, one by Alstom using calcium compounds as an oxygen carrier, and the other by a group at Ohio State University using an iron oxide carrier (Andrus *et al.* 2009). Alstom has developed prototypes, 1 MW$_{th}$ and 3 MW$_{th}$, and is currently testing them in Germany and the USA, respectively (IEAGHG 2011a). It is many years away for any of these novel technologies to reach feasible commercial stage.

9.7   ADVANCED CAPTURE SYSTEMS – COSTS AND DEPLOYMENT PROJECTIONS

It has already been seen that for a technology that has not yet been installed, operated and reproduced at a commercial scale, it would be difficult to project costs due to innate uncertainty in estimates. A technology tends to look cheaper if it is away from the commercial status. Dalton (2008) has illustrated the typical trend in cost estimates for a technology as it advances from the concept level to the stage of commercial deployment. The 'Capital cost per unit of capacity' of a technology begins to rise from concept stage, peaks at the 'demonstration stage', and then start falling in the 'deployment stage' levelling off as it gets matured for commercial status. Several studies (Klara 2006; DOE/NETL 2008a; Rubin *et al.* 2007; Metz *et al.* 2005) have estimated potential future cost reductions from technology innovations both in CO_2 capture processes as well as in other power plant components that influence CO_2 capture cost using two different methods of estimation: (a) the 'bottom up' method that uses engineering analysis and costing to estimate the total cost of a specified advanced power plant design; and (ii) the 'top down' method using learning curves derived from past experience with analogous technologies to estimate the future cost of a new technology based on its projected installed capacity at a future point of time. The latter represents the effect of all factors that influence observed cost reductions including R&D costs, learning-by-doing and learning-by-using.

Cost estimates using 'bottom up' method: The US DOE 2006 analysis has dealt with potential advances in the major CO_2 capture options (Klara 2006) for PC plants such as: SC with ammonia CO_2 scrubbing, SC with econamine scrubbing, SC with Multipollutant ammonia scrubbing, USC with amine scrubbing, USC with advanced amine scrubbing, and RTI regenerable sorbent; and for IGCC plants, the advances such as Selexol, advanced Selexol, Advanced Selexol with co-sequestration, Advanced Selexol with ITM and co-sequestration, WGS membrane and co-sequestration, WGS membrane with ITM and co-sequestration, Chemical looping and co-sequestration. As these advanced technologies are implemented the incremental cost is reduced significantly. On an absolute basis, the total cost of electricity generation falls by 19% for the IGCC systems and by 28% for the PC systems, with largest cost reductions coming in the final stages for each plant type. However, since the technologies including advanced solid sorbents for CO_2 capture, membrane systems for WGS reactors and chemical looping for oxygen transport are still *in the early stages of progress*, the cost estimates for these cases are not definite and are most likely to increase as the technology grows to commercial status.

Recent 2010 DOE analysis of capture cost reductions from continued R&D has revealed that the total cost of a new SC PC plant with CCS declines by 27% while the IGCC plant cost falls by 31%, Figure 9.16 (DOE/NETL 2008a). Consequently the future IGCC plant with CCS may cost 7% less than the current IGCC plant w/o capture. For the PC plant, the CCS cost penalty falls by about half in this analysis. Here again, since many of the components assumed in this DOE analysis are still at the early stages of development, cost estimates are highly uncertain. Nonetheless, these estimates provide an approximate indication of the potential cost savings that might be realized. Several other estimates of similar cost reductions for advanced plant designs with CCS are found elsewhere (e.g., Metz *et al.* 2005).

Costs estimates using 'top down' method: This approach models the future cost of power plants with CCS as a function of the total installed capacity of similar type of plants. The future cost reductions are deduced from an analysis that considered historical learning rates for selected technologies to the components of PC, NGCC, IGCC and oxyfuel-power plants with CO_2 capture, shown in Figure 9.16.

The component costs were then summed to estimate the future cost of the overall power plant as a function of new plant capacity. The analysis also considered uncertainties in key parameters that include potential increases in cost during early stages of commercialization. Deployment of 100 GW worldwide for each of the four types

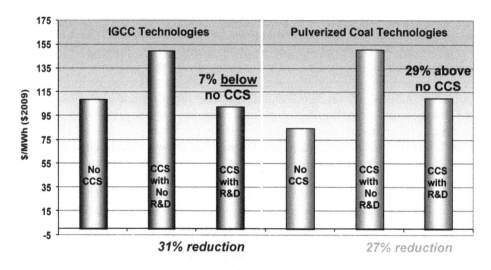

Figure 9.15 Current COE for IGCC and PC power plants with and w/o CCS, plus future costs with advanced technologies from R&D (*Source:* DOE/NETL-2007/1291, 2008a).

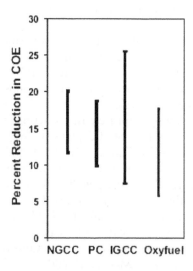

Figure 9.16 Projected cost reductions for four types of power plants with CO_2 capture based on experience curves for major plant components (*Source:* Rubin *et al.* 2007, 2012, reproduced with permission of Prof Rubin, Copyright © 2012, Elsevier).

of power plants is assumed for estimating cost reductions shown in the figure. This is roughly equal to the worldwide capacity of FGD systems two decades after that technology was first introduced to power plants in the USA. The results reveal that each plant type, as well as the CO_2 capture system, has reached a different level of development; and the IGCC plant whose principal cost components are less mature than those of combustion plant, shows the largest potential for overall cost reductions (slightly >25%). The combustion-based plants show a smaller reduction potential since most of their components are well developed and extensively deployed. In all cases, however, the incremental cost of CO_2 capture system falls more rapidly than the cost of the overall plant. The maximum of the COE reduction ranges is similar to DOE's 'bottom up' estimates while the minimum of the ranges, however, is smaller by a factor of 2 to 3. This analysis further indicates a steady rate of cost reductions from continuous advances to capture technologies as CCS is more widely deployed. The reader may also refer to DOE/NETL-2011/1498 publication, 'Cost and performance of PC and IGCC Plants for a range of CO_2 capture', May 27, 2011.

9.8 TECHNOLOGY ROAD MAP FOR CCS

CCS is now established as an important lowest-cost option for the mitigation of greenhouse gases. Hence, many organizations such as USDOE (2010), EPRI, IEA (2009, 2010), Carbon Sequestration Leadership Forum (2009), and Natural Resources Canada (2006) have released 'roadmaps' for the development of efficient, economically viable commercial CCS systems. The IEA and USDOE roadmaps are briefly outlined here.

(i) *The International Energy Agency (2009):* IEA analysis suggests that without CCS, overall costs to reduce emissions to 2005 levels by 2050 increase by 70%. The IEA Roadmap includes an ambitious CCS growth path in order to achieve this GHG mitigation potential, expecting 100 projects globally by 2020 and over 3,000 projects by 2050. To achieve this growth, the following actions are suggested: (a) Development of the magnitude mentioned requires an extra investment of over US$ 2.5–3 trillion from 2010 to 2050, which is about 6% of the overall investment needed to achieve a 50% reduction in GHG emissions by 2050. OECD countries will have to increase funding for CCS demonstration projects to an average of US$ 3.5 to 4 billion/year from 2010 to 2020. Also, systems need to be evolved for taking the technology to commercial stage beyond 2020 in the form of mandates, tax rebates, GHG reduction incentives or other financing mechanisms; (b) The developed countries must lead the CCS effort so that the technology penetrates rapidly into the developing world. This effort will require extended international collaboration and financing for CCS demonstration in developing countries at an average of US$ 1.5 to 2.5 billion per year from 2010 to 2020. To provide this funding, CCS needs to be approved in the Clean Development Mechanism or an alternative financing mechanism as suggested in IPCC Report 2007; recently, CCS was brought under CDM; (c) Since CCS is more than a strategy for 'clean coal', CCS technology must also be adopted by biomass and gas-based power plants, the fuel transformation and gas processing sectors, and emissions-intensive industries like cement, iron and steel, chemicals, and pulp and paper; (d) Costs of CC technology are still high and need to be lowered; more R&D is needed to take it to commercial level, and to address different CO_2 streams from industrial sources and to test biomass and hydrogen production with CCS; (e) CO_2 transport through pipelines has been proven; yet for the future of transportation, long-term strategies for CO_2 source clusters and pipeline networks that optimize source-to-sink transport of CO_2 have to be developed. To deal with this issue, countries have to initiate by region planning and develop incentives for the creation of CO_2 transport hubs; (f) The need to advance the state of global knowledge of CO_2 storage potential is very urgent. The depleted oil and gas fields which offer potential low-cost sites are well mapped. In the case of deep saline formations which are the most viable option for the long-term, only a few regions have adequately mapped their storage potential. It is essential to develop common global methods for CO_2 storage site selection, monitoring and verification, and risk assessment; (g) Considerable progress in developing legal and regulatory frameworks for CCS has been achieved in some regions; most countries still have to address this issue. Near-term regulatory approaches for CCS demonstration are urgently needed; concurrently development of comprehensive approaches for the large-scale commercial deploymentof CCS may have to be undertaken; (h) The genuine concerns of the public about CCS must be addressed. Countries need to develop community-oriented CCS public engagement strategies by providing resources for this activity, and then inform the costs and benefits of planned CCS projects compared to other GHG mitigation options. The investments required are very large; hence, this roadmap's vision will only be possible through extended international collaboration. In particular, providing CCS technology transfer and capacity building to developing countries are very vital and must be ensured. Industrial sectors with a global reach should also expand their CCS collaborative efforts (IEA 2009).

(ii) *USDOE Roadmap:* As part of its Carbon Sequestration Program, the DOE has developed a roadmap displaying the projected timetable for major program elements including CO_2 capture technology development which regularly gets updated (Ciferno 2009). The most recent DOE roadmap was released in 2010 (USDOE 2010a) with the timeline for R&D to commercial deployment of advanced post-combustion capture technologies for existing power plants (Ciferno 2009):

2008–2016: Laboratory & Bench-scale R&D
2010–2020: Pilot-scale & Field testing, 0.5–5.0 MW$_e$
2014–2024: Large-scale Field testing, 5–25 MW$_e$
2016–2030: Large Demonstrations (CCPI), 100 MW$_e$
2020–2030: Commercial Deployment

The timeline in the 2010 DOE roadmap is extended to 2030, roughly a decade longer than the projection in previous (2007) roadmap. DOE's Carbon Sequestration Program released in Februay 2011 (DOE/NETL 2011) comprised of three principal elements: Core R&D, Infrastructure, and Global Collaborations. The technology needs determined by industry and others determine *Core R&D*, and divide the challenges into various focus areas. The *Infrastructure* includes the RCSPs and large-scale field tests where validation of different CCS technology options and their efficacy are established. Under *Global Collaborations*, 'The Carbon Sequestration Program' participates in large-volume tests. The *Core R&D* focuses on developing new CCS technologies to the point of pre-commercial demonstration covering five technical focus areas: (a) Pre-Combustion capture; (b) Geologic storage; (c) Monitoring, Verification, and Accounting (MVA); (d) Simulation and Risk Assessment; and (e) CO_2 utilization – each with specific research goals applicable to each focus area (DOE/NETL 2011). Within each focus area, research plans were determined after identifying specific challenges or uncertainties. The level of research program ranges from laboratory- to pilot-scale activities. Technologies are normally developed in the Core R&D element to a stage where individual companies, utilities, and other business entities are able to design, manufacture, and build the equipment and instrumentation needed to implement or commercialize the processes. The implementation is done through cost-shared cooperative agreements and grants with industry and academic institutions, field work research at other national laboratory complexes, and research at NETL's ORD. The time lines are illustrated for 'Pre-combustion capture': 2015: Development of a comprehensive portfolio of bench-scale technologies which, if combined with other system advances, will enable new power production technology with CO_2 capture (e.g., IGCC) to produce electricity at a cost of no more than 10% above the reference power plant without CO_2 capture; 2015: Start testing pre-combustion promising technologies at 0.1-MW scale; 2018: To initiate the development of second generation pre-combustion technologies which, if combined, will enable production with IGCC facilities at near-zero additional costs. In summary, current roadmaps visualize that improved lowercost capture systems will be generally available for use at power plants and other industrial facilities by 2020. At the same time, both public and private-sector research organizations acknowledge that a sustained R&D effort will be required over the next decade to accomplish that goal especially for many of the promising new processes still in the early stages of development. Further, the extent of future cost reductions is likely to depend on the pace of CCS technology deployment as well as on continued R&D support.

9.9 RETROFITS FOR CO_2 CAPTURE

There are a large numbers of PC combustion units operating for long, and more such units are under construction and also planned for the next few decades. Retrofitting these units for CO_2 capture is vital for managing CO_2 emissions. Retrofit means the addition of a process unit to the back end of the flue-gas system to separate and capture CO_2 from the flue gas, and to dry and compress the CO_2 to a supercritical fluid.

Adding CO_2 capture technology to an existing PC unit is complicated by the choice of options that exist and the number of issues – physical, technical and economic – associated with each. These issues cannot be generalized and are determined by the specifics of each unit. The physical issues include space constraints associated with the unit, and its proximity to a CO_2 sequestration site. The technical issues include technology choice, technology maturity, operability and reliability, impact on efficiency, and retrofit complexity. The economic issues are the investment required (total and $/kW_e$), net output reduction, and change in dispatch order (MIT study 2007). A complete discussion on technical, financial, permitting, legal, and public engagement issues in depth are covered in the 2011 NCC Report (NCC 2011).

For an MEA retrofit of an existing subcritical PC unit, the net electrical output can be derated by over 40%, e.g., from 500 MWe to 294 MWe (Bozuto *et al.* 2001); the efficiency decrease is about 14.5% compared to about 9.2% for purpose-built subcritical PC unit, with no-capture and capture (Table 8.3). With the retrofit, the steam required to regenerate the absorbing solution to recover the CO_2 disturbs the rest of the plant so much that the efficiency is reduced by another 4 to 5%. In the retrofit case, the original boiler is running at full design capacity, but the original steam turbine is operating at about 60% design rating; hence efficiency is far less from optimum. Due to the large power output reduction (41% derating), the retrofit capital cost is estimated to be $1600 per kW_e (Bozuto *et al.* 2001). This was for a specific unit with adequate space, and the retrofit costs vary with location. If the original unit is taken as fully served, the COE after retrofit could be slightly less than that for a newly built PC unit with CO_2 capture. The factors – the residual value for an existing plant, the reduction in unit efficiency and output, increased on-site space requirements and unit downtime – are so complex, they are not fully accounted for in the analysis presented in the MIT report (2007).

Another approach is to 'rebuild' the core of a subcritical PC unit, installing SC or USC technology along with post-combustion CO_2 capture. Although the total capital cost for this approach is higher, the cost/kW_e is about the same as for a subcritical retrofit. The merits are that the resultant plant efficiency is higher, consistent with that of a purpose-built unit with capture, the net power output can effectively be maintained, and the COE is almost the same due to the overall higher efficiency. An ultra-supercritical rebuild with MEA capture is estimated to have an efficiency of 34% and produce electricity for 6.91 ¢/kW_e-h. Hence, the rebuilds including CO_2 capture 'appear more attractive' than retrofits, particularly if low-efficiency PC units are upgraded with high-efficiency technology, including CO_2 capture.

Pilot scale studies indicate that the oxy-fuel method can be retrofitted to existing PC-fired plants. Oxy-fuel is a good option for retrofitting PC and FBC units for capture since the boiler and steam cycle are less affected by an oxy-fuel retrofit. The main issue with retrofitting a subcritical PC plant is the base plant efficiency because about 1/3rd of the plant's gross electricity output is consumed, of which about 20% by ASU, and about 12 to 14% by the CO_2 purification and compression-liquefaction processes. Bozzuto (2001) estimated a 36% derating for an oxy-fuel retrofit against a 41% derating for MEA capture on the same unit. The net plant efficiency with CO_2 capture varies between 23 and 26% (LHV) (Anderson *et al.* 2002). For retrofitting a higher efficiency supercritical PC plant, the energy needs of CO_2 capture is much lower as the energy output of the plant is reduced by about 20% only. The conditions are even more favorable for new SC PC plant with a net efficiency of about 34% (LHV) (Dillon *et al.* 2004).

Studies by B&W and Air Liquide for the DOE (DE-FC26-06NT42747) showed that oxy-fuel is an economically viable retrofit technology for existing boilers. The incremental cost of oxy-fuel for existing boilers varies between 5–7 ¢ a kw-hr, (shown as $29/metric ton and $43/metric ton for a new SC and subcritical retrofit, respectively) which is competitive with other technologies (the incremental cost assumes the plant is fully depreciated; no capital remaining and O&M costs for the pre-retrofit equipment are not included). The efficiency loss for oxy-retrofitting PC ranges between 8–9%; this is considered consistent whether the retrofit is applied to a subcritical or SC base plant (NCC 2012).

Potential of CCS retrofit

Global energy scenarios suggest that CCS retrofit is a necessity for deep CO_2 emission cuts, in addition to constructing new power plants with CCS. Given the complexity and the huge number of units currently in operation worldwide, available information from energy scenarios or global power plant databases are insufficient to provide an in-depth analysis. In reality only a fraction of coal-fired power plants could be technically and economically retrofitted. If other factors are considered, it is likely that the actual potential for CCS retrofitting will be further reduced.

There are several analyses in literature on CCS retrofit. According to IPCC, retrofitting existing plants with CO_2 capture is expected to lead to higher costs and significantly reduced overall efficiencies than for newly built power plants with capture. The cost disadvantages of retrofitting may be reduced in a few relatively new and highly efficient existing plants or where a plant is substantially upgraded or rebuilt (IPCC, 2005). IPCC also lists several other issues with CCS retrofits. In order to minimise site-specific constraints, it has been proposed to build 'CCS ready' plants to reduce these and other disadvantages. This approach is an important option for avoiding 'lock-in' CO_2 emissions from power plants that will be built in the near future. But in the current global coal-fired power fleet, the plants that have already been designed as 'CCS-ready' are very low in number.

Relatively large (300 MW$_e$ or greater), high efficiency coal plants with installed FGD and SCR capability are the best suited for CCS retrofit; and the old, lower efficiency, smaller, subcritical units are not. Rebuilding or repowering is another option depending on significant CO_2 prices being in place (MIT 2009). According to EPRI, cost-effective retrofits for carbon capture are most suitable for boilers that are 300 MW or larger and less than about 35 years old.

IEA's detailed study points out that CCS retrofits to plants with lower efficiencies will tend to have higher generation costs and so are generally less likely to be competitive with new build CCS replacements. For the specific set of parameter assumptions chosen in this study, below a threshold efficiency of about 35% (LHV), retrofit on coal plants would become unattractive (about 33%, HHV), and the Costs of CCS per tonne of CO_2 captured are somewhat higher for retrofitted plants. The study identifies a range of conditions under which costs of electricity may look more favourable for power plants with retrofitted CCS, compared to new-build power plants with CCS (IEAGHG 2011).

USDOE/NETL performed a study on the potential of CCS retrofit to the coal-fired plants under operation (DOE/NETL 2011b). Certain criteria has been used for screening and singling-out power plants that are not amenable to CO_2 capture retrofit: units currently operating; units that have a capacity less than 100 MW; units that have a 2008 reported heat rate greater than 12500Btu/kWh (HHV) or about 29% efficiency (LHV); and units that do not have a defined CO_2 repository within 25 miles (40 kilometres). Bohm et al. (2007) also discusses in detail, the issues specific to retrofitting PC power plants and IGCC plants for CO_2 capture.

Recent Finkenrath, Smith and Volk (OECD/IEA report 2012) discuss key aspects essential for retrofitting existing plants. Using information from the WEPP database and IEA internal statistics, the study discusses the size, regional distribution and characteristics of the globally installed coal-fired fleet of power generation. The profile of the existing fleet is investigated using three case analyses to arrive at the probable number of power plants that have age, size or performance features suitable for CCS retrofitting: (a) In the first case, those plants younger than 30 years which have a power generation capacity above 100 MW; more than 60% of the globally installed fleet, or about 1000 GW, meet these criteria and nearly 60% of these plants are located in China; (b) In the second case, the coal power plants that are younger than 20 years and have a power generation capacity above 300 MW; this applies to 665 GW generation capacity or about 40% of the globally installed fleet, and out of this, 481 GW or 72% are located in China; (c) In the third case, plant of ten years old with 300 MW capacity threshold; this results in 471 GW of existing generation capacity, or about 29% of the globally installed fleet, of which, 390 GW or 83% are located in China. This data presents the magnitude of a specifically suitable fraction of the currently operating plants, but not necessarily the realistic potential for CCS retrofitting. The relevance of China to this study is clear. More detailed regional studies on the retrofit potential in China reveal that about one-fifth of the sites above 1 GW installed capacity appear to be very attractive for CCS retrofitting (Li 2010). From the technology-driven assessment only Asia, USA and Europe will play a key role in the context of retrofit considerations over the coming decades. Since CCS retrofitting technologies are not currently available on a commercial basis, the age of the current installed and assessed coal plant fleet is only a first starting point; with each passing year, the existing plant fleet grows older. While the retrofit potential of the existing fleet shrinks, the interest on future plants and their potential for CCS retrofitting increases. The governments need to motivate the construction of new plants in a way that would allow for economic retrofit of CCS at a later stage, if CCS is not added at the time of original construction. This is particularly vital in order to avoid a future lock-in of high amounts of CO_2 emissions over the long operating lifetimes of coal power plants, which would otherwise come with the expected global capacity additions.

9.10 CAPTURE-READY UNITS

A unit can be considered capture-ready if, at some point in the future, it can be retrofitted for CO_2 capture and sequestration and still be economical to operate. Thus, capture-ready design refers to designing a new unit to reduce the cost of and to facilitate adding CO_2 capture later or at least to not preclude addition of capture later. Capture-ready is not a specific design, but involves several investment and design decisions to be taken during unit design and construction. Even if significant pre-investment for CO_2 capture is typically not economically justified at present (Sekar et al. 2005), some actions make sense. New coal-fired power plants can be designed and built in such a way that the efficiency of the plant can be greater and the CO_2 capture process can be integrated into the plant steam cycle with economic viability. Location of the plants should also consider proximity to geologic storage.

REFERENCES

Abanades, J. (2007): Cost structure of a post-combustion CO2 capture system using CaO, *Environ Sci Technol*, 41(15), 5523–7.

ARPA, Advanced Research Projects Agency, Energy (2010): ARPA-E's 37 projects selected from funding opportunity announcement #1, http://arpa-e.energy.gov/LinkClickaspx?fileticket-b-7jzmW97W0%3d&tabid¼490; April 3, 2010.

ARPA, Advanced Research Projects Agency, Energy (2010a): Energy efficient capture of CO_2 from coal flue gas, http://arpa-e.energy.gov/LinkClickaspx?fileticket¼CoktKdXJd6U%3d&tabid¼490; April 3, 2010.

Allam, R. (2006): Oxyfuel Pathways, in *MIT Carbon Sequestration Forum VII: Pathways to Lower Capture Costs*, MIT, Cambridge, MA.

Allam, R. (2010): CO_2 capture using oxy-fuel systems, *Proc. carbon capture – beyond 2020*, March 4–5, 2010, Gaithersburg, MD: USDOE, Office of Basic Energy Research, 2010.

Amstrong, P.A., Bennett, D.L., Foster, E.P., & Stein, V.S. (2004): ITM oxygen for gasification, *Proc. Of Gasification technologies Conf*, 3–6 Oct. 2004, Washington DC; San fransisco: Gasification Technologies Council, 2004.

Andrus, H.E., Chiu, j.H., Thibeault, P.R., & Brautsh, A. (2009): Alston's CaO chemical looping combustion coal power technology development, *Proc. 34th Intl. Tech. Conf. on Clean coal & Fuel systems*, May 31–June 4, 2009, Clearwater, FL; Pittsburgh:USDOE, 2009.

Andersson, K., Birkestad, H., Maksinen, P., Johnsson, F., Stromberg, L. & Lyngfelt, A., (2003): An 865 MW Lignite Fired CO2 Free Power Plant – A Technical Feasibility Study, *Greenhouse Gas Control Technologies*, (ed.) K. Gale, New York: Elsevier Science Ltd.

Babcox &Wilcox (2006): Demonstration of the Oxy-Combustion Process for CO2 Control at the City of Hamilton, *American Coal Council 2006 Mercury & Multi-Emissions Conf*, 2006.

Battelle Report (2006): Battelle Joint Global Change Research Institute, Carbon Dioxide Capture and Geologic Storage: A Core Element of a Global Energy Technology Strategy to Address Climate Change, 26–27, April 2006.

Beauregard-Tellier, F. (2006): *The Economics of Carbon Capture and Storage*, PRB-05-103E, Parliamentary Information and Research Service, Library of Parliament, Ottawa, 13 March 2006.

Bennion, D.B. & Bachu, S. (2008): Drainage and Imbibition Relative Permeability Relationships for Supercritical CO2/Brine and H2S/Brine Systems in Intergranular Sandstone, Carbonate, Shale, and Anhydrite Rocks, *SPE Reservoir Evaluation & Engineering*, 11, 487–496, doi:10.2118 /99326-PA.

Bensen, S.M.: CCS in underground geologic formations, Workshop at Pew center, available at http://www.c2es.org/docUploads/10-50_Benson.pdf.

Berlin, K. & Sussman, R.M. (2007): Global Warming & The Future of Coal – The Path to CCS, *Centre for American Progress*.

Bhown, A.S., & Freeman, B. (eds) (2009): Assessment of post-combustion CO_2 capture technologies, *Proc 8th annual CCS conference*, May 4–7, 2009, Pittsburgh, PA, Washington: Exchange Monitor Publications, 2009.

Bhown, A.S., & Freeman, B. (2011): Analysis and status of post-combustion CO_2 capture technologies, *Environ. Sci.Technol*, 45(20), 8624–8632.

Bhown, A.S. (2012): CO_2 capture R&D at EPRI, NETL Carbon dioxide Capture Technology Meeting, Pittsburgh, PA, July 9–13, 2013.

Blamey, J., Anthony, E.J., Wang, J., & Fennell, P.S. (2010): The calcium looping cycle for largescale CO_2 capture, *Prog Energy Combust Sci*, 36, 260–79.

Blunt, Martin (2010): *Carbon dioxide Storage, Briefing paper No. 4*, Grantham Institute for Climate Change, Imperial College, London, UK, December 2010.

Bohm, M.C., Herzog, H.J., Parsons, J.E., & Sekar, R.C. (2006): Capture-Ready Coal Plants – Options, Technologies and Economics, *Proc. of 8th Intl. Conf. on Greenhouse Gas Control Technologies (GHGT-8)*, Trondheim, Norway, June 2006; *International J of Greenhouse gas Control*, (2007), 1, 113–120.

Brendan Beck (2009): IEA CCS Roadmap, US RCSP – Pittsburgh, 18th Nov. 2009. Carbon Sequestration Leadership Forum (2009): Carbon sequestration leadership forum technology roadmap: a global response to the challenge of climate change, http://www.cslforum.org/publications/documents/CSLF_Techology_Roadmap.pdf; Nov 30, 2009.

Buhre, B.J.P., Elliot, L.K., Sheng, C.D., Gupta, Wall, T.F. (2005): Oxy-fuel combustion technology for coal-fired power generation, *Progress in Energy and Combustion science*, 31, 283–307.

Carnegie Mellon University (2009): *Integrated environmental control model (IECM) technical documentation, IECM v.6.2.4*, Pittsburgh, PA: Dept of Engineering and Public Policy, available at: http://www.iecm-online.com.

Chapel, D.G., Mariz, C.D., & Ernest, J. (1999): Recovery of CO2 from flue gases: Commercial trends, *Proc Canadian Society of Chemical Engineers Annual Meeting*, October 4–6, 1999, Saskatoon, Saskatchewan, Canada.

Chen, C., & Rubin, E.S. (2009): CO_2 control technology effects on IGCC plant performance and cost, *Energy Policy*, 37(3),915–24.

Ciferno, J.P., DiPietro, P., & Tarka, T. (2005): Economic scoping studies for CO2 capture using aqueous ammonia; Pittsburgh: US DOE/NETL.

Ciferno, J.P. (2009): DOE/NETL's existing plants CO_2 capture R&D program, *Proc. Carbon capture 2020 workshop*, Oct. 5–6, 2009, College Park, MD, DOE/National Energy Technology Center, 2009.

Clean Air Task Force (2010): Coal without carbon: an investment plan for federal action. Boston: Clean Air Task Force; 2010.

Clean Air Task Force & Consortium for Science: Policy and outcomes, innovation policy for climate change, Proc National Commission on Energy Policy, Washington, DC.

Coal Utilization Research Council (CURC) (2009): Clean coal technology roadmap, Washington: Coal Utilization Research Council; 2009.

Crew, J. (2011): GE Gasification project update, *Gasification Technologies Conf.*, October 10, 2011.

Cullinane, J.T., & Rochelle, G.T. (2004): Carbon dioxide absorption with aqueous potassium carbonate promoted by piperazine, *Chem Eng Sci*, 59, 3619–30.

Dalton, S. (2008): CO_2 capture at coal-fired power plants – status & outlook, *Proc. 9th Intl. Conf. on GHG control technologies*, Nov. 16–20, 2008, Washington, DC; London: Elsvier, 2008.

David, J., & Herzog, H.J. (2000): The Cost of Carbon Capture, *Fifth Intl. Conf. on Greenhouse Gas Control Technologies*, August 13–16, at Cairns, Australia.

Dillon, D.J., Panesar, R.S., Wall, R.A., Allam, R.J., White, V., Gibbins, J., & Haines, M.R. (2004): Oxy-combustion Processes for CO2 Capture from Advanced Supercritical PF and NGCC Power Plant, in *Greenhouse Gas Technologies Conference 7*, Vancouver, BC.

Dooley, J., Davidson, C., Dahowski, R., Wise, M., Gupta, N., Kim, S., & Malone, E. (2006): *Carbon Dioxide Capture and Geologic Storage: A key Component of a Global Energy Technology Startegy to Address Climate Change*, Joint Global Change Research Institute, Pacific Northwest National Laboratory, College Park, MD, p. 67.

Eisaman, M.D. (2010): CO_2 concentration using bipolar membrane electrodialysis, *Proc. Gordon Research Conference on Electrochemistry*, January 10–15, 2010, Ventura, CA.

EPRI (Electric Power Research Institute) (2009): Post-combustion CO_2 capture technology development, report no 10117644, technical update; Palo Alto, CA.

EPRI (2011): i. Advanced coal power systems with CO_2 capture: EPRI CoalFleet for tomorrow Vision – A summary of technology status, research, development, and demonstrations, 2011; ii. Progress on technology innovation: Integrated generation technology options, 1022782, technical update, June 2011.

European Technology Platform for Zero Emission FF Power Plants (2010): EU demonstration programme for CO_2 capture and storage, March 5, 2010, at http://www.zeroemissions platform.eu/index; 2008.

Favre, E., Bounaceur, R., & Roizard, D. (2009): Biogas, membranes and carbon dioxide capture, *J Membrane Sci*; 328(1e2), 11–14.

Fernandez, K. (2007): TXU, Buyout Partners Announce Plans for Two Carbon Dioxide Capture Plants, *BNA Daily Environment Report*, Mar. 12, 2007, at A-9.

Figueroa, D., Fout, T., Plasynski, S., McIlvried, H., & Srivastava, R.D. (2008): Advances in CO2 capture technology – the US Department of Energy's carbon sequestration program, *Int J Greenhouse Gas Control*, **2**, 9–20.

Finkenrath, M., Smith, J., & Volk, D. (2012): CCS RETROFIT *Analysis of the Globally Installed Coal-Fired Power Plant Fleet*, IEA, Paris, 2012.

Freeman, S.A., Davis, J., & Rochelle, G.T. (2010): Degradation of aqueous piperazine in carbon dioxide capture, *Int J Greenhouse Gas Control*; 2010.

Gale, J. (2003): Geological Storage of Carbon dioxide: What's known, where are the gaps, and what more needs to be done? *Greenhouse Gas Control Technologies*, **1**, 201–206.

Gale, J. (2004): Using Coal Seams for Carbon dioxide Sequestration, *Geologica Belgica*, **7**, 3–4.

Global CCS Institute (2010): The Global Status of CCS 2010, Canberra, Australia.

Global CCS Institute (2011): Strategic analysis of the global status of carbon capture and storage, Prepared by WorleyParsons, at http://www.globalccsinstitute.com/downloads/Status-of-CCSWorleyParsons-Report-Synthesis.pdf; March 2011.

Global CCS Institute (2012): *The Global Status of CCS*, Canberra, Australia, ISBN 978-0-9871863-1-7.

Global CCS Institute (2012a): *Technology options for CO2 capture*. Canberra, Australia, at http://www.globalccsinstitute.com/publications/technology-options-CO2-capture.

Global CCS Institute (2012b): CO_2 capture technologies: Oxy-fuel combustion capture, January 2012; at http://cdn.globalccsinstitute.com/sites/default/files/publications/29761/CO2-capture-technologies-oxy-combustion.pdf.

Gray, M.L., Champagne, K.J., Fauth, D., Baltrus, J.P., & Pennline, H. (2008): Performance of immobilized tertiary amine solid sorbents for the capture of carbon dioxide, *Int J Greenhouse Gas Control*, **2**(1), 3–8.

Greenhouse Gas R&D Programme IA (2008): Carbon Capture and Storage: Meeting the challenge of climate change, IEA/OECD, Paris.

Group of Eight 2008 (2010): G8 summits Hokkaido official documents-environment and climate, http://www.g7utorontoca/summit/2008hokkaido/2008-climate.html; March 3, 2010.

Gupta, J.C., Fugate, M., & Guha, M.K. (2001): *Engineering Feasibility and Economics of CO2 Capture on an Existing Coal-Fired Power Plant* [Report No PPL-01-CT-09], NETL/USDOE, 2001, Alstom Power Inc.

Guo Yun, Cao Wei-wu, & Huang Zhi-qiang (2010): Advances in Integrated Gasification Combined Cycle system with Carbon Capture and Storage Technology, *Intl. Conf. on Advances in Energy Engineering* 2010.

Habib, M.A., Badr, H.M., Ahmed, S.F., Ben-Monsour, R., Megzhani, K., Imashuku, S. (2011): A review of recent developments in CC utilizing oxy-fuel combustion in conventional and ion transport membrane systems, *International. J. Energy Research*, **35**(9), 741–764.

Hand, E. (2009): The power player, *Nature*; **462**, 978–83.

Harry, J. (2007): IGCC Power, How Far Off, *Gas Turbine World*, **37**, Jan. 2007.

Hashim, S.M., Mohamad, A.R., & Bhatia, S. (2010): Current ststus of ceramic-based membranes for oxygen separation from air, *Adv. Colloid Interface Sci.*, **160**, 88–10.

Hendriks, C. *et al.* (2004): Power and Heat Production: Plant Developments and Grid Losses, Ecofys, Utrecht, Netherlands.

Herzog, H.J., Drake, E., & Adams, E. (1997): CO_2 *Capture, Reuse, and Storage Technologies for Mitigating Global Climate Change*; Cambridge, MA: MIT Energy Laboratory, A White Paper.

Herzog, H.J., & Golomb, D. (2004): Carbon Capture & Storage from Fossil Fuel Use, *Encyclopedia of Energy* at encyclopedia_of_energy_articleCCS-MIT2004.pdf.

Herzog, H.J., & Jacoby, H.D. (2005): *Future Carbon Regulations and Current Investments in Alternative Carbon-Fired Power Plant Designs*, in: MIT Global Climate Change Joint Program Publications. 2005, MIT: Cambridge, MA. p. 19.

Herzog, H.J. (2007): The Future of Coal – Options for a Carbon constrained World, Presentation at *Cambridge-MIT Electricity Policy Conference*, Sept. 27, 2007.

Herzog, H.J. (2009): Carbon dioxide Capture and Storage, Chapter 13 in a Book, pp. 263–283.

Herzog, H.J. (2009a): Capture technologies for Retrofits, *MIT Retrofit Symposium*, Massachussetts Institute of Technology, March 23, 2009.

Herzog, H.J. (2009b): A Research Program for Promising Retrofit Technologies, Prepared for the *MIT Symposium on Retro-fitting of Coal-Fired Power Plants for Carbon Capture*, MIT, March 23, 2009.

Hildebrand, A.N., & Herzog, H.J. (2008): Optimization of Carbon Capture Percentage for Technical and Economic Impact of Near-Term CCS Implementation at Coal-Fired Power Plants, *Energy Procedia*, 2008, GHGT-9.

Holt, N. (2007): *Preliminary Economics of SCPC & IGCC with CO2 Capture and Storage*, in *2nd IGCC and Xtl Conference*, 2007, Freiberg, Germany.

Holt, N. (2007): CO_2 capture & storage, EPRI CoalFleet program, PacificCorp energy IGCC/climate change working group, Palo Alto: Electric Power Res. Inst.

IEA (2008): CO_2 *Capture and Storage: A Key Carbon Abatement Option*, 2008, Paris, France.

IEA (2008b): *Energy Technology Perspectives 2008*, OECD/IEA, Paris.

IEA (2009): *Technology Roadmap – Carbon capture and storage*, at www.iea.org/publications/freepublications/.../CCS_Roadmap.pdf.

IEAGHG R&D Program (2009a): Oxy-fuel combustion network, at http://www.CO2capture andstorage.info/networks/oxyfuel.htm, Dec.15, 2009.

IEA (2010): *Technology roadmap: carbon capture and Storage*, Paris: IEA, at http://www.iea. org/papers/2009/CCS_Roadmap.pdf; April 3, 2010. IEA (2010a): *Energy Technology Perspectives 2010*, OECD/IEA, Paris.

IEA (2010b): CO_2 *Emissions from Fuel Combustion*, OECD/IEA, Paris.

IEAGHG (2011a): Second Oxy-fuel combustion Conf. abstracts, at http://www.ieaghg.org/ docs/General_Docs/OCC2/Abstracts/Abstract/occ2Final00050.pdf, June 25, 2011.

IEA Greenhouse Gas (IEAGHG) R&D Programme (2011): CO2 capture and storage, at http://www.co2captureandstorage.info/co2db.php; March 2011.

IEA (2011): *World Energy Outlook 2011*, OECD/IEA, Paris.

IEAGHG Program (2011): Retrofitting CO_2 Capture to Existing Power Plants, IEAGHG Technical Report (2011/02): IEAGHG, Gloucester, UK, available at www.ieaghg.org/index.php?/ 2009120981/technical-evaluations.html.

IPCC (2005): IPCC Special Report on Carbon Dioxide Capture and Storage, prepared by Working Group III of the Intergovernmental Panel on Climate Change, Figure ..., Cambridge University Press.

Jordal, K, et al. (2004): Oxyfuel Combustion of Coal-Fired Power Gerneration with CO_2 Capture – Opportunities and Challenges, XDQ, 2004.

Kapila (2009): Carbon Capture and Storage in India, at web.geos.ed.ac.uk/carbcap/website/ publications/.../wp-2009-04.pdf.

Kazanc, F., Khatami, R., Manoel-Crnkovic, P., & Levendis, Y.A (2011): Emissions of NO_x and SO_2 from coals of various ranks, bagasse and coal-bagasse blends, Energy and Fuels, 25(7), 2850–2861.

Khatami,R., Stivers,C., Joshi, K., Levendis, Y.A., & Sarofin, A.F (2012): Combustion behavior of single particles from three different coal ranks and from sugar cane bagasse in O_2/N_2 and O_2/CO_2 atmospheres, Combustion and Flame, 159, 1253–1271.

Khatami, R., Stivers, C., & Levendis, Y.A (2012): Ignition characteristics of single coal particles from three different ranks in O_2/N_2 and O_2/CO_2 atmospheres, Combustion and Flame, 159, 3554–3568.

Katzer, J.R. (2007): The Future of Coal-based Power generation, MIT Energy Initiative, Presented at *UN Sustainable development and CCS Meeting*, September 10–11, 2007.

Klara, S.M. (2006): CO_2 capturedevelopments, Presentation at *Strategic initiatives for coal*, Queenstown, MD; Pittsburgh: NETL, Dec. 2006.

Klara, J.M., & Plunkett, J.E. (2010): The potential of advanced technologies to reduce carbon capture costs in future IGCC plants, *Intl. J. Greenhouse gas Control*, 2010 (Spl. Issue).

Knuutila, H., Svendsen, H.F., & Juliussen, O. (2009): Kinetics of carbonate based CO_2 capture systems, *Energy Procedia*, 1, 1011–8.

Kolbitsch, P., Pröll, T., Bolhar-Nordenkampf, J., & Hofbauer, H. (2009): Operating experience with chemical looping combustion in a 120kW dual circulating fluidized bed (DCFB) unit, *Energy Procedia*, 1, 1465–72.

Kotowicz, J., Chmielniak, T., & Janusz-Szymanska, K. (2010): The influence of membrane CO_2 separation on the efficiency of a coal-fired power plant, *Energy*; 35, 841–50.

Kuuskraa, V.A. (2007): A program to accelerate the deployment of CCS: rationale, objectives and costs, Paper prepared for the 'Coal initiative reports' series of the Pew Center on global climate change, Arlington.

Kubek, D., Higman, C., Holt, N., & Schoff, R. (2007): CO_2 Capture Retrofit Issues, Presented at *Gasification Technologies 2007*, San Francisco, October 2007.

Lackner, K. (2003): A Guide to CO_2 Sequestration, *Science*, 300, 13 June 2003.

Lake, L.W. (1989): *Enhanced Oil Recovery*, Prentice-Hall, New Jersey.

Leonardo Technologies, Inc. (2009): CO_2 capture technology sheets. Pittsburgh: DOE, National Energy Technology Laboratory; 2009.

Liang, D.P., Harrison, R.P., Gupta, D.A., Green, W.J., & Michael, M.C. (2004): CO_2 capture using dry sodium-based sorbents, *Energy Fuels*, 18(2), 569–75.

Li, J. (2010): Options for Introducing CO2 Capture and Capture Readiness for Coal-fired Power Plants in China, PhD Thesis, Imperial College London, UK.

Li, X., Sun Yongbin, & Li Huimin (2009): Development of IGCC projects abroad – Overview, *Power Design*, 6, Mar. 2009, 2S-33.

Lupin, M. (2011): CIUDEN $20\,MW_{th}$ PC oxycombustion system: first operating experiences, Proc. of CCT2011, *Fifth Intl. Conf. on Clean coal technologies*, Zargoza, Spain, 8–12 May 2011; London: ICA Clean coal centre.

Manchao He, Luis, S., Rita, S., Ana, G., Varagas Jr, E., & Na Zhang (2011): Risk assessment of CO_2 injection processes and storsge in carboniferous formations: A review: *Journal of Rock mechanics & Geotechnical engineering*, 3(1), 39–56.

Marion, J., Mohn, N., Liljedahl, G., Nsakala, N.y, Morin, J.-X., & Henriksen, P.-P. (2004): Technology options for controlling CO_2 emissions from fossil-fuelled power plants, *Proc. of 3rd Annual Conf. on Carbon Sequestration*, Alexandria, VA.

McCormick, M. (2012): A GHG Accounting Framework for CCS Projects, Center for Climate and Energy Solutions (C2ES), February 2012.

MIT Study (2007): *The Future of Coal: Options for a Carbon-Constrained World*, Massachusetts Institute of Technology, Cambridge, MA, August 2007.

MIT Energy Initiative (2011): Carbon capture and sequestration technologies at MIT, at http://sequestration.mit.edu/.

MIT (2012): at http://sequestration.mit.edu/tools/projects/index_capture.html.

MIT Energy Initiative (2013): Schwarze Pumpe Fact Sheet: Carbon Dioxide Capture and Storage Project, CCS Technologies@MIT, modified March 13, 2013, at http://sequestration.mit.edu/tools/projects/vattenfall_oxyfuel.html.

Morrison, G. (2008): Roadmaps for Clean Coal Technologies-*IGCC, Supercritical and CCS, Energy Technology Roadmaps Workshop*, 15–16 May 2008, IEA, Paris.

NASA (2009): Definition of technology readiness levels, at http://estonasa.gov/files/TRL_definitions.pdf; Nov. 15, 2009.

Natural Resources Canada (2006): Canada's carbon dioxide capture and storage technology roadmap, Ottawa: Natural Resources Canada; 2006.

National Coal Council (2011): Expedited CCS Development: Challenges & Opportunities, March 18, 2011, National Coal Council, Washington DC.

National Coal Council (2012): Harnessing Coal's carbon content to Advance the Economy, Environment and Energy Security, June 22, 2012; Study chair: Richard Bajura, National Coal Council, Washington DC.

OECD/IEA and UN Industrial Development Organization (2011): Technology roadmap carbon capture and storage in industrial applications, Paris: IEA, p. 46.

Oexmann, J., Hensel, C., & Kather, A. (2008): Post-combustion CO_2-capture from coal-fired power plants: preliminary evaluation of an integrated chemical absorption Process with piperazine-promoted potassium carbonate, *International J Greenhouse Gas Control*; 2, 539–52.

Orr, F.M. Jr. (2004): Storage of Carbon Dioxide in Geological Formations, *J. Petroleum Technology*, 56(3), pp. 90–97.

Pacala, S., & Socolow, R. (2004): Stabilization Wedges: Solving the Climate problem for the next 50 Years with Current Technologies, *Science*, 305, 13, 2004.

Pennline, H.W., Granite, E.J., Luebke, D.R., Kitchin, J.R., Landon, J., & Weiland, L.M. (2010): Separation of CO_2 from flue gas using electrochemical cells, *Fuel*, 89, 1307–14.

Phillips, J. (2007): Response of EPRI's Coal Fleet for Tomorrow program to recent market developments, *Gasification Technologies Conf.*, 2007, San Francisco, October 2007.

Platts (2010): *World Electric Power Plants Data Base (WEPP)*, Platts, New York.

Plaza, M.G., Pevida, C., Arenillas, A., Rubiera, F., & Pis, J.J. (2007): CO2 capture by adsorption with nitrogen enriched carbons, *Fuel*; 86, 2204–12.

Powell, C.E., & Qiao, G.G. (2006): Polymeric CO2/N2 gas separation membranes for the capture of carbon dioxide from power plant flue gases, *J Membrane Sci*; 279, 1–49.

PowerClean, T.N. (2004): *Fossil Fuel Power Generation State-of-the-Art*, P.T. Network, Editor, University of Ulster: Coleraine, UK, pp. 9–10.

Press Releases:

(1) 'AEP to Install Carbon Capture on Two Existing Power Plants, Company Will Be First to Move Technology to Commercial Scale', Press Release (March 15, 2007), *available at* http://www.aep.com/newsroom/newsreleases/default.asp?dbcommand=displayrelease& ID=1351.

(2) 'BP and Edison Mission Group Plan Major Hydrogen Power Project for California', Press Release, February 10, 2006, available at http://www.bp.com/genericarticle.do?categoryId= 2012968& contentId= 7014858. 33–37.

(3) 'Xcel Energy Increases Commitment to IGCC', Press Release, August 15, 2006, http://www.xcelenergy.com/XLWEB/CDA/0, 3080,1-1-1_15531_34200-28427-0_0_0-0,00. html.

Rao, A.B., & Rubin, E.S. (2002): A Technical, Economic and Environmental assessment of amine-based CO_2 capture technology for power plant greenhouse gas control, *Environmental Science and Technology*, 36(20), 4467–75.

Radosz, M., Hu X., Krutkramelis, K., & Shen Y. (2008): Flue-gas carbon capture on carbonaceous sorbents: toward a low-cost multifunctional carbon filter for 'green' energy producers, *Industrial Engineering Chem Res*, 47, 3784–94.

Reuters-News-Service (2005): *Vattenfall Plans CO2-Free Power Plant in Germany*, May 20, 2005, Available from: www.planetard.com.

Rochelle, G. (2009): Amine scrubbing for CO_2 capture, *Science*, 325, 1652–4.

Rubin, E.S., Hounshell, D.A., Yeh, S., Taylor, M., Schrattenholzer, L., Riahi, K., *et al.* (2004): The effect of government actions on environment technology innovation: applications to the integrated assessment of carbon sequestration technologies, Report from Carnegie Mellon University, Pittsburgh, PA, to USDOE, German town, MD: Office of Biological and Environmental Research, January 2004.

Rubin, E.S. (2005): The Government role in Technology innovation: Lessons for the Climate change policy Agenda, *Proc 10th Biennial conference on transportation energy and environmental policy*. Davis: Dept. of Transportation Studies, University of California.

Rubin, E.S., Yeh, S., Antes, M. & Berkenpas, M. (2007): Use of experience curves to estimate the future cost of power plants with CO_2 capture, *Intl. J Greenhousegas Control.*

Rubin, E.S., Chen, C., & Rao, A.B. (2007): Cost and performance of fossil fuel power plants with CO_2 capture and storage, *Energy Policy*, 35(9), 4444–54.

Rubin, E.S. (2008): CO_2 Capture and Transport, *Elements*, 4, pp. 311–317.

Rubin, E.S. (2010): Will carbon capture and storage be available in time? *Proc. AAAS annual meeting*, San Diego, CA. Washington: American Academy for the Advancement of Science 2010; 18–22 February 2010.

Rubin, E.S., Hari, M., Marks, A., Versteeg, P., & Kitchin, J. (2012): The Outlook for improved carbon Capture technology, *Progress in Energy & Combustion Science*, 38 (5), 630–671, doi: 10.1016/j.pecs.2012.03.003.

Scheffknecht, G., Al-Makhadmeh, L., Schnell, U., &Maier, J. (2011): Oxy-fuel combustion: A review of the current state-of-the-art, *International J. GHG Control*, 5(1), S16–S35.

Simbeck, D. (2008): The Carbon capture technology landscape, *Proc Energy frontiers – International emerging energy technology forum*, Mountain View: SFA Pacific, Inc.

Simbeck, D.R. (2001): CO_2 Mitigation Economics for Existing Coal-Fired Power Plants, *Pittsburgh Coal Conference*, December 4, at Newcastle, NSW, Australia.

Simbolotti, G. (2010): CO_2 Capture & Storage, *IEA/ETSAP, Technology Brief E-14*, October 2010, G. Tosato, Project Coordinator.

Sjostrom, S., & Krutka, H. (2010): Evaluation of solid sorbents as a retrofit technology for CO_2 capture, *Fuel*; 89, 1298–306.

Socolow, R. (2005): Can We Bury Global Warming?, *Scientific American*, 293(1), 52, July 2005 (estimating that coal plants accounted for 542 billion tons of CO_2 emissions from 1751–2002 and will account for 501 billion tons of CO_2 from 2002–2030).

Sourcewatch (2010): Southern Company abandons carbon capture and storage project, at http://www.sourcewatch.org/indexphp?title=Southern_Company# cite_note-11.

Specker, S. *et al.* (2009): The Potential Growing Role of Post-Combustion CO_2 Capture Retrofits in Early Commercial Applications of CCS to Coal-Fired Power Plants, *MIT Coal Retrofit Symposium*, Cambridge, MA, available at http://web.mit.edu/ mitei/docs/reports/speckerretrofits.pdf.

Thijs, P., Synnove-Saeverud, S.M. & Rune, B. (2009): Development of thin Pd-23%Ag/SS composite membranes for application in WGS membrane reactors: *Ninth Intl. Conf. on catalysis in membrane reactors*, Lyon, France, June 28

Turkenburg, W.C., & Hendriks, C.A. (1999): Fossil fuels in a sustainable energy supply: the significance of CO_2 removal, Utrecht, The Netherlands: Ministry of Economic Affairs, The Hague.

US DOE/NETL (2007): Cost and performance baseline for fossil energy plants, In: *Bituminous coal and natural gas to electricity final report*, Pittsburgh: National Energy Technology Laboratory; 2007.

US DOE/NETL (2008): Carbon dioxide capture from flue gas using dry regenerable sorbents, Project facts, Pittsburgh: National Energy Technology Laboratory.

USDOE/NETL (2008a): PC oxycombustion power plants, Vol 1, bituminous coal to electricity, *report no. DOE/NETL-2007/1291;* Pittsburgh: NETL, 2008.

US DOE (2009): Assessment of power plants that meet proposed greenhouse gas emission performance standards, *DOE/NETL-401/110509*, Pittsburgh: NETL.

US DOE (2010a): Carbon sequestration technology roadmap and program plan, Pittsburgh: National Energy Technology Laboratory; 2010.

US DOE/NETL (2010): Carbon dioxide Capture R&D annual technology update, draft. Pittsburgh: National Energy Technology Laboratory, 2010.

USDOE/NETL (2011): Carbon Sequestration Program: Technology Program Plan – Enhancing the Success of CCTs – *Applied Research and Development from Lab- to Large-Field Scale'*, DOE/NETL-2011/1464, Feb. 2011, at www.netl.doe.gov.

USDOE/NETL (2011a): 'Cost and Performance of PC and IGCC Plants for a Range of Carbon Dioxide Capture', DOE/NETL-2011/1498, May 2011.

USDOE/NETL (2011b): Coal-Fired Power Plants in the United States: Examination of the Costs of Retrofitting with CO_2 Capture Technology, Pittsburgh: NETL, at www.netl.doe.gov/ energy-analyses/pubs/GIS _CCS_ retrofit.pdf.

US DOE/ NETL (2011): Carbon capture and storage database, at http://www.netl.doe.gov/ technologies/carbon_seq/database/indexhtml;March 2011.

USDOE/NETL (2012): Advancing Oxycombustion Technology for Bituminous Coal Power Plants: AnR&D Guide, DOE/NETL-2010/1405, February 2012.

USEPA (2005): *Continuous Emissions Monitoring System (CEMS) Data Base of 2005 Power Plant Emissions Data*, 2005, EPA.

van Dijk, H., Walspurger, S., Cobden, P.D., Jansen, D., van den Brink, R.W. & de Vos, F. (2009): Performance of WGS catalysts under sorption-enhanced WGS conditions, *Energy Procedia*, 1, 639–46.

van Selow, E.R., Cobden, P.D., van den Brink, R.W., Wright, A., White, V., *et al.* (2009): Pilot scale development of the sorption enhanced water-gas shift process, in: Eide Il (ed.), CO_2 *capture for storage in deep geologic formations*, Berks: CPL Press, pp. 157–80.

Varagani, R.K., Chatel-Pelage, F., Pranda, P., Rostam-Abadi, M., Lu, Y., & Bose, A.C. (2005): Performance Simulation and Cost Assessment of Oxy-Combustion Process for CO2 Capture from Coal-Fired Power Plants, in *Fourth Annual Conference on Carbon Sequestration*, 2005, Alexandria, VA.

Wall, T., Stanger, R., & Santos, S. (2011): The current State of Oxy-fuel technology: demonstrations & technical barriers, *2nd Oxy-fuel Combustion Conf.*, Queensland, Australia, September 12–16, 2011.

Wappel, D., Khan, A., Shallcross, D., Joswig, S., Kentish, S., & Stevens, G. (2009): The effect of SO_2 on CO_2 absorption in an aqueous potassium carbonate solvent, *Energy Procedia*, 1(1), 125–31.

Weiss (2011): A New HP Version of Lurgi's FBDB Gasifier in bringing value to clients, *Gasification technologies Conf.*, October 10, 2011.

WEO (2006): note 5, p. 493 (Nov. 2006); Yamagata, note 7, estimating 1400 gigawatts based on data from the IEA's and Platt's database.

White, K., Torrente-Murciano, L., Sturgeon, D., & Chadwick, D. (2010): Purification of oxy-fuel derived CO_2, *Intl. J GHG control*, 4(2), 137–42.

Williams, R.H. (2005): Climate-Compatible Synthetic Liquid Fuels from Coal and Biomass with CO_2 Capture and Storage, Princeton Environmental Inst., Princeton University, Dec. 19, 2005, at http://www.climatechange.ca.gov/ documents/2005-12-19_ WILLIAMS.PDF.

Williams, R.H. (2006): Climate-Friendly, Rural Economy-Boosting Synfuels from Coal and Biomass, Presentation to Brian Schweitzer (Governor of Montana). Helena, Montana. November 15, 2006.

Williams, T. (2006): Carbon, Capture and Storage: Technology, Capacity and Limitations; Science and Technology Division, Library of Parliament, Parliament of Canada, 10 March 2006, at www2.parl.gc.ca/content/lop/researchpublications /prb0589-e.html.

WRI: CCS Overview, Available at www.wri,org/project/carbon-dioxide-capture-storage/ccs-basics.

Zhao, L., Ernst, R., Ludger, B., & Detlef, S. (2010): Multi-stage gas separation membrane Processes used in post-combustion capture: Energetic and economic analyses, *J Membrane Science,* **359**(1e2), 160–72.

Zeng, Y., Acharya, D.R., Tamhankar, S.S., Ramprasad, N., Ramachandran, R., Fitch, F.R., MacLean, D.L., Lin, J.Y.S., & Clarke, R.H. (2003): Oxy-fuel combustion process, *US Patent Application Publication No US 2003/0138747 A1.*

Zheng, L. (ed.) (2011): *Oxy-fuel combustion for power generation and CO_2 capture,* Woodhead Publishing Series in Energy No. 17, Cambridge, UK: Woodhead Publishing Ltd. February 2011.

Websites:

Carbon Sequestration Leadership Forum: http://www.cslforum.org/;

CO_2 Capture and Storage Association: http://www.ccsassociation.org.uk/;

CO_2 CRC: http://www.co2crc.com.au/;

CO2GeoNet: http://www.co2geonet.com/;

EU Zero Emissions Technology Platform: http://www.zero-emissionplatform.eu/;

Global CCS Institute: www.globalccsinstitute.com;

International Energy Agency: http://www.iea.org;

Greenpeace: http://www.greenpeace.org/international/press/reports/technical-brifing-ccs;

IEA Clean Coal Centre: http://www.iea-coal.co.uk/site/index.htm;

IEA GHG R&D Program: http://www.ieagreen.org.uk/;

Intergovernmental Panel on Climate Change: www.ipcc.ch;

Massachusetts Institute of Technology: http://sequestration.mit.edu/index.html;

IMO: http://www.imo.org/includes/blastdataonly.asp/data_id=17361/7.pdf;

Natural Resources Canada: http://www.nrcan.gc.ca/es/etb/cetc/combustion/co2trm/htmldocs/technical_reports_e.html;

NOVEM: http://www.cleanfuels.novem.nl/projects/international.asp;

Pew Center on Global Climate Change: http://www.pewclimate.org/technology-solutions;

The Carbon Trust: http://www.carbontrust.co.uk/default.ct;

UN FCCC: http://unfccc.int;

US NETL: http://www.netl.doe.gov/technologies/carbon_seq/index.html; http://www.netl.doe.gov/technologies/coalpower/cctc/;

US DOE Carbon Sequestration Website: http://carbonsequestration.us/;

US EPA: http://www.epa.gov/climatechange/emissions/co2_geosequest.html;

World Business Council for Sustainable Development: http://www.wbscd.org/;

World Coal Institute: http://www.worldcoal.org/;

World Energy Council: http://www.worldenergy.org/;

World ResourcesInstitute: http://www.wri.org/project/carbon-capture-sequestration

Chapter 10

Coal-to-liquid fuels

Introduction

The liquefaction of coal, viz, converting coal into liquid fuels which can substitute petroleum products is generally referred to as Coal-to-liquids (CTL) process. These liquid fuels (also called synthetic fuels) are accomplished using three basic approaches: pyrolysis, direct coal liquefaction (DCL) and indirect coal liquefaction (ICL) (Bridgwater & Anders 1994), of which the latter two are in practice. The syntheticic fuels can be prepared almost identical to conventional petroleum fuels in their properties.

Coal liquefaction processes were first developed in early 20th century but were later held up by the relatively low price and large-scale availability of crude oil and natural gas. Large scale applications existed in Germany during World War II, providing 92% of Germany's air fuel and over 50% of their petroleum supply in the 1940s (DOE). Similarly, South Africa developed CTL technology in the 1950s during an oil crisis which now plays a vital role in the country's economy, providing over 30% of their fuel demand (Sasol website). The oil crises of the 1970s and the threat of depletion of conventional oil supplies incited a renewed interest in the production of oil substitutes from coal during the 1980s. However, as was the case for coal gasification, the wide availability of cheap oil and natural gas supplies in the 1990s has effectively slowed down the commercial prospects of these technologies. And now, the R&D has increased in these processes.

10.1 CHEMISTRY OF COAL-TO-LIQUID SYNTHESIS

Basically, coal-to-liquid synthsis depends on a carbon source combining chemically with a hydrogen source. The Fischer-Tropsch (F-T) process is a chemical reaction in which CO and H_2 present in the syngas (produced through gasification of fuel) react in the presence of a suitable catalyst producing liquid hydrocarbons of various forms (synthetic fuels). Iron and cobalt-based materials are used as typical catalysts. The utility of the process is primarily in its role in producing fluid hydrocarbons or hydrogen from a solid feedstock, such as coal or various carbon-containing solid wastes.

Generally, non-oxidative pyrolysis of the solid carbonaceous material produces syngas which can be used directly as a fuel. If liquid petroleum-like fuel (or lubricant, or wax) is required, the F-T process can be applied. If hydrogen production is to be maximized, water gas shift reaction can be done, generating only CO_2 and H_2 and leaving no hydrocarbons in the product stream (Bowen & Irwin 2006). Generally, the F-T process gives two types of products, described by two different chemical processes:

$$nCO + 2nH_2 \rightarrow nH_2O + C_nH_{2n} \qquad \text{(olefins)} \qquad (10.1)$$

$$nCO + (2n + 1)H_2 \rightarrow nH_2O + C_nH_{2n+2} \quad \text{(paraffins)} \qquad (10.2)$$

The resulting products depend on the catalysts and the operating conditions of the reactor. Olefin-rich products with 'n' in the range 5–10 (naphtha) can be used for making synthetic gasoline and chemicals in high-temperature F-T processes. Paraffin-rich products with 'n' in the range of 12–19 are suitable for making synthetic diesel and waxes in low-temperature F-T processes.

In the Bergius process, splitting coal into shorter hydrocarbons, resembling ordinary crude oil is done by adding hydrogen under high pressure and temperature, thus eliminating the need for a gaseous middle stage.

$$nC + (n + 1)H_2 \rightarrow C_n + H_{2n+2} \qquad (10.3)$$

10.2 TECHNOLOGY CHOICES

Several technological pathways are available for producing liquid fuels from coal, although only a few have proved commercially feasible. Technological infrastructure available from conventional crude oil processing and gas liquefaction along with modern equipment can provide synergistic effects, and be able to comply to a large extent with the environmental and economic constraints.

Pyrolysis at high temperatures is the oldest method for obtaining liquids from coal. Typically, coal is heated to around 950°C in a closed container, and the heat causes decomposition; the volatile matter is removed, increasing carbon content. This is similar to the coke-making process and the resulting tar-like liquid is mostly a byproduct. The liquid yields are very low, the process is inherently low efficient, and upgrading costs are relatively high. Furthermore, the liquids require further treatment before they can be used in vehicles. Hence, this process is not pursued to the commercial scale.

Direct coal liquefaction (DCL) requires breaking coal down in a solvent at elevated temperature and pressure, followed by interaction with hydrogen gas and a catalyst. Direct coal liquefaction was developed as a commercial process in Germany based on research to convert lignite into synthetic oil by a Chemist, Friedrich Bergius (Collings 2002). Most of the direct processes developed in the 1980s were modifications or extensions of Bergius's original concept. However, none of the industrialized countries have pursued DCL technology to meet their own liquid fuel needs as modern DCL technology is not proven at the commercial scale. The largest scale at which there has been experience with DCL in the USA is a Process Development Unit at the Hydrocarbon Technology, Inc. (HTI) R&D facility that consumes 3 tonnes of coal per day (Williams and Larson 2003). Most interest in the world in finding alternatives to crude oil is focused on gas-to-liquids (GTL) technology, which aims to exploit low-cost 'stranded' natural gas resources in various parts of the world.

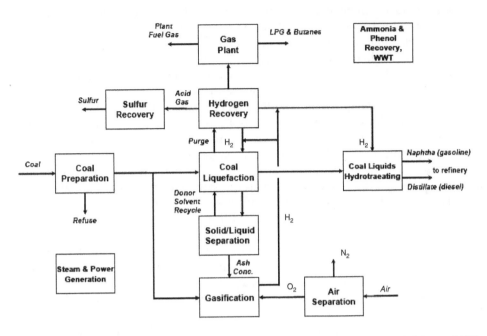

Figure 10.1 Direct liquefaction process Flow diagram (*Source:* DOE/NETL, Winslow & Schmetz 2009).

Indirect liquefaction (F-T synthesis) involves first gasifying coal and then making synthetic fuels from the syngas. Despite established scientific principles, only a small number of commercial installations generating liquid fuels from coal have come up, most of them based on ICL technology. The most successful operation in the world is Sasol in South Africa (Collings 2002). Indirect liquefaction processes were developed in Germany at the same time as direct processes. In the early 1920s, Franz Fischer and Hans Tropsch patented a process to produce a mixture of alcohols, aldehydes, fatty acids and hydrocarbons known as synthol from syngas (Fischer & Tropsch 1930).

10.2.1 Direct coal liquefaction

Direct liquefaction processes add hydrogen to the hydrogen-deficient structure of the coal (to increase H/C ratio), breaking it down only as far as is necessary to produce distillable liquids. The coal is ground so that it can be mixed into a coal-derived recycle-solvent to form a coal-oil slurry feed. The slurry containing 30%–50% coal is then heated to about 450°C in a hydrogen atmosphere at a pressure of about 1500 to 3000 psi for about one hour (Figure 10.1). The coal fragments are further hydrocracked to produce a synthetic crude oil. The liquids produced have molecular structures similar to those found in aromatic compounds. A variety of catalysts are used to improve the rates of conversion to liquid products. Hydrogen is needed in the DCL process both to make synthetic crude oil (represented simply as $CH_{1.6}$) and to reduce the oxygen, sulfur, and nitrogen in the coal feedstock. These elements are removed from the liquid fuel products in the forms of H_2O, H_2S, and NH_3. The oxygen is removed so that hydrocarbon fuels can be obtained. The nitrogen and sulfur compounds are removed to avoid poisoning the cracking catalysts in the refining operations downstream of the DCL plant. H_2 might be obtained from natural gas via steam reforming or from coal via gasification (Williams & Larson 2003).

One tonne of coal yields about one-half tonne of liquids and are much aromatic. Liquid yields can be in excess of 70% of the dry weight coal, with overall thermal efficiencies of 60–70% (Couch 2008; Benito et al. 1994). The resulting liquids are of much higher quality, compared with pyrolysis, and can be used unblended in power generation or other chemical processes as a synthetic crude oil (syncrude). However, further upgrading and hydrotreating is needed before they are usable as a transport fuel and refining stages are needed in the full process chain. Refining can be done directly at the CTL facility or by sending the synthetic crude oil to a conventional refinery. A mix of many gasoline-like and diesel-like products, as well as propane, butane and other products can be recovered from the refined syncrude. DCL naphtha can be used to make very high octane gasoline component; however, aromatics content of reformulated gasolines is now limited by US EPA. DCL distillate is a poor diesel blending component due to high aromatics which results in low cetane. Raw DCL liquids contain contaminants such as sulfur, nitrogen, oxygen, possibly metals and require extensive hydrotreatment to meet clean fuels specifications (Winslow & Schmetz 2009).

There are two variants of DCL: A single-stage direct liquefaction process gives distillates via one primary reactor. Such processes may include an integrated on-line hydrotreating reactor, which is intended to upgrade the primary distillates without directly increasing the overall conversion. A two-stage direct liquefaction processis designed to give distillate products via two reactor stages in series. The primary function of the first stage is coal dissolution and is operated either without a catalyst or with only a low-activity disposable catalyst. The heavy coal liquids produced in this way are hydrotreated in the second stage in the presence of a high-activity catalyst to produce additional distillate (Winslow & Schmetz 2009).

The range of liquid products produced by direct liquefaction depends on four major variables: the nature of the coal being processed; the solvent or solvent mix used; the process conditions, including temperature, pressure, residence time and catalyst; and the number of reactor stages used, and the subsequent refining of the initial products. Processes have been developed to use coals from low rank lignites to high volatile bituminous coals. Higher-rank coals are less reactive and anthracites are essentially non-reactive.

Benefits of DCL: (a) Direct liquefaction efficiency may be higher than indirect technology. One ton of a high volatile bituminous coal can be converted into approximately three barrels of high quality distillate syncrude for refinery upgrading and blending; (b) Direct Liquefaction provides high octane, low sulfur gasoline and a distillate that will require upgrading to make an acceptable diesel blending stock; (c) Development of direct liquefaction technology could lead to hybrid (direct/indirect) processes producing high quality gasoline and diesel, and (d) direct liquefaction may have a better carbon footprint than indirect technology.

There are, however, challenges to address such as high capital costs, investment risk, environmental challenges that include dealing with CO_2 and criteria pollutants and water usage, and major technical issues apart from oil price fluctuations. The main

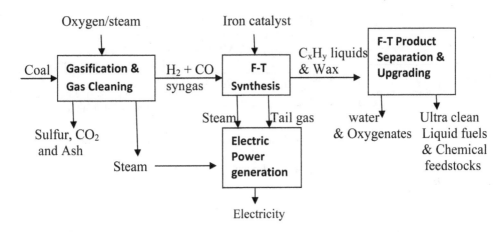

Figure 10.2 Representation of Fischer-Tropsch indirect liquefaction process (Redrawn from National
Mining Association, Liquid fuels from US Coal, at www.nma.org/pdf/liquid_coal_fuels_
100505.pdf).

technical issues for research: (i) first commercial unit is established in China (Shenhua);
more first-of-kind large scale operations with carbon management are needed to verify
baselines and economics; (ii) R&D activity should focus on process issues such as
further improvement in efficiency, product cost and quality, reliability of materials
and components, and data to better define carbon life cycle; (iii) the timelines for
demonstration and development of direct liquefaction technology, and (iv) integrating
and demonstrating CCS.

At commercial level, there has not been much progress. In 2002, the Shenhua
Group Corporation, the largest stateowned mining company in China took up
the designing and construction of the world's first DCL commercial plant in Inner
Mongolia Autonomous Region (Fletcher *et al.* 2004; Wu Xiuzhang 2010), which
recently became operational.

10.2.2   Indirect-coal-liquefaction

Indirect liquefaction is a multi-step process that first requires the gasification of coal
to produce a syngas which is then converted to liquid fuel via two methods: the
Fischer-Tropsch process or the Mobil process (also called methanol synthesis). In
the Fischer-Tropsch process, which is more common, the syngas is cleansed of impurities and is reacted via chemical reactions Eqn 10.1 and Eqn 10.2 given above, over an
appropriate F-T catalyst to form predominantly paraffinic liquid hydrocarbons having
wide molecular weight (Figure 10.2). The initial syngas can be derived from coal alone,
or from a coal/biomass mixture. The process is the same when biomass is included, but
the amount of CO_2 emitted during the process decreases as the proportion of biomass
increases. In the less-common Mobil process, the syngas can be converted to methanol,
which is subsequently converted to gasoline via a dehydration sequence (methanol-
to-gasoline, MTG). Indirect liquefaction of coal during Fischer-Tropsch produces a

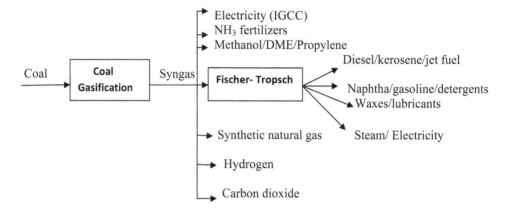

Figure 10.3 Coal-derived products (Redrawn from GTI Gasification Symposium, Dec. 2, 2004, available at http://www.gastechnology.org/webroot/downloads/en/1ResearchCap/1_8Gasification andGasProcessing/LepinskiSmall.pdf).

significant amount of CO_2 that is removed from the fuel as a necessary step during the final stages of the process (AAAS Policy Brief 2009). However, recent research has suggested a modified Fischer-Tropsch method that could significantly reduce CO_2 emissions during liquefaction (Hildebrandt *et al.* 2009).

There are two main operating modes for F-T synthesis: (a) High temperature (HTFT) technology, which operates between 300–350°C to maximize the gasoline fraction; and (b) Lower temperature (LTFT) technology, which operates in the range 200–250°C and maximizes the production of diesel. HTFT produces a more complex product line up than LTFT; and the gasoline produced contains significant amounts of benzene and its derivatives, regarded as being detrimental in transport fuels. HTFT offers more promise for production of chemicals, and the vast amounts of light hydrocarbons produced can be used to make a substitute/synthetic natural gas (SNG). The product distribution from F-T synthesis can be estimated from thermodynamic considerations, but in practice there is a strong degree of empiricism as there are many competing reactions (IEA Clean coal Center 2009).

South Africa has used this technology to produce motor fuels and petrochemical feedstocks since 1960s. The ICL has a lower thermal efficiency, but yields a large number of byproducts (Figure 10.3). Fuels that can be produced include methanol, dimethyl ether (DME), Fischer-Tropsch diesel- or gasoline-like fuels, and hydrogen. The availability of CO and H_2 as molecular building-blocks at an ICL plant also facilitates production of chemicals. Using modern technology, indirect liquefaction produces environmentally compatible zero-sulfur liquid fuels that are cleaner than required under current emissions laws and regulations.

DME can be reformed into hydrogen as easily as methanol, and thus is potentially suitable for future use as a hydrogen source for stationary or vehicle fuel cells. DME can also be used as an LPG substitute in cooking, where it burns with a clean blue flame over a wide range of air/fuel ratios (Fleisch *et al.* 1995; ICC 2003). DME is relatively inert, non-corrosive, non-carcinogenic, almost non-toxic, and does not form peroxides by prolonged exposure to air (Hansen *et al.* 1995). DME is produced globally today at a rate of about 150,000 tonnes/year in small-scale facilities by dehydration of MeOH (Naqvi 2002). More details on DME production are found in the literature (e.g., Larson & Tingjing 2003).

Catalysts have important role in aiding chemical processes in both DCL and ICL. Catalysts commonly used are transition metals – iron, ruthenium or cobalt. Transition metal sulphides, amorphous zeolite and similar compounds are also utilized. These are grouped as 'supported catalysts (Co/Mo, Co/Ni), dispersed catalyst (Fe, Mo), and HTI proprietary catalyst (GelCat – iron-based)'.

In general, catalysts have a large impact on process efficiency as well as influence over the resulting products. Many catalysts are highly sensitive to sulfur compounds and other substances, requiring special treatment and separation techniques to avoid catalyst poisoning. For detailed discussions on F-T synthesis, catalysts and so on, the reader may particularly refer to Bridgwater & Anders 1994, Dry 2002, Davis & Occelli 2006, Yang *et al.* 2006, Duvenhage & Coville 2006, Longwell *et al.* 1995, Bacaud *et al.* 1994, Speight 2008, Tarka *et al.* 2009. Considerable research has been done on different catalysts for CTL processes, and finding the right choice of catalysts and optimizing their performance has been a great challenge.

R&D programs

USDOE has supported DCL programs through sponsored projects during 1968–1995. Several DCL processes have emerged from this research which are classified as Phase I processes: SRC-II, Exxon Donor Solvent (EDS), and H-Coal; and Phase II process: Lummus Integrated Two-Stage Liquefaction (ITSL), Wilsonville Two-Stage Liquefaction, HRI/HTI Catalytic Multi-Stage Liquefaction (CMSL), and U. Ky./HTI/CONSOL/Sandia/LDP Advanced Liquefaction Concepts (ALC). Details on these processes are available in literature. Winslow and Schmetz (2009) have discussed these programs in great detail, and the general findings are summerized:

(a) The H-Coal and EDS programs (Phase I) demonstrated the technical and engineering feasibility of direct coal liquefaction, although many issues were not adequately resolved, including those of process yield, selectivity, product quality, and, ultimately, economic potential;

(b) Improvements in distillate yields and quality were shown in HTI bench-scale program with dispersed catalysts; low sulfur and nitrogen were achieved with in-line hydrotreating;

(c) Since direct liquefaction is capital-intensive, increasing liquid yields greatly reduce the capital cost component of the process on dollars/barrel/stream day basis. Liquid fuel yields were increased from 45% to 50% (MAF coal basis) for Phase I processes to about 75% (more than 4.5 bbl/t of MAF coal) for Phase II processes, while the yields of less valuable gaseous and non-distillate fuels were reduced proportionately for Illinois Basin coal;

(d) The liquids made in the Phase I were planned to replace crude oil, but they were unstable, highly aromatic, and had high heteroatom (sulfur, nitrogen, oxygen) contents, resulting in concern about refinability, storage stability, and human health, principally related to carcinogenicity. In the Phase II work, considerable improvement to liquid fuel quality could be achieved: the fuels contain no resid, no metals, and low levels of heteroatoms and can be refined in conventional refineries to meet current specifications for motor and turbine fuels. Product quality evaluations ensured that acceptable transportation fuels can be produced by direct coal liquefaction;

(e) The Phase ll processes could make quality naphtha that can be processed in conventional refineries into high-quality gasoline;

(f) No undesirable blending interaction with conventional gasolines and naphthas is expected. Direct coal liquefaction middle distillates can serve as blend stocks for the production of diesel fuel and kerosene;

(g) The low heteroatom content with accompanying higher hydrogen contents of Phase II process products alleviate the carcinogenicity concerns related to Phase I process products;

(h) The Phase I work demonstrated successful continuous operation of plants as large as 200 tpd of coal feed (Ashland Synthetic Fuels, Inc., Catlettsburg, KY);

(i) The Phase II processes are sufficiently similar to the Phase I processes, in terms of process equipment and unit operations; in addition, some of the key process equipment, such as the ebullated bed reactor, is used in petroleum refineries around the world;

(j) Materials of construction and equipment designs were found to overcome corrosion, erosion, and fouling problems experienced in Phase I plants; these new materials and designs were demonstrated to be suitable; As a result, the scale-up of the Phase II processes to commercial scale can be done with reasonable confidence;

(k) In the Phase II work, emphasis was also to apply direct liquefaction to low-rank coals; it proved that the inexpensive lignite, subbituminous, and bituminous coals were suitable feedstocks;

(l) The Phase II work showed that direct liquefaction is a flexible process for subbituminous and other low rank coals; and also applicable to a mixed feedstock containing coal and petroleum residues, heavy oil, or bitumen, and to coal and waste polymers, allowing a single plant to operate with the most economical feedstock available at a given place and time;

(m) Issues that were significantly problematic initially were moderated by improved materials, equipment, or process design during the development program, which include overall plant reliability, deashing, and product compatibility with conventional fuels, let-down valve erosion, preheater coking, and corrosion in distillation columns.

Sasol in South Africa, the first commercial-scale developer of ICL, has developed a number of different versions of ICL technologies; the ones used from 1950s to late 1980s were the oldest; advanced technologies that include Sasol Advanced Synthol High-Temperature F-T synthesis and the Sasol Slurry Phase Distillate Low-Temperature F-Tsynthesis are utilized since 1990s (Collings 2002) and has gained a lot of operational experience with its ICL plants currently in operation (Sasol website). So far, over 1.5 billion barrels of synthetic oil has been produced (WCI 2006). The details of FTS process development and commercialization activities are summarized in Table 10.1. The US DOE undertook a project to demonstrate advanced

Table 10.1 FTS Process development and commercial activity.

Country	Participant	Technology	Scale	Stage	Period
South Africa	Sasol	Lurgi/Sasol	7 million t/a	commercial	1955–
New Zealand	Methanex	Mobil (MTG)	0.75 million t/a	commercial	1983–
Malaysia	Bitutu	Shell (SMDS)	0.5 million t/a	commercial	1993–
Japan	Mitsubishi	AMSTG	1 barrel/d	Pilot	1986–90
Denmark	Topsoe	TIGAS	1 t/d	Pilot	1984–86
USA	Exxon	AGC 21	200 barrels/d	Pilot	1990–96
USA	Syntroleum	Syntroleum	2 barrels/d	Pilot	1994–97

(Data taken from Zhang Kai & Wu Xuehui 2006)

Fischer-Tropsch synthesis technology to convert anthracite wastes (residues of coal mining) into 5000 barrels per day (b/d) of synthetic diesel, naphtha, and kerosene fuels, while co-producing 35MW of electricity in the state of Pennsylvania (Larson & Tingjin 2003).

In addition to the USA, several countries including Canada, Japan, China, Australia, and South Africa, and European countries (UK, Germany, Poland, and Estonia) have launched R&D programs in CTL technologies. A report by IEA Clean Coal Center released in 2009 discusses these programs in detail.

The Mobil process (Methanol synthesis) is one of the technically well-developed commercial processes. Most methanol process technologies are offered by Lurgi, ICI, Mitsubishi, Linde, and Toyo corporations. ICI process differs with the Lurgi process basically in the type of reactor used: ICI uses multi-quench reactors while Lurgi uses multi-tubular reactors. Modern methanol plants can yield about 1 kg of methanol per liter of catalyst per hour. Although the commercial processes use a fixed-bed reactor in a gas recycle loop, new developments in methanol technology include use of liquid-phase slurry and fluidized-bed reactors. The liquid-phase slurry reactor offers an improved control of temperature and is of great interest for both methanol and FTS hydrocarbon production. Z Processes have been developed to use coals from low rank lignites to high volatile bituminous coals. Higher-rank coals are less reactive and anthracites are essentially non-reactive (Zhang Kai & Wu Xuehui 2006).

In China the production of methanol (primarily for use as chemical feedstock) by ICL processes is commercially established. China produced 3.3 million tonnes of methanol in 2001 (the bulk from coal) and has announced at least three new coal-to-methanol projects. Yankuang Group (Shangdong) built and put into operation a demonstration unit (10,000 t/a) in 2004 (Zhang Kai & Wu Xuehui 2006). China has an estimated 10 to 15 modern coal gasification facilities in operation to make hydrogen for ammonia production. Also, there is considerable interest in Shanxi Province to use methanol as a vehicle fuel (Niu 2003). There is also interest in pursuing ICL to produce dimethyl ether from coal (Larson & Tingjin 2003).

10.2.3 Comparison between DCL and ICL

Several studies on comprehensive analysis on DCL and ICL are found in literature (e.g., Hook & Aleklett 2009; Couch 2008; Yu *et al.* 2007; Williams & Larson 2003).

While DCL provides unrefined syncrude, ICL usually results in final products. ICL has a long history of commercial performance, while DCL does not.

In DCL, coal liquefaction and refining is nearly similar to ordinary crude oil processing, and creats a synthetic crude oil. If the complete breakdown of coal is avoided, efficiency can be slightly improved and the total liquefaction equipment can be reduced. In ICL, on the contrary, a set of criteria for the desired fuel are set up and pursued, using products resulting from F-T synthesis. Many of the various processes will yield hydrocarbon fuels superior to conventional oil-derived products. Harmful materials in coals inherently present have to be eliminated to protect the synthesis reactor catalysts. All ICL-derived products are better than the petroleum-derived products in terms of energy content or other characteristics; ICL fuels are inherently clean and virtually free from nitrogen, sulfur and aromatics, and generally release lower emissions when combusted (Durbin *et al.* 2000; Szybist *et al.* 2005).

Technology requirements: Heat, energy, catalysts and a few other chemicals are necessary in addition to coal. Water is a vital requirement of the process, in the form of hot steam or feedstock for hydrogen production and for cooling. Water consumption is large and approximately equal for DCL and ICL. The water consumption for a 50,000 barrel/day facility with American coal would be in the region of 40,000–50,000m^3/day (DOE/NETL 2006a). Grinding of coal and mixing it with water are other process steps that consume energy and water. Therefore, water availability is an essential issue while planning to install CTL facilities.

The DCL system requires hydrogen, the most costly component of the DCL system, to crack the coal into syncrude. High-efficiency designs often acquire hydrogen from steam reforming of natural gas, but DCL systems can also be modified to produce hydrogen from coal by the water-gas shift reactions. Necessary heat process for obtaining syncrude is usually provided by coal.

ICL utilizes huge amounts of steam to break down coal into syngas, requiring substantial energy input. Treatment and purification of the syngas is necessary for protecting the catalysts which involves gas cooling and different separation stages, all necessitating extra energy. However, some of this energy can be produced from sulfur and other compounds separated out from the syngas in the recycling processes. Some ICL configurations actually generate more electricity than they consume by converting excess heat into electricity (Williams & Larson 2003a).

DCL or ICL refining and product upgrading requires additional heat, energy and hydrogen. This extra energy requirement is up to 10% of the energy content of the syncrude which can also be provided by coal. Additional energy must be also provided to reduce GHG and other emissions to avoid environmental concerns.

System efficiency: The estimated overall efficiency of the DCL process is 73% (Comolli *et al.* 1999). Other estimates give the thermal efficiency between 60 and 70% (WCI 2006; Williams & Larson 2003a). SHELL estimated the theoretical maximum thermal efficiency of ICL as 60% (van den Burgt *et al.* 1980; Eilers *et al.* 1990). The overall efficiencies of ICL (making methanol or di-methyl-ether) is 58.3 and 55.1% (Williams & Larson 2003a). Tijmensen *et al.* (2002) has shown an overall energy efficiency of ICL of about 33–50% using various biomass-blends. Typical overall efficiencies for ICL are around 50%. van Vliet *et al.* (2009) has performed detailed analysis of energy flows for ICL diesel.

However, efficiency comparisons must be viewed with caution, because DCL efficiencies are estimated generally for making unrefined syncrude which requires more refining before utilization; and ICL efficiencies are often for deriving final products. If the refining of DCL products is considered, some ICL-derived fuels can be produced with higher final end-use efficiency than their DCL-counterparts (Williams & Larson 2003a). It is also sometimes unclear, whether the extra energy needed for process heat, hydrogen production, and process power is included in the analyses, making efficiency comparisons even more delicate.

Emission Profiles: Emissions and combustion characteristics of DCL and ICL fuels differ. The sulfur content of CTL products is low compared to petroleum-derived fuels for both DCL and ICL. Comprehensive analysis of emission characteristics of synthetic and conventional fuels are reported by Huang *et al.* 2008 and Hori *et al.* 1997. DCL products are typically rich in polycyclic aromatics and heteroatoms (Mzinyati 2007; Farcasiu *et al.* 1977; Jones *et al.* 1980; Leckel 2009), while ICL has lower aromatics content. High-temperature F-T synthesis yields branched products and contains aromatics, which are virtually absent in low-temperature F-T synthesis (Lipinski 2005). Recent environmental regulation trends in the United States have begun to limit the aromatic content in transportation fuels (Williams & Larson 2003a), giving the advantage to ICL fuels.

Toxic trace metals and inorganic compounds, such as cadmium, selenium, arsenic, lead and mercury, can be passed on to the final fuel product in both DCL and ICL processes. Removal of mercury and other metals is generally insignificant and inexpensive in ICL (Williams & Larson 2003a) whereas it is more complicated and more costly, but not impossible in DCL. Cetane and octane numbers resulting from the chemical properties of the various products also differ. High quality diesel is produced in ICL mostly due to the dominance of straight-chain products. However, low densities are a problem for ICL products, but this can be mitigated by blending (Leckel 2009). Both DCL and ICL fuels emit large amounts of carbon dioxide compared with conventional petroleum-derived fuels. However, methods for reducing or even neutralizing emissions without significant increase in production costs are available. There are substantial differences between DCL and ICL technologies with regard to the potential and cost of mitigation of GHG emissions (Williams & Larson 2003a).

DCL generates about 90% more CO_2 than conventional fuel on a well-to-wheel basis (Vallentin 2008). This is in agreement with other studies, but if reduction measures are implemented, the emissions could be reduced to not more than 30% extra compared with conventional petroleum fuels. ICL technology generates around 80–110% more CO_2 emissions compared with conventional fuels, if the CO_2 is vented (Williams & Larson 2003a; Vallentin 2008). However, there are ICL system configurations where $H_2S + CO_2$ co-capture/co-storage can reduce emissions (Williams & Larson 2003a). Well-to-wheel analysis has shown that even with CCS, CTL production chain emissions are higher than for petroleum-derived fuels, mostly coming from mining (van Vliet *et al.* 2009). Typical properties for specific ICL distillates can be found in Leckel (2009). Properties of final products in both DCL and ICL are given in Table 10.2. In general, CTL fuels can improve emission characteristics and reduce transportation emissions of sulfur, aromatics, NOx and particulates compared to conventional fuels (Huang *et al.* 2008). However, the potential for CO_2

Table 10.2 Typical properties of DCL and ICL final products.

	DCL	ICL
Distillable product mix	65% diesel, 35% naphtha	80% diesel, 20% naphtha
Diesel Cetane number	42–47	70–75
Diesel sulfur content	<5 ppm	<1 ppm
Diesel aromatics	4.8%	<4%
Diesel specific gravity	0.865	0.780
Naphtha octane number	>100	45–75
Naphtha sulfur content	<0.5 ppm	Nil
Naphtha aromatics	5%	2%
Naphtha specific gravity	0.764	0.673

(Source: Hook & Aleklett 2009)

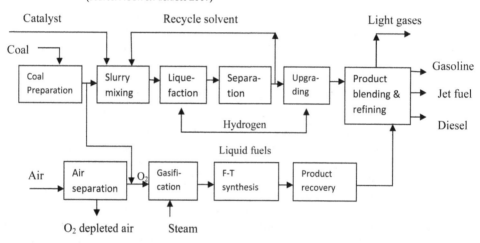

Figure 10.4 Hybrid concept of integrating DCL and ICL (Redrawn from USDOE-NETL 2008).

emission reductions if the full supply chain is analysed for either DCL or ICL, is not promising.

Hybrid liquefaction process

The hybrid process concept combines technologies from both the indirect and direct liquefaction processes. There are two key features of this hybrid process: (a) the hydrogen required for the direct process can be manufactured in the indirect process; and (b) the hybrid process yields both a high-quality diesel fuel from the indirect process and a highquality gasoline from the direct process. The hybrid process flow diagram is shown in Figure 10.4.

10.2.4   Costs of DCL & ICL

The capital cost of a CTL facility is usually the largest cost, with O&M costs being the next. The front-end expenditure is high – a 10,000 barrel-a-day plant could cost

Table 10.3 Estimated costs of CTL in the USA.

Plant capacity	Capital investment
20,000 b/d plant	$1.5–4.0 billions
80,000 b/d plant	$6–24 billions
One Mb/d industry	$60–160 billions

(*Source:* Drawn from Northern Plains Resource Council 2005)

$600–700 million or more to construct. The refinement process is three to four times more expensive than refining an equivalent amount of oil. When biomass is mixed with coal, the process becomes even more expensive, and is only viable with oil prices above $90 per barrel, according to DOE. If the cost of sequestrating the captured CO_2 is included, the price of the end product would increase by a projected $5 a barrel (AAAS Policy Brief 2009). The recent RAND study estimated that 'CTL production plus carbon storage' could produce fuel at a cost from $1.40 to $2.20 per gallon or more by 2025 (RAND 2008). CTL as an economically attractive alternative is dependent on the prevailing oil prices.

Depending on the local supply, quality and so on, the coal costs are usually around 10–20%. ICL plants in the USA provide break-even crude oil-prices in the range, US$ 25–40 per barrel, depending on whether measures such as CO_2 capture are included (Williams & Larson 2003a). Older and more modest studies claim break-even crude oil prices around US$35 per barrel (Lumpkin 1988). Liquid fuel costs for a Chinese DCL facility have been estimated at around US$24/barrel (Fletcher & Sun 2005). The development of coal prices and the economic situation in recent years has influenced break-even prices. The most recent study of CTL costs suggests a break-even price of US$48–75 per barrel (Vallentin 2008). Expected costs for ICL and DCL do not seem to differ much and can be assumed to be virtually identical. Table 10.3 lists some estimated costs for construction of three different capacities in the USA. The DCL facility in Inner Mongolia in China has an overall cost of approximately US$ 4 billion (Tingting 2009).

Sasol and China were planning two additional 80,000 b/d ICL plants in Shaanxi (650 km west from Beijing) and Ningxia (1000 km west from Beijing), with US$5 billion as estimated capital cost per plant (Sasol 2006). Currently, only two projects are approved and the Chinese National Development and Reform Commission suspended all other CTL projects in September 2008 (Tingting 2009).

10.2.5   Outlook for CTL

There is currently a great deal of international interest in converting coal to liquid fuels, especially for transport. This is due to (a) the ongoing uncertainties in the supply and price of oil when faced with the continuing increases in demand – particularly in the developing economies, (b) oil reserves not expanding to meet this increase in demand, (c) the fact that global coal reserves being extremely large and widely distributed, the coal supplies are likely to be more dependable, and the price less prone to wide fluctuations. Hence, the production of liquid fuels from coal would (a) greatly increase the availability and lifetime of liquid transport fuels, (b) increase the opportunities to use indigenous fuel reserves, and (c) send a signal to oil producing countries regarding the maximum tolerable price ceiling (IEA Clean Coal Center 2009).

Industrialized countries have, however, focused more on ICL or GTL for making synthetic fuels and virtually ignored DCL technologies. As a result, DCL systems generally require new technology, with the important exception of the H_2-production (Williams and Larson 2003a). However, DCL can provide essential new feedstock for refineries, if conventional crude oil becomes scarce.

On the contrary, the existing support infrastructure for ICL seems stronger than for DCL because commercialization of ICL technology began quite early (Leckel 2009), and has gone through several designs and developments; most of the required components are already established. ICL fuels, especially ICL diesel, are generally cleaner and can surpass many DCL fuels and conventional fuels in terms of emissions. It is also easier to implement CCS and GHG emission reduction in ICL plants compared to DCL. In essence, ICL technology will generally put the coal energy system on a track more dedicated to environmental concern, while DCL does not offer this possibility to the same extent. ICL offers more variable systems, capable of producing many more products than DCL systems, especially in polygeneration designs. Further, F-T fuel is increasing its share (USDOE 2004) although gasification technologies are commonly used globally for producing chemicals. Since similarities exist with GTL technologies and the chemical industries based on syngas derived from gasification (Leckel 2009), it is easier for expansion into new ICL projects. However, realizing this expansion and scaling up the existing capacity could still be a major challenge.

Although coal-to-liquid technology is a well established process, making the greenhouse gas emissions of synthetic fuels derived from coal comparable to those of oil requires further research into emissions from CTL and CBTL production and large-scale carbon sequestration.

Shenhua direct coal liquefaction project

The Shenhua DCL project, from R&D to commercialization, is discussed by Wu Xiuzhang at a conference held on May 26th, 2010. With thorough understanding of the property of Shenhua coal, liquefaction process and catalyst, ShenhuaGroup has developed its proprietary DCL process (Figure 10.5).

The general process is: Coal first goes through solvent extraction; then, with the presence of catalyst at high temperature and high pressure, coal slurry is hydrogenated through which the complex organic molecular structure in coal will be changed to increase H/C ratio. In this way, coal is directly converted into liquids. The end products are LPG, naphtha, diesel, and phenol. The successful commissioning of Shenhua DCL Demonstration Plant in 2008 (starting the plant work in 2004) establishes that China has become the only country in the world in possession of the core technology of direct coal liquefaction enabling a commercial plant with a capacity over 1 MMt/annum (Wu Xiuzhang 2010).

Coproduction of synthetic fuels and electricity: CTL systems can also be designed to provide electricity as a major coproduct (Figure 10.3). In a CTL co-production plant, syngas which is not converted to liquid fuels after the synthesis process is burned in the gas turbine combustor of a combined cycle power plant. In such a plant configuration, typically 25–35% of energy output (electricity plus diesel and jet fuel plus gasoline or

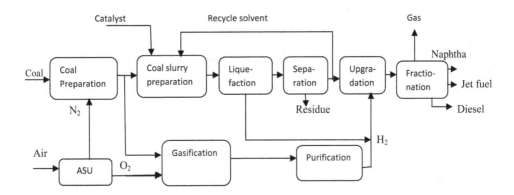

Figure 10.5 Schematic of Shenhua DCL Process (Redrawn from Wu Xiuzhang 2010).

naptha) is electricity. As in the CTL, CO_2 has to be removed from syngas before it is synthesized in the presence of a catalyst which can be compressed at low cost for sending through the pipeline. Also, when an iron F-T synthesis catalyst is used, additional CO_2 generated in synthesis can be captured at high partial pressure downstream of synthesis. The relative profitability of investments on 'CTL plants with and without co-production' depends on electricity and oil prices. Modest scale coproduction systems with CO_2 capture assessed as power generators have proved economically attractive in the context of high oil prices and low CO_2 selling prices. As a result, they might be preferred for synthetic fuels' production in regions where new electricity supplies are needed, if the institutional issues arising by coproduction route can be solved. Evaluated as power generators, such coproduction plants could be built or considered as rebuild options for older coal power plants.

In case large quantities of coal need to be converted to synthetic fuels under a possible eventual carbon mitigation policy, it may be necessary to pursue not only CCS but also the coprocessing of biomass with coal. For such systems that capture CO_2, storing photosynthetic CO_2 underground represents negative emissions that can be used to offset positive CO_2 emissions from coal. For example, a coproduction plant cogasifying coal with 5% biomass (energy basis) with CCS has a greenhouse gas intensity of 0.5 (NCC 2012).

REFERENCES

AAAS Policy Brief (2009): Coal-to-Liquid Technology, at www.aaas.org/spp/cstc/briefs/coalto liquid/index.shtm

Bacaud, R., Jamond, M., Diack, M., & Gruber, R. (1994): Development and evaluation of iron based catalysts for the hydroliquefaction of coal, *Intl J of Energy Research*, **18**(2), pp. 167–176, doi: 10.1002/er.4440180214.

Benito, A., Cebolla, V., Fernandez, I., Martinez, M.T., Miranda, J.L., Oelert, H., & Prado, I.G. (1994): Transport fuels from two-stage coal liquefaction, *Intl. Journal of Energy Research*, **18**(2), pp. 257–265; doi: 10.1002/er.4440180225.

Bridgwater, A.V. & Anders, M. (1994): Production costs of liquid fuels by indirect coal liquefaction, *Intl J of Energy Research*, **18**(2), 97–108, doi: 10.1002/er.4440180207.

Bowen, B.H., & Irwin, M.W. (2006): CCTR Basic Facts File #1, The Energy Center at Discovery Park, Purdue University, West Lafayette, IN, July 2006.

GTI Gasification Symposium, Dec. 2, 2004, at http://www.gastechnology.org/webroot/down loads/en/1ResearchCap/1_8GasificationandGasProcessing/LepinskiSmall.pdf).

Collings, J. (2002): *Mind over Matter – The Sasol Story: A Half-Century of Technological Innovation*, 2002; Available at http://sasol.investoreports.com/ sasol_mm_2006/index.php.

Comolli, A.G., Lee, L.K., Pradhan, V.R., Stalzer, R.H., Karolkiewicz, W.F., & Pablacio, R.M. (1999): *Direct Liquefaction Proof-of-Concept Program*, final report prepared by Hydrocarbon Technologies Inc. (Lawrenceville, NJ) and Kerr-McGee Corporation (Oklahoma, OK) for the Pittsburgh Energy Technology Center, US DOE, DE-92148-TOP-02, available at http://www.osti.gov/bridge/servlets/purl/772402Us5qhE/webviewable/772402.pdf.

Couch, G.R. (2008): *Coal-to-liquids*, IEA Clean Coal Centre, Publ. CCC/132; available at http://www.coalonline.info/site/coalonline/content/ browser/81994/coal-to-liquids.

Davis, B.H., & Occelli, M.L. (2006): *Fischer-Tropsch Synthesis, Catalysts and Catalysis*; Elsevier: Amsterdam, p. 42026.

Dry, M. (2002): The Fischer-Tropsch process 1950–2000, *Catalysis Today*, **71**(3–4), pp. 227–241.

Durbin, T.D., Collins, J.R., Norbeck, J.M., & Smith, M.R. (2000): Effects of biodiesel, biodiesel blends, and a synthetic diesel on emissions from light heavy-duty diesel vehicles, *Environmental Science and Technology*, **34**(3), 349–355, DOI: 10.1021/es 990543c.

Duvenhage, D., & Coville, N.J. (2006): Deactivation of a precipitated iron Fischer-Tropsch catalyst – a pilot plant study, *Applied Catalysis A: General*, **298**, pp. 211–216, DOI: 10.1016/j.apcata.2005.10.009.

Eilers, J., Posthuma, S.A., & Sie, S.T. (1990): The shell middle distillate synthesis process (SMDS), *Catalysis Letters*, **7**, pp. 253–269; DOI: 10.1007/BF00764507.

Fanning, L.M. (1950): Our Oil Resources, McGraw-Hill: New York, 1950.

Hook, M., & Aleklett, K., 2009: A review on Coal-to-liquid fuels and its coal consumption, *Int. J. Energy Research*, published online in Wiley InterScience (www.interscience.wiley.com), DOI: 10.1002/er.1596.

Farcasiu, M. (1977): Fractionation and structural characterization of coal liquids, *Fuel*, **56**(1), 9–14; DOI: 10.1016/0016-2361(77)90034-5.

Fischer, F., & Tropsch, H. (1930): U.S. Patent 1,746,464. Feb. 11, 1930.

Fletcher, J.J., & Sun, Q. (2005): Comparative analysis of costs of alternative coal liquefaction processes, *Energy and Fuels*, **19**, pp.1160–1164, doi: 10.1021/ef049 859i.

Fletcher, J.J., Sun, Q., Bajura, R.A., Zhang, Y., & Ren, X. (2004): Coal to clean fuel – the Shenhua investment in direct coal liquefaction, Third US-China Clean Energy Conference, Morgantown, U.S.A., 18–19 October 2004; Available at: http://www.nrcce.wvu.edu/conferences/2004/China.

Hildebrandt, D., Glasser, D., Hausberger, B., Patel, B., & Glasser, B.J. (2009): Producing Transportation Fuels with Less Work, *Science*, **323**, p. 1680.

Hori, S., Sato, T., & Narusawa, K. (1997): Effects of diesel fuel composition on SOF and PAH exhaust emissions, *JSAE Review*, **18**, 255–261; doi: 10.1016/S0389-4304(97)00022-2.

Huang, Y., Wang, S., & Zhou, L. (2008): Effects of F-T diesel fuel on combustion and emissions of direct injection diesel engine, *Frontiers of Energy and Power Engineering in China*; **2**, 261–267; doi: 10.1007/s11708-008-0062-x.44.

IEA Clean Coal Center (2009): Review of worldwide coal to liquids Research, D&D activities and the need for further initiatives within Europe.

Jones, D.G., Rottendorf, H., Wilson, M., & Collin, P., (1980): Hydrogenation of Liddell coal – Yields and mean chem. structures of the products, *Fuel*, **59**(1), 19–26.

Larson, E.D. & Tingjing, R. (2003): Synthetic fuel production through Indirect coal liquefaction, *Energy for Sustainable Development*, VII (4), pp. 79–102.

Leckel, D. (2009): Diesel production from FischerTropsch: the past, the present, and new concepts, *Energy and Fuels*, 23(5), 2342–2358, doi: 10.1021/ef900064c.

Lipinski, J.A. (2005): Overview of coal liquefaction. U.S.–India Coal Working Group Meeting, Washington, DC, 18 November 2005; Available at http://www.fe.doe.gov/international/Publications/cwg_nov05_ctl_lepinski.pdf.

Longwell, J.P., Rubin, E.S., & Wilson, J. (1995): Coal: energy for the future, *Progress in Energy and Combustion Science*, 21(4), pp. 269–360, doi: 10.1016/ 0360-1285(95)00007-0.

Lumpkin, R.E. (1988): Recent progress in direct liquefaction of coal, *Science*, 239, pp. 873–877; doi: 10.1126/science.239.4842.873.

Mzinyati, A.B. (2007): Fuel-blending stocks from the hydrotreatment of a distillate formed by direct coal liquefaction, *Energy and Fuels*, 21(5), pp. 2751–2761; DOI: 10.1021/ef060622r.

National Coal Council (2012): Harnessing Coal's Carbon content to Advance the Economy, Envi-ronment and Energy Security, June 22, 2012; Study chair: Richard Bajura, National Coal Council, Washington DC.

Niu, J. (2003): Demonstration of fuel methanol and methanol vehicle in Shanxi Province, in *Proc. of the Workshop on Coal Gasification for Clean and Secure Energy for China*, Task Force on Energy Strategies and Technologies, China Council for International Cooperation on Environment and Development, Tsinghua University, Beijing, 25–26 August 2003.

Northern Plains Resource Council (2005): Montana's Energy Future, 2005; available at http://www.northernplains.org/ourwork/coaltodiesel.

RAND Technical Report (2008): Unconventional Fossil-Based Fuels Economic and Environmental Trade-Offs, sponsored by the National Commission on Energy Policy, 2008.

Sasol: Unlocking the Potential Wealth of Coal – Information brochure; available at www.sasol.com/sasol_internet/downloads/CTL_ Brochure_1125921891488. pdf.

Speight, J. (2008): Synthetic Fuels handbook: properties, processes & Performance, McGraw Hill.

Szybist, J.P., Kirby, S.R., & Boehman, A.L. (2005): NOx emissions of alternative diesel fuels: a comparative analysis of biodiesel and F-T diesel, *Energy and Fuels*, 19(4), 1484–1492, doi: 10.1021/ef049702q.

Tarka, T.J., Wimer, J.G., Balash, P.C., Skone, T.J., Kern, K.C. *et al.* (2009): Affordable low-carbon diesel from domestic coal & biomass, DOE/NETL.

Tingting, S. (2009): Shenhua plans to triple capacity of its direct coal-to-liquids plant, China Daily, 8 January, 2009; 14; available at http://www.chinadaily.com.cn/cndy/200901/08/content_7376581.htm.

Tijmensen, M., Faaij, A., Hamelinck, C., & van Hardeveld, M. (2002): Exploration of the possibilities for production of Fischer-Tropsch liquids and power via biomass gasification, *Biomass and Bioenergy*, 23, pp. 129–152; doi: 10.1016/S0961-9534(02)00037-5.

USDOE (2004): Current Industry Perspective Gasification, Office of Fossil Fuels information brochure from 2004; 28, Available at http://gasification.org/Docs/News/2005/Gasification_ Brochure.pdf.

US DOE/NETL (2006a): Emerging Issues for Fossil Energy and Water, 49; available at http://www.netl.doe.gov/technologies/oilgas/ publications/AP/ IssuesforFEandWater.pdf.

USDOE: Early Days in Coal Research, available at http://fossil.energy.gov/aboutus/history/syntheticfuels_history.html.

USDOE/NETL (2008): Coal-to-Liquids technology, clean liquid fuels from coal, March 2008, Program contacts: Miller, C.L., Ackiewicz, M., & Cicero, D.C; DOE: Office of fuel energy, Washington DC.

USDOE/NETL Report (2009): January 2009.

Vallentin, D. (2008): Policy drivers and barriers for coal-to-liquids (CTL) technologies in the United States, *Energy Policy*, 36, pp. 3198–3211, doi:10.1016/j.enpol.2008.04.032.

Van den Burgt, M., Van Klinken, J., Sie, S.T. (1985): The shell middle distillate synthesis process. Paper presented at the *Fifth Synfuels Worldwide Symposium*, Washington, D.C., 11–13 November, 1985.

van Vliet, O.P.R., Faaij, A.P.C., & Turkenberg, W.C. (2009): Fischer-Tropsch diesel production in a well-to-wheel perspective: a carbon, energy flow and cost analysis, *Energy Conversion and Management*, 50(4), pp. 855–876, doi: 10.1016/j.enconman 2009.01.008.

Winslow, J., & Schmetz, E. (2009): Direct Coal Liquifaction Overview, Presented to *NETL, Leonardo Technologies, Inc.* March 23, 2009.

Williams, R.H., & Larson, E.D. (2003): A comparison of direct and indiredt liquefaction technologies for making fluid fuels from Coal, *Energy for Sustainable Development*, **VII**(4l), pp. 103–129, December 2003.

Williams, R., & Larson, E. (2003a): A comparison of direct and indirect liquefaction technologies for making fluid fuels from coal, *Energy for Sustainable Development*, 7, 79–102.

Wise, J., & Silvestri, J. (1976): Mobil Process For the Conversion of Methanol to Gasoline, Presented at the 3rd Intl. Conf. on Coal Gasification and Liquefaction, University of Pittsburgh: Pittsburgh, PA, 1976; 15.

World Coal Institute (2006): Coal: Liquid Fuels, available at http://www.worldcoal.org/assets_cm/files/PDF/wci_coal_liquid_fuels.pdf.

Wu Xiuzhang (2010): Coal Liquefaction Project: From R&D to Commercial Demonstration, Shenhua Group Co. Ltd, May 26th, 2010, PDF.

Yang, J., Sun, Y., Tang, Y., Liu, Y., Wang, H., Tian, L., Wang, H., Zhang, Z., Xiang, H., & Li, H. (2006): Effect of magnesium promoter on iron-based catalyst for Fischer–Tropsch synthesis, *Journal of Molecular Catalysis A: Chemical*, **245** (1–2), pp. 26–36; doi: 10.1016/j.molcata.2005.08.051.

Yu, Z., Wu, L., & Li, K. (2007): Development of alternative energy and coal-to-liquids in China, Presented at the 24th Intl. Pittsburgh Coal Conf, Johannesburg, South Africa, 10–14 September 2007.

Zhang Kai & Wu Xuehui (2006): Progress in Coal Liquefaction technologies, *Petroleum Science*, 3(4), pp. 90–99.

Chapter 11

CCTs in developing countries

11.1 STATE OF AFFAIRS

Global demand for coal has increased steadily since the 1970s despite the increasing environmental regulations on coal use; and as seen earlier, the IEA World Energy Outlook 2011 projects that the demand for coal will increase by another 25% between 2009 and 2035. The IEA in its annual *Medium-Term Coal Market Report (MCMR 2012)* states that coal's share of the global energy mix will continue to rise, and by 2017 coal will come close to surpassing oil as the world's top energy source. Although the growth rate of coal will slow from the early pace of the first decade of the 21st century, global coal consumption will by 2017 stand at 4.32 billion tonnes of oil equivalent (btoe), versus around 4.40 btoe for oil. It is expected that coal demand will increase in every region of the world except in the USA, where coal is being replaced by shale gas. The world will burn around 1.2 billion more tonnes of coal per year by 2017 compared to today – equivalent to the current coal consumption of Russia and the USA combined. Coal's share of the global energy mix continues to grow each year, and if the current policies continue, coal will catch up with oil within a decade according to the Report. China and India will lead the growth in coal consumption over the next five years, and China will surpass the rest of the world in coal demand during the outlook period, while India will become the largest coal importer and second-largest consumer, exceeding the USA. The report's forecasts are based on an assumption, that carbon capture and sequestration will *not be available* during the outlook period. That is, CCS technologies are not taking off as once expected, and CO_2 emissions will keep growing substantially (IEA's *MCMR* 2012). This is slightly in contrast to the CCS development schedules given in IEA Roadmap.

The problems between the use of coal for power generation, economic development and the increasing demand for environmental security are mounting but are considered manageable. The development and implementation of CCTs are central to reduce these tensions. CCTs, which cover technologies ranging from the preparation of coal through combustion and the clean up of waste gases to carbon capture and storage, will reduce the pollution emission intensity of coal and make coal usage cleaner. But implementation of CCTs incurs huge costs though the economic and environmental benefits follow such investments. For example, IGCC power plant technology can increase efficiencies by 20–30% compared with conventional coal-fired power plants; the captured CO_2 from CCS power plants can be injected into oil fields to increase the oil recovery rate by 4–18%; and carbon storage technologies, such as the creation of bio-charcoal, can improve soil fertility, agricultural productivity and water quality. CCTs can also bring new export opportunities for developing countries. Low-rank coal such as brown coal in Indonesia which has had no previous market, if upgraded, may extend export opportunities. However, despite these economic and environmental benefits, many developing countries do not use CCTs. Instead, coal-fired power plants with little application of CCTs are concentrated in, or even relocated to, other countries. For example, the Electricity Generating Authority of Thailand built coal-fired power plants in neighbouring countries after facing internal public resistance to coal-fired power plants. Such relocations usually worsen the emissions performance of these plants: if the power plants were built in Thailand, the Thai policy and regulatory regime would have led to the installation of better and more CCTs (Shi & Jacobs 2012).

In developing countries, there is a lack of adequate public awareness and policy regulation; and the institutional, technical and economic capacities and legal framework are not yet ready for full implementation of CCTs. The governments in developing countries often place a higher priority on improving income and life style of people and economic development rather than investment in reducing pollution. The trend, however, in China and India in recent years has been changing. Developing countries are essential and the most cost effective players to improving the environmental performance of coal through the promotion of CCTs. As already seen, it is expected that more than 96% of growth in demand for coal between 2009 and 2035 will be from non-OECD countries, most of it from non-OECD-Asian countries. But their low-level uptake of CCTs implies that the net marginal benefits of promoting CCTs would be higher. Thus, making coal use cleaner in developing countries is environmentally significant and economically effective. To make this happen, an international approach is very essential. Many developing countries require international finance generated from a cooperative effort such as financial aid, technology transfers and cooperation mechanisms like the Clean Development Mechanism to invest in CCTs. The transfer of CCTs, including low carbon technologies, could play a critical role in encouraging countries such as China and India to enter into a post-Kyoto Protocol agreement. Japan has proposed a post-Kyoto Protocol emission-reduction credit mechanism which is the 'bilateral offset crediting mechanism'. This will provide support and incentives to the Japanese private sector to export emission reduction and low carbon technologies to developing counties such as India, Indonesia and Vietnam. Through this mechanism, Japan expects to benefit from emission reduction credits, while the countries receiving the technologies will get high-end, efficient technology on competitive terms. Such a win–win outcome is likely to improve global environmental outcomes for coal power generation in a cost effective way (Shi & Jacobs 2012).

11.2   CCS IN DEVELOPING COUNTRIES

It has been clear that despite often-aggressive programs to promote energy efficiency and deploy nuclear, renewable, and other low-carbon energy sources, many developing countries will rely heavily on fossil fuel energy, especially coal-based power, for decades to come. There is therefore a need for developing countries to create strategies that address fossil fuel emissions in a way that minimizes the costs of such efforts, and consequently minimizes impacts to their national development goals. CCS is currently the only near-commercial technology proven to directly remove CO_2 emissions from fossil fuel use at scale. Its deployment could allow developing countries with relatively little disruption to their long-term development strategies to remove CO_2 emissions. If deployed now as an interim measure, it could allow time for other alternative low-carbon technologies to be developed and deployed, permitting fossil fuels to be gradually phased out. This strategy could assist developing countries to transition to a low-carbon economy in the next 3 to 4 decades. While CCS is potentially attractive to some developing countries, there has been limited activity to reach the stage of deploying demonstration projects due mainly to their high cost in the absence of significant carbon financing.

Out of an estimated 1000 GW or more of new coal-fired power plants due for construction through 2035, half of the new installations are expected in China alone. This new addition represents 73% of the present Chinese fleet. Nearly three-quarters of operating coal-fired power plants in China are comparably young and large, and hence potentially attractive for CCS retrofitting. Studies indicate one-fifth of these plants could be very attractive for retrofitting, but significant uncertainties remain. Given the size and profile of the Chinese fleet, further analysis, including site-specific assessments, would provide more details on the technical and economic potential. South Africa would experience a similar growth rate during this period, which would make the country the fourth largest generator of power from coal. The highest growth rate is anticipated for India, with an additional installation of 225 GW, which is more than double the current installed capacity. However, factors such as the general acceptance of CCS as part of the energy mix, have to improve in order to raise India's retrofit potential. The fleets in Japan and Korea are young, large and modern, but factors such as storage availability limit the potential for CCS.

The IEA estimates the total cost for a new average-sized coal-fired power plant that captures up to 90% of its CO_2 emissions to be US$1 billion over 10 years. While the financing mechanisms at hand are grossly insufficient for CCS to enable demonstration projects, the current carbon offset mechanisms are inadequate to drive CCS deployment in developing countries. Sources currently available for partial financing of CCS development in developing countries are very few (Table 4 of Almendre *et al.* 2011). Overall existing CCS financing mechanisms help grow capacity, but their support is insufficient to influence enough funding from capital markets to implement projects. The developed countries need to finance urgently to develop local capacity in developing countries and to lay the base for realizing potential benefits of large-scale CCS deployment in the future.

Under the UNFCCC, the Annex 1 parties have agreed to assist developing countries to undertake mitigation of carbon dioxide emissions. Since 2009, a number of governments and organisations that include the EU, the Global CCS Institute and the Norwegian, UK and US governments have collectively contributed or allocated hundreds of millions of dollars to current and future activities to support CCS capacity and project development in developing countries. These contributors have directly supported by financing specific activities, as well as through CCS capacity development funding mechanisms managed by ADB, APEC, CSLF, and World Bank. The most significant funding contribution of £60 million to support CCS in developing countries has come from the UK in 2012 (GCCSI 2012).

The IEA estimates that 70% of CCS deployment will need to happen in non-OECD countries to achieve global emission reduction targets by 2050 (IEA 2012b). The MEF Technology Action Plan for CCS recommended that 4 of the 10 commercial-scale projects should be in developing countries (IEA 2009). Estimates from some developing countries themselves have assigned a similar importance to CCS. Models produced by the Chinese show that China can stabilize its annual carbon dioxide emissions by 2030 through energy efficiency and fuel substitution, but a future decrease in annual carbon dioxide emissions after the 2030 peak will only occur if CCS is implemented in the next few years and then reaches significant commercial-scale deployment post-2030 (Development Research Centre of the State Council, 2009 & others).

Persistent concerns such as the high cost of CCS, access to energy, and permanence of storage, and other challenges are prompting a go-slow approach in some developing countries. If countries identify that CCS is a relevant technology for their low-emission strategies, then they must start right away the enabling, preinvestment, and demonstration activities to place themselves in a beneficial position from emission reductions from CCS in the coming decades. Many of these enabling and pre-investment activities will need to address country-specific requirements, and include: (a) developing geologic storage assessment; (b) developing legal and regulatory frameworks; (c) understanding the technology and project development framework through pre-feasibility and feasibility studies; (d) understanding funding and commercial issues; and (e) good practices for public engagement. Some of these activities can take a number of years to develop. For instance, storage characterisation from the basin level down to the site-specific level can take 3–6 years or longer, depending on how much information is already available. Developing appropriate legislative and regulatory frameworks can take considerable time, depending on the individual conditions of each country or region. Some countries have undertaken dedicated CCS scoping studies to investigate their CCS potential, such as the country's emissions profile (to ascertain whether there is a high degree of emissions from fossil fuel based power generation and industrial processes suited to CCS), its storage potential, and the feasibility of transporting CO_2 to likely storage sites.

Current Status: There are at least 19 developing countries engaged in CCS activities, ranging from capacity development, pre-investment, and planning activities, and in two cases it involves the operation of a CCS project. Most of these 19 countries are at an early stage of carefully assessing the opportunities and potential for CCS. The growing awareness of CCS as a potential mitigation technology within developing countries that heavily rely on fossil fuel based energy has facilitated the inclusion of CCS in the UNFCCC's CDM.

According to 'CCS development lifecycle', developed by GCSSI to help conceptualize different stages of CCS development, most developing countries are still at the early 'scoping' stage, although there are some developing countries which are further along the development lifecycle, notably Algeria, Brazil, Mexico, South Africa, the UAE, and China. China is now clearly transitioning from purely focusing on CCS R&D to taking steps towards creating an enabling environment for the demonstration and deployment of CCS. China's *12th Five-Year Plan (2011–15)* issued in 2011 emphasizes the country's plans to develop fully integrated CCS demo projects with the captured CO_2 to be used for EOR or for geologic storage (GCCSI 2012). There has been a preliminary analysis of legal and regulatory issues and/or review in the majority of developing countries that have an interest in CCS, including Brazil, China, Botswana, India, Indonesia, Jordan, Kosovo, Malaysia, Philippines, South Africa, Saudi Arabia, Thailand, Trinidad and Tobago, and Vietnam. Most of these preliminary analyses can be found in studies funded through APEC, the ADB, the CSLF, the Global CCS Institute, and the World Bank while some studies are still being finalised. The depth of analysis, however, differs between studies. In Latin America, CCS is seen as a crucial component in the region's efforts to combat climate change, particularly for emerging oil-based economies such as Mexico, Brazil, and Venezuela. The Latin American Thematic Network on Carbon Dioxide Capture and Storage was formed to help facilitate the development of CCS by promoting collaboration and to integrate CCS activities by scientists, research centres, and other agencies. Regarding storage assessment, Brazil, Mexico, and South Africa have already undertaken. Very preliminary storage assessments have been undertaken (or are currently being undertaken) as part of broader CCS scoping studies in a number of other developing countries, including Botswana, Indonesia, Jordan, Kosovo, the Maghreb region, Malaysia, Philippines, Thailand, and Vietnam.

Pilot and demonstration projects provide a focus for CCS associated activities, and their learning-by-doing highlight the importance, at least in the short term, of funding for enabling and pre-investment activities in these countries. In the medium term, more significant funding is needed for the 'extra' CCS costs associated with construction and operation of at least 5–10 demonstration projects in these countries. A key factor behind the interest in CCS in a number of developing countries is the link with EOR and/or gas processing. Given that EOR can help make CCS projects commercially viable, developing countries with EOR potential are well placed to take further CCS steps in the future (e.g. Indonesia, Malaysia, the Middle East, and North African countries).

The status of implementation of CCTs in China and India is briefly discussed here for two main reasons: one, they are the greatest examples of fast growing economies in the developing world with coal attracting the largest share in the energy-mix; two, they are currently the highest CO_2 emitters along with the USA in the world.

11.3 CHINA

China is by far the world's largest coal producer; its coal use amounts to nearly a third of worldwide coal consumption and accounts for about two-thirds of the country's primary energy supply (e.g., Chen & Xu 2010). China more than tripled its share of global coal-based generation from 11% in 1990 (471 TWh) to 37% in 2011 (3170 TWh) (Clemente 2012). China's per capita carbon emissions increased from 1980 to 2005 by about 4%; and driven mainly by population expansion, economic growth, urbanization and a booming transportation sector, China's future carbon emissions will continue to increase (Chen 2005; Chen et al. 2007). A sustainable energy development strategy was proved imperative for China (Chen & Wu 2004). Coal being the main energy source, sustainable development required the development and deployment of clean coal technologies such as supercritical and ultra-supercritical boilers, circulating fluidized bed combustion, and integrated gasification combined cycles. In addition, R&D and demonstration on carbon capture and storage technologies should be encouraged (Chen *et al.* 2006). Though China has announced its intention to reduce the country's reliance on coal, for the foreseeable future it will remain a dominant fuel, and will very likely still account for more than half of the country's primary energy supplies in the year 2030. The largest contributor to future increase in the demand for coal would be the electric power sector whose recent growth has been dramatic (Wei Yiming et al. 2006). Although Chinese energy statistics reported by different sources (e.g. IEA-WEO 2011, EIA-IEO 2011) exhibit substantial variation and several inconsistencies, there is no dispute about coal remaining a major player.

Presently China has an installed coal-fired capacity of approximately 400 GW which is growing by 50–100 GW per year (World Bank 2008). These coal-fired power plants are mostly responsible for the rapid increase in its greenhouse gas emissions, and currently China is the world's largest emitter.

In the eleventh Five-year Plan for NESD (2006–2010), China started a program to replace inefficient small units with large units, and to install FGD systems on all new coal power projects and accelerated desulphurization retrofits to all coal-based generating units larger than 135 MW. Phasing out of the small units, and replacement with 300 MW class subcritical units would reduce coal usage by 90 Mtce and cut SO_2 emissions by 1.8 Mt and CO_2 by 220 Mt (Chen & Xu 2010). The initiation in 2007 of China's dual programs of Large Substituting for Small (LSS) and Energy Conservation Power generation (ECPG) is expected to result in decommissioning of 114 GW of small inefficient plants and the addition of 112 GW of more efficient SC units. China's future growth in power generation centers on moving away from 300 MW and 600 MW subcritical boilers to larger and more efficient SC/USC boilers of 600 to 1000 MW size. These are expected to reduce emissions by as much as 40%. In the first year of the LSS program, more than 500 small inefficient thermal units (of total capacity, 14.4 GW) were decommissioned. During 2006–2011, the average coal consumption nation-wide dropped from around 366 gce/kWh to 330 gce/kWh indicating that the set goal of 320 gce/kWh for 2020 would be easily achieved (Clemente 2012).

China's 12th Five-year Plan (2011–2015) aims to cut carbon intensity (the ratio between changes in CO_2 emissions and gross domestic product) by 17%. The country is now installing its most advanced coal units using SC and USC steam conditions and modern SO_2/NO_x and particulates control systems. Asian Development Bank (ADB) observes that these steps have resulted in significant reductions in coal consumption, GHG and pollutant emissions, and an impressive improvement in energy efficiency (Clemente 2012).

SC/USC technology implementation: China started using supercritical technology in the 1990s with the procurement of 10 units (4 × 320 MW; 4 × 500 MW; and 2 × 800 MW) from Russia. The steam conditions of these units: 23.5 MPa/540°C/540–570°C. The first plant utilizing western technology was the Shi Dong Kou, commissioned in 1992 and consisting of 2 × 600 MW units with 25.4 MPa/538°C/565°C steam conditions. The second plant utilizing western technology was the Waigaoqiao plant in Shanghai (next to the Shi Dong Kou), which consists of two 900 MW units with steam conditions: 24.7 MPa/538°C/565°C. The project was financed by the World Bank (World Bank 2008).

Research on supercritical and ultra-supercritical systems in China, though started relatively late, has developed rapidly in recent years. In late 2004, the first domestically fabricated 600 MW supercritical unit was put into operation at China Huaneng

Yuhuan Power plant at Taizhou, Zhejiang province: 1000 MW ultra-supercritical boilers; generation efficiency of 45% (*Source*: www.panoramio.com/photo/27957989, by zhoubin 1972).

Group's Qinbei Power Plant in Henan Province. Due to localization of the technology the cost is reduced significantly with Qinbei's unit costing about 4000 RMB/kW.

The Huaneng's Yuhuan power plant (4 × 1000 MW), in Zhejiang Province is the country's first commercially operated power plant using indigenously-made 1000 MW *ultra-supercritical* pressure boilers. Units 1 & 2 went online in 2006, unit 3 in 2007, and unit 4 in 2008. Unit 4 generates 22 billion kWh of electricity per year. The generation efficiency of the Yuhuan power plant is 45% with the coal consumption per kWh being 285.6 gce/kWh which is 80.4 gce/kWh less than the national average in 2006. These units are touted as the 'world's cleanest, most efficient and most advanced' PC units.

At the same time, two 1000 MW *ultra-supercritical* units went into operation at China Huadian Corporation's Zouxian power plant in Shandong Province in 2006 and 2007, respectively. The power generation efficiency is 45% and the gross coal consumption rate is 272.9, 66.1 gce/kWh, less than the national average of 339 gce/kWh in 2006. The two plants both utilize high-efficiency dust removal and desulphurization with operating conditions identical to international standards (Chen & Xu 2010). The first two units of Yuhuan plant and the Zouxian plant utilize steam conditions: 26.2 MPa/605°C/605°C (World Bank 2008).

At of the end of 2006, China had supercritical plants in operation, around 30 GWs of installed capacity; most of them have been designed for 24.7 MPa/565°C/565–593°C, but two have steam conditions: 24.7 MPa/600°C/600°C. By the end of 2007, approximately 120 GWs of installed capacity was expected to be utilizing supercritical conditions (Mao 2007).

Between 2004 and 2007, about 123.6 GW of 600 MW supercritical units were installed (Chen & Xu 2010). In terms of number of plants, more than 150 supercritical or ultra-supercritical units of 600 MW size or more have been put into operation/under construction (Deng 2008).

Bobcock & Wilcox (B&W) announced in September 2010 that their local company (B&W Beijing Co.) would build two 1000 MW USC coal-fired boilers for a large power plant in Zheijiang province. The two Spiral Wound Universal Pressure (SWUP) boilers which are unique in design will use one of B&W's most advanced and efficient boiler design.

Another development announced was installing the Ovation expert control system by Emerson Process management at two new 1000 MW USC coal-fired generating units under construction in Anhui province. The Ovation system, now being used in many of the 1000 MW units in China, will perform data acquisition, as well as monitor and control all plant's major components including the boiler, turbine and generator. The system also manages the modulating control system, sequence control system, electrical control systems, furnace safety supervisory system, feedwater turbine control system and flue gas desulfurization system (Clemente 2012).

Chinese manufacturers have developed joint ventures and licensing agreements, so the majority of the equipment for SC and USC plants is already manufactured in China. More specifically, Shanghai Boiler Works has teamed up with Alstom and Siemens; Harbin Boiler Group works with Mitsubishi; and Dongfang Boiler Industrial Group has a joint venture with Hitachi. While there is a clear commitment to supercritical technology, subcritical plants, including very small ones (well below 100 MW), continue to be built in China for a variety of reasons: (a) lack of adequate manufacturing capacity for the state-of-the-art SC and USC plants to satisfy rapidly increasing demand; (b) not practical to close down all manufacturing facilities in China, which have been producing smaller units (up to 300 MW) or to convert them to manufacture state-of-the-art plants in a short period of time; in fact, some of these facilities, in addition to satisfying domestic demand, are targeting exports (typical sizes being exported are in the 100–300 MW range) to other countries (World Bank 2008).

Looking forward, China's coal-based power generation will be increasingly clean. The NDRC uses policy instruments and economic incentives to prioritize the starting of cleaner and larger coal plants. It was expected that between 2010 and 2020, all new power plants – 600 MW and higher unit capacity – would be required to be supercritical and about half of the newly built power generating units, ultra-supercritical. As a result, supercritical units would account over 15% of the total power capacity by 2010 and 30% by 2020 (Huang 2008; Chen & Xu 2010). The average efficiency of China's coal-fired power plants is projected to improve from the present 30% to above 40% in 2030.

Circulating fluidized bed technology is another preferred option in China for mid-sized (300–450 MW) and larger (400–600 MW) utility units because of its fuel flexibility, low emissions capability and the ease of scale-up. There are at least 200 (possibly up to 500) CFB units in operation, ranging in size from 3 MW to 300 MW. The total installed capacity is estimated at 10,000 MW. Approximately 2,500 small bubbling AFBC boilers have also been constructed, but there are no accurate statistics regarding to their operating status (World Bank 2008). In 2006, the first 210 MW CFB 1025 t/h boiler was successfully put into commercial operation in Jiangxi. The

Baima project, the China's first 300 MW CFB demo plant, was setup by Alstom in cooperation with Dongfang Boiler of China. The engineering of the unit was done in France and the manufacturing was shared between Alstom and the Dongfang (Morin 2003). Several institutes and companies including Tsinghua University, Institute of Engineering Thermophysics, Chinese Academy of Sciences, Sichuan Boiler Works, Hangzhou Boiler Works, Jinan Boiler Works and Wuxi Boiler Works so on are involved in research and development of CFB technology. Many CFB units, 300 MW capacity, are in operation, that include the China Huaneng Group's Kaiyuan power plant consisting of two lignite-fueled 300 MW CFBs, Honghe power plant and Xunjiansi power plant. Many more CFB projects of same capacity are planned for implementation on a co-production or license basis. A 600 MW CFB plant is known to be under construction and, if successful, it would be the largest CFB plant in the world (Chen & Xu 2010).

Having developed capability to manufacture desulfurization equipment, coal-fired power generation units with FGD systems in China accounted for 30% of the total installed thermal capacity at the end of 2006. Since then about 40% of the new units have been fitted with FGD systems. In 2007, the total capacity of contracted projects by desulphurization companies was 374 GW, of which 208 GW had gone into operation. According to the government's SO_2 Pollution Control Plan, the country's existing coal-fired power plants are required to install fluegas desulphurization units with a total capacity of 137 GW which would reduce SO_2 emissions by 4.9 million tons (China Desulphurization Industry Report, 2007–2008). At present, the capital and operating costs of high-efficiency NOx removal technologies is relatively high. Several selective catalytic reduction (SCR) demonstration projects are in operation, and about 6000 MW of SCR units are under construction. Given the development trends, it was estimated that the SCR installation rate for power generation would reach 30% in 2020 and 50% in 2030, while the installation rate of FGD systems would touch 80% in 2010, 90% in 2020 and 95% in 2030 (Chen & Xu 2010).

Gasification technologies: The primary gasification technologies being developed in China are ash agglomerating fluidized bed coal gasification, non-slag and slag two-stage entrained flow bed coal gasification, two-stage dry feed entrained flow bed coal gasification, coal-water slurry gasification with opposed multi-burners, and coal-water slurry gasification with multiple materials. The multi-nozzle striking flow coal slurry-feed gasifier was successfully developed by Huadong Science and Technology University, with a 1150 t/d gasifier put into operation in 2005. An intermediate test facility with a 36 t/d 2-stage dry-feed pressurized gasification plant has been built by the Xi'an Thermal Power Research Institute, with a preliminary design of a 1000 t/d gasifier now completed. Nineteen Texaco gasifiers are in service in China and 12 Shell gasifiers were under construction or put into service in 2006.

Liquefaction: China has made great progress on coal liquefaction technology in recent years. The China Coal Research Institute, Shenhua Group Cooperation Ltd, Shanxi Coal Chemistry Institute of the Chinese Academy of Sciences, and the Yankuang Group have developed direct/indirect coal liquefaction technologies and are ready for industrialization. Demonstration projects, integrating CCS into CTL (coal-to-liquids) by the Shenhua and Yitai BIHI-TECH Company in Ordos, Inner Mongolia have been working well since 2004, and the commercial operation has begun in 2008. CTL is not a major source of CO_2 that could considerably add to global warming; but the project owners are likely to pursue these capital-intensive CTL-CCS projects only where EOR opportunities exist to provide stable revenue over longer periods (Morse *et al.* 2010). In 2006, total methanol consumption in China was around 8.86 Mt, of which around 65% was produced from coal. About 0.6 Mt/year of DME is produced from methanol and such demonstration plants for MTO/MTP are planned (Chen *et al.* 2008a).

IGCC: IGCC was included in 2006 for future power generation in the National Program for Medium-to-Long-Term Scientific and Technological Development (2006–2010), announced by the State Council of China. The overall goal is to form clusters of these advanced coal technologies and support the development of China's energy equipment manufacturing industry to achieve efficient, clean, and affordable use of coal. China aims to develop domestic IGCC plants by importing, understanding and using foreign technology, with increased localization. 300 MW/400 MW IGCC power units are preferred based on the current gas turbine capacity. Eleven research institutes have been studying since mid-1990s, the technical feasibility of the technology, and are now collaborating with Texaco/General Electric, Shell, and the Asian Development Bank to develop this technology. Domestic electric power design institutes have gained the ability to produce general designs of IGCC power projects, Combined Cycle islands, and the balance of the plant control and instrumentation system. China has imported manufacturing for E class and F class gas turbines from GE, MHI (Mitsubishi Heavy Industries Ltd.) and Siemens for over a decade. F class gas turbine power plants locally designed have been put into operation now. Local manufacturers can make the outside shell of the gasifier with fabrication facilities for 2000 t/d gasifiers, with internal components being imported. China has started research and development of low heating value gas turbines. The 5×104 Nm^3/h air separation units can be made domestically with a 6×104 Nm^3/h ASU being designed and fabricated. The Huaneng Group launched the GreenGen IGCC power plant with CCS in Tianjin. The project has three stages: In the first stage (2006 to 2009), to develop a 250 MW IGCC power plant and investigate the polygeneration options, especially pilot testing of hydrogen production and fuel cell power generation; the main project areas being the coal gasifier with a processing capacity of 2000 tons per day, the gas and steam turbine combined cycle consisting of 171 MW gas turbine and 110 MW steam turbine, a waste heat boiler, and a syngas purification system with a processing capacity of 160,739 m^3/h; the second stage from 2010–2012: improve the IGCC and poly-generation systems, build a 400 MW IGCC plant, optimize the gasification technology and develop H$_2$ and CO$_2$ separation system; the third stage from 2013–2015: build a 400 MW GreenGen demonstration plant including H$_2$ production, fuel cell and H$_2$ power generation with CCS, and operating the plant with near zero carbon emission to prove the economic viability of the GreenGen plant in order to move to the stage of commercialization (Chen & Xu 2010). CCS technology integrated with IGCC, as visualized by GreenGen, is now the major focus of state-supported CCS for power plants in China. The impetus for making IGCC power plants a state priority are the energy security concerns, as well as co-benefits from reduced local pollution and synergies with chemicals production. The direct benefits to the country's energy security through higher energy efficiency and the development of potentially rewarding domestic intellectual property are clearly the reasons for the Chinese government to fund GreenGen. The outcome from GreenGen especially regarding IGCC costs and associated CO$_2$ capture will be crucial as China prepares the roadmap of its power generation arrangement beyond 2020 (Morse *et al.* 2010).

Table 11.1 Costs of different coal-fired power systems in China.

	PC + FGD	SC + FGD	USC + FGD	IGCC	CFB
Capacity (MW)	300	600	1000	400 class	300
Unit cost (RMB/kW)	4596	3919	3924	7751	4799
Ref. electricity price (RMB/MWh)	352.7	321.0	310.7	504	354.8

(*Source*: Electric Power Planning and Design Institute, China, 2006)

Yankuang Company, with indigenous R&D support put up in April 2006, the demonstration of the first coal gasification-based poly-generation system that provides an output of 60 MW and 240 thousand tons of methanol per year. This unit has provided the learning-experience for a long-term development of IGCC and polygeneration systems in China. Few more plants are planned which include a 200 MW Banshan IGCC power plant in Hangzhou, and a 200 MW IGCC power plant and a retrofit to 120 MW combined cycle power plant to use IGCC in Dongguan, Guangdong province (Chen & Xu 2010).

Costs: The costs of the various coal-fired power generation technologies in China are listed in Table 11.1 (Electric Power Planning and Design Institute, 2006). Here, the standard coal price is assumed to be RMB 430/t and the annual operating time is 5000 h. The IGCC unit cost is much higher compared to other technologies though IGCC offers same efficiency as USC plants. The IGCC offers better air pollution control, lower water demand, reduced solidwaste, the possibility for polygeneration, and lower costs to add CO_2 capture (to new plants).

Carbon Capture & Storage: There are three major national science and technology programs in China sponsored by the Ministry of Science and Technology (MOST): (i) National Key Technology R&D Program, (ii) National Basic Research Program (973 Program), and (iii) National High-tech R&D Program (863 Program). These Programs study different aspects of CCS.

Some Chinese companies have also started their own R&D. For example, PetroChina, the largest oil and gas producer and distributor in China, has been operating China's first CO_2 storage and usage project at the Jilin Oil Field since 2006. In 2008, the China Huaneng Group installed a pilot-scale post-combustion carbon capture facility at their Beijing thermal power plant, capable of recovering more than 85% of the carbon dioxide from the flue gases. The project was designed and developed by the Xi'anThermal Power Research Institute, with all of its equipment locally made. Further, Huaneng Group and Shanghai Electric Co. have set up a research center which has plans to build 2×660 MW USC units with CO_2 capture at Shanghai Shidongkou Power Plant.

International collaboration on CCS: At the EU-China Summit 2005, China and EU agreed to cooperate to develop and demonstrate advanced, near-zero emissions coal technology through carbon capture and storage in China and the EU by 2020. The collaboration agreed with UK and EU spreads over three phases: phase i – exploring the options for near-zero emissions coal technology through CCS in China during 2006–2009; phase ii – defining and designing a demonstration project from 2010 to 2011; and phase iii – constructing and operating a demonstration project from 2012 to 2015. As a follow up, the Cooperation Action within the CCS China-EU

(COACH) project and the Near Zero Emissions Coal (NZEC) initiative involving European and Chinese partners are initiated. Several Chinese research institutes and industry have tied up many collaborative projects related to CCS, with EU, since 2006. China-Australia Geological Storage project with Australia was another one undertaken during 2008–2010.

Though China is leveraging international support for developing CCS projects, they are advancing at a slower rate. All of these projects certainly represent useful research efforts, but they do not represent a level of investment comparable to the Shenhua and GreenGen projects. China's interests are not aimed at installing post-combustion capture on the existing 600 GW fleet of conventional coal plants which are its biggest source of emissions. The implementation of CCS in the present-day Chinese power sector presents special obstacles. The electricity prices in China are tightly controlled in order to meet dominating political priorities; as a result, the Chinese power market cannot internalize increased costs, making it nearly impossible to set up a commercially viable CCS model on its own. In 2008, much of the Chinese power market could not even bear the cost fluctuations of its coal. Moreover, CCS costs extend beyond the power sector to the entire coal value chain; and adding CO_2 capture reduces generation efficiency by 20–30%. It is estimated that CCS at scale in China, as prescribed by the IEA Blue scenario, would demand about 200–300 million tons of additional coal production per year. Beyond obvious added cost increases for generators, increasing coal production to these levels would require new mining capacity, rail infrastructure, port expansions and shipping capacity that call for massive investments, to maintain the chief objective of supplying cheap and reliable electricity. Costs would be well in excess of 100 million RMB (15 billion US$).

CCS would also likely come at the expense of some investments in wind, solar and nuclear power which enhance China's diversity of energy supply. Secondly, other key factors such as technological and regulatory uncertainty, high costs and the lack of clear carbon policies that could provide a steady income for capital-intensive CO_2 capture deter global investments in CCS. Since China's highest priorities are security and diversity of fuel supplies, cheap and reliable electricity, and development of domestic intellectual property for key energy technologies, its involvement in CCS projects should be viewed in that context, rather than climate change concerns. Hence, China's plans are likely to promote the development of China's CCS demonstration efforts but they do not translate into incentives to deploy CCS at scale in power plants where they are most needed. Fundamental and interrelated country's interests – energy security, economic growth and macro-economic stability – come in the way of large-scale implementation of CCS in China, unless foreign funding can entirely support such an implementation (Morse *et al.* 2010). Major CCS projects under operation/construction/planning in China are given in Table 11.2.

US–China Clean Energy Research Centre: Abundant coal resources and widespread use in the USA and China present challenges as well as opportunities for the two countries in environmental performance and commercial development. Advanced Coal Technology consortium was created to address technology and practices for advanced coal utilization. Joint research would be undertaken in the areas of advanced power generation, clean coal conversion technology, pre-combustion capture, post-combustion capture, oxy-combustion capture, CO_2 sequestration, CO_2 utilization, simulation and assessment, and communication and integration under a

Table 11.2 Major CCS projects in China.

CCS Projects	Technology	Partnership	Finances sharing	Status
GreenGen Corporation	IGCC Pre-combustion Decarbonization Gasification or partial oxidation Shift plus CO_2 separation	Huaneng with seven other state-owned Companies; Peabody Energy	Registered capital RMB 300 million (US$ 44 million) Huaneng 51%; 7 other companies 7% each Total investment will reach RMB 7 billion	Under construction
Shenhua CTL	Coal to synfuels (DCL)	Shenhua Group, Sasol, WesVirginia University	US$ 1.4 billion	CTL operational
Huaneng Beijing Thermal Power	Pot combustion	Huaneng; Australia CSIRO	US$ 4 million research project by CSIRO	Operational since 2008
Near Zero Emission Coal	Research, Development & Demonstration (R&DO)	UK; China Ministry of Science & Technology	US$ 5.6 million equivalent from the UK Govt's Dept of Energy & Climate Change	Under planning
COACH Project	R&DO	COACH project groups: 20 partners (R&D, manufacturers, Oil & gas companies etc.)	Partly funded by European Union	Under planning
Shanghai Huaneng Shidon ghou	Post combustion	Huaneng	Corporation investment	Under construction

(Taken from Morse *et al.* 2010)

five-year *Joint Work Plan* to significantly advance technology in the area of clean coal in both China and the United States[1].

11.4 INDIA

Coal usage in India, especially in power generation, was briefly explained in Chapter 2. With one of the largest hard coal reserves in the world (currently third highest), coal-fired power has been the foundation of India's growing electricity sector for the past three decades. The power sector consumes nearly three quarters of all coal used in India, with most of the remainder used by heavy industries such as iron, steelmaking, cement and fertiliser production. Over the last five years, the demand for coal has been growing at an average rate of 8–9% annually as compared to a 5–6% increase in domestic

---

[1] Available at http://www.us-china-cerc.org/Advanced_Coal_Technology&CER_coal_JWP_english_OCR_18_Jan_2011.pdf

production (BP statistical world energy review 2012). This has widened the demand-supply gap, leading to growing dependence on imported coal. In 2011–12, the country imported around 100 million tons of coal (including thermal and coking coal) (Coal Imports, Press Information Bureau, 14th May 2012). By 2015, Coal India Ltd.(CIL) is expected to see a shortfall of around 80 million tonnes, thus limiting CIL's ability to meet growing demand. The shortage of coal is not only affecting operational plants but is also raising concerns around the viability of future power projects. The lack of coal linkages is making it incrementally difficult for power-generation companies to raise capital for their proposed thermal plants. Further, as per recent reports, the government of India is likely to lower the country's power capacity addition target for the Twelfth Five Year Plan from 1,00,000 MW to 75,000 MW (Government may lower power generation target for 12th Plan, Livemint, 7 February, 2012) as a result of fuel shortage (India – KPMG-Research 2013).

Nearly all the coal-fired plants in India are subcritical pulverized coal plants. Standardized designs have been used with the following plant sizes: 60 MW, 110/125 MW, 200/210/250 MW and 500 MW. According to the Central Electricity Authority, 200/210/250 MW and 500 MW units form the backbone of the Indian power industry and together constitute about 60% of the total thermal capacity (CEA 2003). Technology for these plants was originally acquired from the Czech Republic and Russia and, in the 1970s, from Combustion Engineering Inc., USA (presently part of Alstom). These plants consumed about 279 million tons of coal and 25 million tons of lignite in 2004–2005 (CEA 2006a), making the power sector the largest consumer of domestic coal, about 80% of coal produced in 2004–2005 was used by power plants (Ministry of Coal, 2006). This domination of coal in the power sector is likely to continue in the future. According to the Working Group for the 11th five-year Plan (CEA 2007b), about 46.6 GW of new coal-based capacity is expected to be installed by 2012 (which is about 68% of total planned addition of 69GW) for which the coal requirement is projected to about 545 MT. Central Electricity Authority states that 98 GW of coal-based generation is under construction for commissioning during 2012–2017 (12th Plan period). Long-term scenarios from the Planning Commission (2006) suggest that annual coal consumption by the power sector might range between 1 to 2 billion tons by 2030.

Over the years, India has made significant progress in manufacturing major component systems for coal-fired power plants. Its manufacturing facilities are state of the art and the organizations (e.g., BHEL) involved have demonstrated excellent expertise and achieved significant technological progress (World Bank 2008).

Indian coal is abundant, but its quality and accessibility is very problematic. It is generally high in ash content, low in calorific value and full of mineral sediments. Only an estimated 6% of India's domestic coal reserves are in the prime coking coal category suitable for efficient and high-quality industrial purposes. Lower quality hard coal can be washed, dried and freed of deadrock, but this involves lot of effort and expenditure. India has thus for many years imported high-quality coking coal for its iron and steel sector, some of which is mixed with domestic coal to reduce the latter's ash content. India's steam coal is generally more suitable for its intended use than domestic coking coal, but high ash content can still reduce boiler efficiency and severely hamper efforts to meet emission standards. Such resulting load limitations are known to contribute to interruptions in electricity supply (Pakiam 2010). The industry, however, has successfully addressed the issues associated with the erosive nature of the coal and the high amount of ash.

Of the 200 thermal power plants of different size and capacity, about 40% are more than than 20 years old. A few of the existing power plants in India have efficiencies in the range, 35–40%, though most of them have an efficiency ranging from 20–30%. These efficiencies are low relative to OECD countries (the average efficiency in OECD countries is 36%, HHV-basis, though the newly built power plants demonstrate 37–42%). In India, the most commonly used design is the 500 MW plant utilizing subcritical steam conditions: 16.9 MPa/538°C/538°C (Central Electricity Authority 2003; Mott McDonald 2006; EPDC of Japan 1999).

Clean Coal Technologies: The clean coal research in India has started with clear objectives: (a) to improve the quality of non-coking coal at the pre-combustion stage to provide value addition for power generation, (b) to adopt new coal combustion and conversion technologies for utilizing coal with high efficiency, and (c) to reduce carbon dioxide and other emissions in the environment through renovation and modernization of plants.

The country is still years away from developing supercritical coal technologies cost-effectively. Nevertheless there are already initiatives underway to commission more efficient coal units for domestic use (see Table 11.3). Considering the variable quality and high ash content of domestic coal stocks, import of the steam coal will be needed to create fuel mixtures suitable for both sub- and super critical power plant configurations.

While assessing the CCTs for India, there are key drivers to be considered: (a) Maximizing performing efficiency of the coal resource, both domestic and imported, is of strategic importance for India which is driven by energy security considerations, as well as economics and the need to reduce environmental impacts; (b) Since India needs reliable and affordable electricity, the highest efficiency technology may not be necessarily the best choice, if it is not reliable; a trade-off between reliability and efficiency may be needed; (c) India has not been having adequate power supply to meet the demand for many years, and closing this gap is of overriding importance for the country; (d) While the power industry in India has good experience in burning the domestic coal which has very high ash content, some of the new technologies may need a gestation period before they achieve adequate reliability; and (e) Physical constraints such as lack of land and water, and the inability of the power plant suppliers to meet tight delivery schedules.

Nonetheless, Supercritical plants are being introduced at unit sizes of 660 MW and 800 MW. Six 660 MW units are under construction at Sipat and North Karanpura plants since 2011. Also, supercritical plants have been specified for five ultra-megapower plants (4,000 MW each) which are preparing for implementation (Malti Goel 2010).

The Jhajjar Power Plant, 1,320 MW (2 × 660 MW), is one of India's first supercritical coal-fired plants using domestic coal located at Village Khanpur, Dist. Jhajjar, Haryana (see photo above). Developed by a private producer, CLP Ltd., both the units of the plant have been successfully synchronized and have achieved Commercial Operation Date (COD) in 2012. The two units will provide about 270 lakh additional units of electricity daily to the state when operating at rated capacity. Central Electricity Authority of India has developed guidelines on the introduction of large

Jhajjar Power Plant in India, 2 × 660 MW, Supercritical technology, fuel: HSD & coal (*Source*: CLP India Pvt Ltd website).

supercritical plants (CEA 2003) as the country set a goal to use supercritical technology for 60% of the new coal-fired plants built during the 12th Five-Year Plant (2012–2016). CEA recommends that the new plants should be of 600–1,000 MW size and utilize steam conditions in the range, 565 to 593°C for about 10 to 15 units, and higher steam temperatures for the next group depending on site-specific techno-economic considerations.

The ultra-mega projects being implemented in India provide an excellent framework with regard to institutional capacity to finance, plan, construct and operate such plants. It is suggested that introduction of SC technology in India should be accompanied by an institutional capacity-building program which includes training on plant operation and maintenance, water chemistry control, etc. directing especially at the State Electricity Boards (SEB). As supercritical and ultra-supercritical plants are introduced in the country, the subcritical plants will continue being manufactured and used. Several factors such as shorter lead times, capacity to produce them indigenously and familiarity by the electric utilities of the country make subcritical units attractive, at least for the short-term (next 10 years). It is, however, suggested these plants should be as large as possible (e.g., 500 MW) and be designed with the aim of high efficiency, with steam conditions preferably, 16.9 MPa/538°C/565°C (World Bank 2008). India has nearly 50 GWs of installed capacity represented by plants, 11 to 30 years old, which have reduced reliability, output and efficiency relative to design conditions. The government has placed them into three categories: units to retire, to rehabilitate and to replace with new state-of-the-art units. Rehabilitation of existing units encounter some problems that include lack of adequate financing, lack of interest for such projects by the large power plant suppliers who are overbooked with orders for new power plants, and difficulty in guaranteeing the performance of the rehabilitated plants. The World Bank is working to provide resources to overcome these hurdles with support from Global Environment Facility (GEF) Program.

CFB technology has been used successfully in India because country's low rank coal is a most suitable fuel for CFB boilers. There are more than 36 CFB units in operation representing 1,200 MW of installed capacity; most of them are relatively small size (2–40 MW) with the largest unit being 136 MW. Two 250 MW lignite-fired CFB units are under construction by BHEL, India; the first CFBC boiler of 175t/h capacity was commissioned at Sinarmas Pulp and Paper (India) Ltd., in Pune. If SO_2 emission regulations are introduced in the future, it is likely that this technology will be utilized more. World Bank suggests that CFB as well as pulverized coal plants, could utilize biomass co-firing, an option considered CO_2-neutral.

The first coal gasification test facility in India was installed by BHEL with a 10 MW captive power plant using gas from 'coal washery rejects' at the TISCO Jamadova Colliery in 1987. Fertilizer plants at Ramagundum and Talcher installed entrained bed gasifiers for utilizing indigenous coal for production of ammonia. Although these plants have encountered some practical problems, the experience gained has helped to install two CFB gasifiers of 390t/h capacity successively at Surat Lignite Power Plant. The Neyveli Lignite Corporation Ltd. has also installed a Circulating fluidized bed gasifier based on the Wrinkler process to demonstrate the use of lignite. Coal India and Gas Authority India Ltd. have jointly started work in 2010 to set up a large surface coal gasification plant. Research on IGCC technology was started in India; and India was the first Asian country to construct a 6.2 MW IGCC demonstration plant in 1986. This closed loop combined cycle unit with air-blown fluidized-bed gasifier installed by BHEL was unique in terms of testing coal with up to 40% ash at temperatures of 960°C and 1050°C at 0.8 MPa. Gujarat Sanghi Steam Works and Ahmedabad Electric Co. are other institutions in the country that developed IGCC technology (Malti Goel 2010). Also, a comprehensive study funded by USAID was completed in 2006 assessing the feasibility of using IGCC in India. This study recommends a demonstration project of approximately 100 MW utilizing fluidized bed gasification technology. GTI's U-Gas technology was identified as most suitable. BHEL, India is developing its own fluidized bed gasification technology at a 6.2 MW pilot plant in Trichy; this could be utilized for the demonstration at 100 MW scale size. Implementing IGCC with imported coal will be the country's strategic decision to participate in the advancement of this technology and preparing better if CCS is required in the future. Although CCS is not yet pursued seriously in India, under the National Programme on CO_2 Sequestration (NPCS), research was started in 2006 by the Department of Science & Technology, Government of India, identifying the following thrust areas: (a) CO_2 sequestration through micro-algae bio-fixation, (b) carbon capture process development, (c) terrestrial agro-forestry sequestration modelling network, and (d) policy development studies. Further, India considers assessing its sequestration potential (geological) and monitoring CCS related developments.

Underground gasification is also being explored in India. Coal India Ltd., GAIL, ONGC and Reliant Industries Ltd. are surveying sites to test the technology. The Mehsana and Gondwana coal producing areas have been identified as most suitable. There is, however, no adequate information to determine whether this option is feasible and cost-effective. R&D activity is being carried out in various universities and institutes.

Steps are being taken to utilize more washed coal. The Ministry of Environment and Forestry stipulates that coal shipped more than 1,000 km from the mine should be washed and contain less than 34% ash. Advanced coal cleaning processes to suit the indigenous coal for power generation have been developed which are cost-effective and can reduce the ash content by about 10% or more. Also, Coal India announced that the coal in all new coal mines would be washed (Coal Age 2007). As a result, it was projected that coal washing would reach 55 million tons in 2007 and 163 million tons by 2012 (Sanyal 2007). This would certainly reduce transportation costs and improve plant reliability and potentially efficiency. The overall cost-effectiveness of coal washing requires site-specific assessment as it is impacted significantly by factors such as the characteristics of the coal (not all coals are washed easily), the distance between the mine and the plant and the design of the plant. Hence, an integrated analysis which includes coal mining, coal transport and energy conversion (power plants) need to be carried out, and the cost-effectiveness of options such as upstream coal beneficiation (cleaning) and clean coal conversion technologies in an integrated manner must be addressed.

Proposed CCT roadmap for India

At a CCT Workshop organized by DST and BHEL in October 2006, a time-bound Road map was drawn to establish the required R&D infrastructure for implementation with an ultimate goal of achieving emission-free-electricity-generation.

Ongoing and near-term (up to 2012): (i) improved coal recovery, coal benefaction, reduction in cost, (ii) more emphasis on FBC; SC power plant boilers; IGCC demonstration, (iii) enhanced energy recovery from coal: CBM, CMM etc, (iv) pilot-scalestudiesonliquefaction;

Medium-term (2012–2017): (i) IGCC, PFBC, USC power plants, (ii) enhanced energy recovery from coal, (iii) commercial scale coal liquefaction, (iv) zero emission technologies pilot-scale, (v) carbon sequestration pilot-scale;

Long-term (2017 and beyond): (i) zero emission technologies commercialization, (ii) carbon sequestration demonstration plant, (iii) IGFC and production of hydrogen fuels from coal.

There have been several barriers and constraints, technical and non-technical, for the advancement of CCTs in India, and India is not able to keep pace with international developments. The research and development community and policy analysts have suggested that India should first launch programmes to raise the average combustion efficiency rates of coal from 30 to 40% over two decades, with international financial and technological assistance where necessary to support the accelerated deployment of supercritical units. If CCS is being seriously considered, it should only be part of a long-term innovation strategy, supported with domestic institutional capacity building. Coal gasification and advanced combustion technologies are also potentially important long-term future options, but are presently subject to considerable uncertainties in terms of technical and cost trajectories, their suitability for India's conditions as well as the timing of India's greenhouse gas reduction commitments. In the short term, measures to limit transmission and distribution losses would also help to reduce to some extent, the need for additional generation capacity. This would provide time to evolve more considered supply-side technological decisions in the longer term (Pakiam 2010). Despite these differing views and issues, the coal-based power generation capacity has

Table 11.3 Supercritical projects currently under development in India.

Developer	Location	Capacity (MW)	Remarks
Adani Power	Mundra, Gujarat	4,620	4 phases; 990 MW is currently operational. Phase 3, approved by the UN to earn annual carbon offsets for one decade
Adani Power	Tirora, Maharashtra	3,300	3 phases under construction; Seeking carbon revenues
Tata Power	Gujarat	4,000	Under construction; application for carbon revenues was rejected
Reliance Power	Andhra Pradesh Madhya Pradesh Jharkhand	3,960 × 3	Under planning and construction
CLP Power India	Jhajjar, Haryana	1,320 (2 × 660)	Achieved COD in 2012*
Indiabulls Power Ltd	Nandgaonpeth, Maharashtra	1,320	Under planning & construction
GMR Energy	Chattisgarh	1,370	Under planning & construction
Total capacity		**27,810 MW**	

Source: Reuters, 'FACTBOX – India Coal Plants Seeking UN Carbon Offsets', 12th August 2010 at (www.alertnet.org/thenews/newsdesk/SGE67B04A.htm (21 Sept. 2010);
*at www.clpindia.in/operations_jhajjar.html

been on the rise, and during April–December 2012, 9505 MW was installed (Power Ministry website).

China and India in broad comparison: India's population is almost as large as that of China (around 1.2 billion compared with 1.3 billion) and has a similarly rapid rate of economic growth. Like China, India has extensive coal reserves and it is the world's third largest coal producer after China and the USA. Coal use in India is growing rapidly, with the electric power sector accounting for a large share of new demand. However, India's per capita electricity consumption of 660 kW$_e$h/year is around 35–38% of China's, and its current rate of coal consumption is about a fifth of that of China. In India, as in China, self generation by industry is also a significant source of coal demand. A large share of future growth in the electricity sector will be coal-based. Current government plans project growth in coal consumption of about 6%/year (Government of India 2005). At this rate, India's coal use would reach the current level of the United States coal consumption by about 2020, and would match current Chinese usage by about 2030. This suggests that there may be time to introduce cleaner, more efficient generating technologies before the greatest growth in coal use in the Indian power sector occurs.

REFERENCES

Almendra, F., West, L., Zheng, L., & Forbes, S. (2011): CCS Demonstration in developing countries: Priorities for a Financing Mechanism for Carbon Dioxide Capture and Storage, WRI Working Paper, World Resources Institute, Washington DC; Available online at www.wri.org/publication/ccs-demonstration-in-developing-countries.

CCT Initiative – Road Map for future development, India (2006): Clean Coal Technology DST-BHEL Workshop, October 26–27, 2006.

Chen, W.Y., & Wu, Z.X. (2004): Current status, challenges, and future sustainable development strategies for China energy, *Tsinghua Science and Technology*, 9 (4), pp. 460–467.

Chen, W.Y. (2005): The costs of mitigating carbon emissions in China: findings from China MARKAL-MACRO modeling, *Energy Policy*, 33(7), pp. 885–896.

Chen, W.Y., Wu, Z.X., & Wang, W.Z. (2006): Carbon capture and storage (CCS) and its potential role to mitigate carbon emission in China, *Environmental Science*, 28 (6), pp. 1178–1182.

Chen, W.Y., Liu, J., Ma, L.W., Ulanowsky, D., & Burnard, G.K. (2008a): Role for carbon capture and storage in China, In: Ninth Intl. Conf. on Greenhouse Gas Control Technologies, Washington, DC.

Chen, W., & Xu, R. (2010): *Energy Policy*, 38, pp. 2123–2130.

China Securities News (2007): China's coal to oil industry has beginning to take shape, November 13, 2007.

China Environment Protection Agency (2008): China Environment Protection Statistical Yearbook 2007, Beijing: China Environment Protection Publishing House.

China Statistics Bureau (2008): China Statistical Yearbook 2007, China Statistics Press, Beijing.

Coal Age Magazine (2007): MISC. NEWS, May 2007, p. 8.

CEA (2003): *Report of the Committee to recommend next higher size of coal-fired thermal power stations*, November, 2003, Central Electricity Authority, Government of India.

CEA (2006a): All India electricity Statistics: General Review 2006, Central Electricity Authority, Government of India.

CEA (2007b): Report of the Working Group on Power for 11th Plan, CEA, Govt of India, at http://cea.nic.in/planning/WG%2021.3.07%20pdf/03%20contents.pdf.

Clemente, Jude (2012): China leads the Global Race to Cleaner Coal, POWER magazine, Dec 1, 2012, available at www.powermag.com/coal/china-leads-the-Global-Race-to-cleaner-coal_5192.html.

Deng, J. (2008): Adopting clean and high-efficiency power generation technologies vigorously to promote the sustainable development of electric power industry, *Huadian Technology*, 30(1), pp. 1–4.

Development Research Center of the State Council, the Energy Research Institute under NDRC, and the Tsinghua University Nuclear and New Energy Research Institute (2009); 2050 China Energy and CO_2 Emissions Report (Chinese); UNDP China and Renmin University (2009); China Human Development Report (2009/10): China and a sustainable future: towards a low carbon economy and society; Online at: http://hdr.undp.org/en/reports/nationalreports/asiathepacific/china/nhdr_China_2010_en.pdf.

Energy Bureau, National Development and Reform Commission, November 28, 2005: 'A Comparison of World and Chinese Energy Statistics', Shijie yu Zhongguo de nengyuan shuju bijiao; at http://nyj.ndrc.gov.cn/sjtj/t20051128_51344.Htm, & http://www.sp-china.com/news/powernews/200605110002.htm

Electric Power Planning and Design Institute (2006): Design Reference Cost Index of Thermal Power Plant since 2005, China Electric Power Press, Beijing.

Electric Power Development Corp of Japan (1999): Adoption of supercritical technology for Sipat super thermal power plant, January 1999.

Global CCS Institute (2012): The Global Status of CCS – 2012, Canberra, Australia.

GOI (2005): 'Draft Report of the Expert Committee on Integrated Energy Policy', Planning Commission, Government of India, New Delhi, December 2005, at http://plannning commission.nic.in/reports/genrep/intengpol.pdf.

Huang, Q.L. (2008): Clean and highly effective coal-fired power generation technology in China, *HuadianTechnology*, 30(3), pp. 1–8.

IEA (2011): International Energy Agency – World Energy Outlook 2011, Paris.

IEA (2009): Major Economies Forum (MEF) 2009; Technology action plan carbon capture, use and storage. Online at: http://www.majoreconomiesforum.org/the-global-partnership/carbon-capture-use-a-storage.html.

IEA's *MCMR* (2012): News release, 17 Dec. 2012, available at http://www.iea.org/newsroom andevents/pressreleases/2012/december/name,34441,en.html.

Jhajjar Power Ltd. (2012): 1320MW Supercritical power plant, built by CLP India Pvt Ltd., details at www.clpindia.in/operations_jhajjar.html.

Kapila, R.V. (2009): Investigating Prospects for CCS technologies in India, School of Geosciences, University of Edinburgh.

Ma Kai (ed.) (2005): *Strategic Research on the Eleventh Five-Year Plan*. Beijing: Beijing China Science Technology Press. October 2005; Shiyiwu guihua: Zhanlueyanjiu. Beijing: Beijing kexiejishu chubanshe.

Malti Goel (2010): Implementing Clean Coal Technology: Barriers and Prospects, In '*India Infrastructure Report 2010*', chapter 13, pp. 209–221.

Mao, Jianxiong (2007): Electrical Power Sector and Supercritical Units in China, presented at the *Workshop on Design of Efficient Coal Power Plants*, Vietnam, October 15–16, 2007.

Ministry of Coal (2006): Annual report 2005–2006, Ministry of Coal, Govt of India.

Mott MacDonald (2006): India's Ultra Mega Power Projects/Exploring the use of carbon financing, October 2006.

Morin, J. (2003): Recent ALSTOM power large CFB and scaleup aspects including steps to supercritical, In: *47th IEA Workshop on Large Scale CFB*, Poland.

Morse, R., Rai, V., & He, G. (2010): The Real Drivers of CCS in China, *ESI Bulletin*, November 2010, 3–6.

National Bureau of Statistics, 2004, China Electric Power Yearbook, Beijing: China Electric Power Press, p. 671.

National Coal Council (2012): Harnessing Coal's carbon content to advance the Economy, Environment and Energy Security, June 22, 2012; Study chair: Richard Bajura, National Coal Council, Washington DC.

NDRC (2007): Special Plan for Mid-and Long-Term Energy Conservation, National Development and Reform Commission, Beijing.

Pakiam, Geoffrey (2010): The role of Coal in India's Energy sector, *ESI Bulletin*, 3, Issue 2, November 2010; at http://www.esi.nus.edu.sg/docs/esi-bulletins/esi-bulletin-vol-3-issue-2-nov-2010084D65E6EEC2.pdf.

Planning Commission (2006): Integrated Energy Policy: Report of the Expert Committee, Planning Commission, Government of India.

Sanyal, B. (2007): Coal India's profits for 2006–07 may dip, *The Hindu Business Line*, March 15, available at http://www.thehindubusinessline.com/2007 /03/15 /18hdline.htm

Shi, X. and Jacobs, B. (2012): Clean coal technologies in developing countries, September 25th, 2012, at http://www.eastasiaforum.org/2012/09/ 25/ clean-coal-technologies-in-developing-countries/.

US-China Coal Energy Research Center (2011): at http://www.us-china-cerc.org /Advanced_ Coal_Technology & CERC_coal_JWP_english_OCR_18_ Jan_2011. pdf).

Vincent, C., Dai, S.F., Chen, W.Y., Zeng, R.S., Ding, G.S., Xu, R.N., Vangkilde-Pedersen, T., & Dalhoff, F. (2008): Carbondioxide storage options for the COACH project in the BohaiBasin, China, In: *Ninth Intl. Conf. on Greenhouse Gas Control Technologies*, Washington, DC.

Wei Yiming, Han Zhiyong, Fan Ying, Wu Gang (eds.) (2006): *China Energy Report* (2006), Beijing, China: Science Press (Zhongguo Nengyuan Baogao: Zhanlue yu zhengceyanjiu (2006), Beijing: Kexue Chubanshe), p. 12.

World Bank (2008): Clean Coal Power generation Technology Review: World-wide experience and Implications for India, Background paper: India – Starategies for low carbon growth.

Chapter 12

Outlook for clean coal technologies

Preamble

Mitigating climate change and achieving stabilization of greenhouse gas concentrations in the atmosphere is the objective of the United Nations Framework Convention on Climate Change. This would require deep reductions in global energy related carbon dioxide emissions. The consequences of increasing CO_2 levels in the atmosphere on the environment and life are being witnessed globally (IPCC 2007). IEA Technology Perspectives 2008 shows a scenario where global GHG emissions peak between 2020 and 2030 and will be halved, if appropriate and immediate actions are taken, by 2050. Developing new or improved low-carbon energy technologies, especially clean coal technologies, thus become very essential. Large capacity and higher efficiency coal-fired power generating technologies facilitate reduction of carbon dioxide emissions. Such technologies have been developed and demonstrated. Large-scale R&D efforts are being continued to enhance the performance and bring down the costs for wider implementation. Adequate policies, regulations, legislations and financial mechanisms specifically designed to promote technology change are required. Since the 'technology change' plays a critical role in making available affordable new low-carbon energy technologies, well-meaning and committed international collaborations and knowledge-sharing arrangements have to be fully developed.

Recently, the carbon capture, utilization and storage (CCUS) initiative where the captured carbon dioxide is viewed as a commodity has been vigorously pursued (such as, CO_2 enhancing oil recovery). This approach provides high potential for the economic growth, the environmental safety and energy security. The 2012 National Coal Council's Report (with reference to the US scenario) discussed the economic and environmental potential of captured CO_2 from power plants, and highlighted the confronting challenges to convert them into viable CCUS technologies.

Current status

To summarize what has been discussed in the pages so far, Coal is the most abundant and economical energy source in the world, and also the fossil fuel with the highest carbon content per unit of energy. Yet, due to the rising demand for large amounts of affordable and reliable electricity globally, especially in the developing world, having plenty of reserves, coal has remained a prominent energy source for electricity generation. Ten countries represent 84% of the global operating coal-fired power generation capacity (OECD/IEA 2012). These countries also represent all together more than 85% of the world's total CO_2 emissions from electricity and heat generation using coal and peat – about 8.5 Gt of CO_2 are emitted from China, the USA, India, Germany, Russia, Japan, South Africa, Australia, Korea and Poland.

The world-wide coal-fired power plants have been operating with subcritical PC technology, and the global plant efficiency averages around 33% on LHV basis. Enhancing plant efficiency was considered essential to considerably reduce CO_2 emissions as well as the coal consumption. Replacement of about 30 GW, and retrofitting around 200 GW of existing subcritical coal-fired plants with advanced coal-combustion (SC and USC) technologies are suggested while simultaneously applying the state-of-the-art for the new power plants to be planned and constructed (OECD/IEA 2010). The new SC and USC PC plants can reach thermal efficiencies of 44% and above. The supercritical PC plants have been installed in large numbers successfully in the USA since early 1990s, and in Canada, Australia, Europe, and so on in the last decade. These provide a lot of operational experience. China has launched an ambitious program to move to larger and more efficient SC and USC boilers ranging in size from 600 MW to 1000 MW. According to IEA, deployment of the clean coal technologies globally is expected to reach over 100 GW by 2025; and in the commercialization phase, USC coal plants would be applied with a capacity of 550 GW by 2050 which facilitates heavy reduction of emissions (IEA 2008).

FBC and IGCC offer more promise to the conventional emission abatement measures. The FBC which is relatively low-cost, clean and efficient with low emissions, is particularly suited to *poor quality fuels* including *a range of coals*; and has potential to suit to oxy-firing which make the technology most acceptable.

IGCC systems are among the cleanest and most efficient of the emerging clean coal technologies, though costly; pollutants are removed before the gas is burned in the gas turbine and thermal efficiencies of over 50% are likely to reach in the future. Currently, R&D efforts are on full swing to develop alternative component systems to bring down the cost of electricity.

Renewable energy technologies (RETs) – particularly wind and solar – are recognized globally as the best alternative to fosil fuel sources, and many countries focus on increasing the share of RETs in the energy-mix. Their consumption is projected to grow rapidly in the next two decades (e.g., USEIA, IEA, and BP Energy Outlooks). Even so, a major shift towards alternative sources has been difficult, as the technologies are still too expensive and their production volumes are not to the required scale to replace fossil fuel sources. Very recently natural gas, whose emission levels are relatively low compared to coal when burned, has been emerging as an important energy source as technically recoverable shale gas reserves are largely identified in the USA and in other countries in Asia, South America, Africa, and EU. But, the unknowns relating to shale gas are plenty. Supply, long-term environmental impacts, deliverability, cost, and price stability remain unanswered. Predictions of future supply and price of natural gas have a high level of uncertainty. Factors increasing the demand for gas further cloud the future – LNG export facilities are being built, the chemical industry is restoring, gas vehicles are flowing into the market, and gas-based generation capacity is growing (NCC 2012).

CCS is recognized as the most critical technology for capturing streams of CO_2 from coal-fired power plants and sequester into geological formations. But retrofitting the existing small and low efficient conventional PC plants with CO_2 capture facility is a useless proposition; there is large cost and efficiency penalty. High efficiency and larger capacity plants alone have technical and economic viability for adding CCS. The designs of PC power plants planned now must be 'capture-ready' to retrofit CCS when commercially ready at some future point of time, and must be of capacity >600 MW$_e$. Coupling higher efficiency coal-fired power technologies with CCS will unravel the full value of coal in an environment-friendly way. A few demonstration plants of CCS are operating globally providing experience to move to commercial level. Captured CO_2 utilizing for EOR was established globally as economically beneficial and environmentally safe activity. Several other economic applications for CO_2 are under development.

Polygeneration wherein gasification of coal possibly with other fuels (biomass or petroleum residues) provides heat, power and synthetic fuels; and Underground coal gasification (UCG), the gasification of coal operated *in situ* underground, are other choices vigorously pursued recently. These technologies have some merits and are more suited to certain regions of the world.

If these measures are fully implemented, the expected reductions in global CO_2 emissions may reach upto 1.7 Gt/year, and in coal consumption upto 0.5 Gt/year (OECD/IEA 2010).

Issues for large-scale implementation

The United States has taken the lead in clean coal research and development, and deployment of CCTs. China, among the developing world, is moving faster in RD&D and implementation (Clemente 2012). Between 2006 and 2010, around 295 gigawatts (GW) of clean coal capacity was installed globally. Considering the growth of total installed capacity of coal-fired power plants worldwide from 1,263 GW in 2005 to an estimated 1,700 GW in 2011, the progress in the deployment of CCTs is not *significant*. In the future too, coal-fired installed capacity will continue witnessing growth, but probably at a slower rate.

Many existing subcritical PC plants, although upgraded to reduce emissions of several pollutants, the efficiency gains are minimal as the heat rates can be improved by 3–5% at best. This is because the heat rate is primarily dependent on unit design, specific fuel type, and capacity factor; and the plant design cannot be changed once built. Replacing the subcritical PC plants by SC and USC PC technologies is considered expensive and in some regions unsuitable for local coals. But the recent experience proves otherwise except in the case of high ash coals where there is not enough operational experience. It is true that the boiler and steam turbine costs can be as much as 40–50% higher for a USC plant than for a sub-critical plant (ETP 2008); but the balance-of-plant cost can be 13–16% lower because of reduced coal consumption, coal handling and flue gas handling. Even so, compared to sub-critical PC, the investment costs for USC steam cycle can be 12–15% higher (Burnard & Bhattacharya IEA 2011). In addition, the manufacturing base for high power boilers and associated components is inadequate to meet the demand.

As outlined in Chapter 8 – Section A, further improvement in achievable efficiency by higher USC steam parameters is dependent on the availability of new, high temperature alloys for boiler membrane wall, superheater and reheater tubes, thick walled heaters and steam turbines. The Thermie Project and COMTES700 project of European Commission, and the 'Advanced materials for Ultra-Supercritical boiler systems' in the US are the major programs aimed at development of higher steam parameters. The EC projects are aimed to develop 375 bar, 700°C/720°C (5439 psi, 1292°F/1328°F) steam conditions, whereas the US Program to develop 379 bar, 730°C/760°C (5500 psi, 1346°F/1400°F) (Gierschner 2008; Dalton 2006; Weitzel 2004). These steam conditions should increase generating efficiency to 44–46% (HHV) or over 50% (LHV) for bituminous coal, but require further materials advances, particularly for manufacturing, field construction, and repair. It may also be possible to raise efficiency over 50% (LHV, net) with integrated, pre-drying for high moisture lignite coals. The recent successful demonstration of low temperature pre-drying process using waste heat at Coal Creek Station, North Dakota (USDOE: CCPI June 2012) offers promise for utilizing high moisture lignite coals.

FBC technology operates in a number of versions, but the Circulating fluidised bed combustion (CFBC) is the one gaining most market penetration. As mentioned earlier, the first super critical CFBC unit, 460 MW$_e$, 282 bar, 563°C/582°C, designed by Foster-Wheeler is operating in Poland since 2009, with Polish lignite coal and has a design efficiency of 43.3% (LHV, net). A second unit of 330 MW$_e$ will be installed in Russia at the Novocherkasskaya GRES facility (Jantti et al. 2009). Since SC/USCs require much higher than 600°C superheat or reheat temperatures, the designs need to be improved considerably, as the CFBC technology operates at somewhat low temperatures. To progress to higher steam conditions, some areas in CFBC need investigations: (a) development of materials with higher temperature and pressure resistance; (b) improving the manufacturing technology using these materials; and (c) hastening demonstration of large SC units. These studies will better the prospects for reaching efficiency of over 45% (LHV net) or 43% (HHVnet) for hard coal. The BFBC technology is considered the best economic solution for 'distributive' energy generation, and hence highly appropriate for developing countries like India. The BFBC boilers and furnaces, 0.5 MW$_{th}$–500 MW$_{th}$ which is a suitable unit size for distributive power generation can be locally produced and maintained.

IGCC technology is currently expensive; oxygen production is a major part of the energy consumption and capital cost. IGCC demonstration plants under operation or construction or at an advanced stage of planning are based on long-established designs. New cycles and systems are required for advancing efficiency in the future gasification processes. For example, the future plants may be requiring components of better quality such as larger and more efficient gas turbines, higher duty steam cycles, more efficient oxygen separation processes including ion-membrane technology and solid sorbents in the longer term, and improvements to ancillary components such as solids pumps (Henderson, 2008; Minchener, 2005; Barnes 2011). These areas have to be investigated for wider implementation of this high potential technology well suited for pre-combustion CO_2 capture. Projects are being developed by several firms: American Electric Power, Duke Energy, Texas Clean Energy Project, and Southern Company in the USA; by ZAK/PKE, Centrica (UK), Nuon Magnum (Netherlands), and E.ON and RWE (Germany) in Europe; and GreenGen and Dongguan Taiyangzhou in China, Wadoan Power in Australia, and Osaki CoolGen in Japan (Wikipedia – Free encyclopedia, Burnard & Bhattacharya IEA 2011). The vital areas of research for improvement are discussed in detail in Chapter 8.

Efficiency improvements are possible also through advanced solid oxide fuel systems (SOFC) power production. Recent analysis of SOFC/gas turbine systems projects an averall efficiency of nearly 60% together with a carbon capture of over 90% (excluding losses for compression of CO_2 for transport and storage). To develop such systems with effective control is challenging. The design of gasifiers also needs improvement for reducing the capital cost relative to throughput and to reduce the effects of corrosion. These can be tackled using widely-used 'extensive computational modeling and analysis', for which data on fundamental process understanding and process rate parameters for gasification processes at high temperatures and pressures (upto $1600°C$ and 80 bar) are lacking. For accurate model development and validation, more information under well-defined flow and boundary conditions are essential. This is a vital and productive area for research (Shaddix 2012).

For reducing emissions from *existing* coal-based plants, they have to be *retrofitted* with CCS or to be closed, to meet the emissions targets. *Replacing* or *repowering* an old, inefficient plant with a new, efficient unit with CO_2 capture can provide a net efficiency gain that decreases all emissions from the plant and fuel use. The net impact of the CO_2 capture energy penalty must be assessed in the context of a particular situation or strategy for reducing CO_2 emissions. As higher plant efficiency reduces energy penalty and associated impacts, innovations that raise the efficiency of power generation also can reduce the impacts and cost of carbon capture. IEA analysis suggests that without CCS, overall costs to reduce emissions to 2005 levels by 2050 increase by 70%. IEA's Roadmap includes an ambitious CCS growth path in order to achieve this GHG mitigation potential which calls for gigantic R&D activity covering several technological issues. RD&D program for developing retrofit technologies for the *operating* power plants was suggested. Oxygen production, boiler modification and flue gas purification are identified as the broad areas of research which are under intense study (e.g., Herzog 2009; DOE/NETL 2012; NCC 2011, 2012). Shaddix of Sandia National laboratory (Shaddix 2012) discusses critical areas of study to be undertaken in oxy-combustion for carbon capture. In chapter 9, several ongoing research areas are outlined that include chemical loop combustion, ITM and OTM technologies, development of solid sorbents for oxygen production. The outcome from these research and development efforts would make possible the implementation of commercial level oxy-fuel projects.

In CCS, another major issue involving costs is the expansion of transportation infrastructure to move the large amounts of captured carbon, and identification of storage sites of huge capacity. Global collaborations would yield better results in these areas.

Another new area of thermal conversion technology is 'direct carbon fuel cell technology (DCFC)'. A DCFC generates electrical power directly through electrochemical oxidation and has a maximum theoretical efficiency of 100%. Another benefit is the small loss in system efficiency for small-scale power generation, thereby allowing distributed energy generation and reduced distribution losses. Since a solid fuel is utilized, gas leakage is not an issue. However, several major challenges have to be overcome in this novel technology which has the potential to revolutionize power generation from coal, spectacularly improving efficiency, reducing emissions, and allowing carbon capture (Shaddix 2012).

Coal-to-liquid technology and Underground coal gasification are attractive from emissions point of view. Alongwith CTL technology, UCG potentially provides an effective route for energy extraction. Integration of UCG into an ICL plant offers the prospect of providing gaseous and liquid fuels in a clean and effective way that allow stable prices, and significantly enhanced security of fuel supplies. The IEA Report (IEA Clean Coal Center 2009) argues a case for UCG demonstration in Europe, coupled with the production of liquid fuels by Fischer-Tropsch conversion to utilize vast unexploited coal reserves in Europe. This approach can be pursued in all regions of the world wherever relevant, via collaboration with active demonstration programmes, which are currently operating in China.

CO_2 as a commodity

The newly emerged concept of carbon dioxide as a commodity would unravel the potential of CCTs for improving the economy as well as the environment. For example, developing large-scale CO_2-EOR would benefit enormously to unearth 'stranded oil' as well as to exploit extensive Residual Oil Zone resources (ROZ) globally. This approach is in contrast to capturing and simply storing CO_2 in deep geological formations. As coal is projected to remain the main source of power generation globally, the new clean coal-fired power plants and CTL plants with carbon capture will be a massive source of affordable and reliable CO_2 for the large-scale EOR. This approach will achieve deep emissions reductions, widespread economic advantage, and many other benefits and opportunities.

Beyond EOR, several potential pathways to CCUS are available for countries with no oil fields that could offer an economic advantage. A few examples: (a) Cement production, (b) Algae ponds to produce biofuels and dry biomass for animal feed by locating these ponds near major coal-fired power plants, and (c) Supercritical CO_2 (S-CO_2) power generation cycles for potential applications for closed cycle, high efficiency, coal-fired, and nuclear power plants to generate electric power in size ranges up to 200 MW_{th} (studies by Sandia Laboratories). Research to further develop these novel economical alternatives into commercially viable technologies is very important for effective management of carbon emissions.

Actions & Way Forward

The developed countries and industries have initiated steps to transfer efficient technologies or equipment, more specific to clean coal, to developing countries, through several programs, independently and combinedly. These programs are made possible through (a) bilateral cooperation, and (b) more collective efforts through regional cooperative frameworks such as the Asia Pacific Economic Cooperation (APEC), the development banks like Asian Development Bank, the World Bank and the Global Environment Facility (GEF). A few such cooperative programs are mentioned in Chapter 11 under 'China'. The scale of these efforts is inadequate and has to be intensified and accelerated.

Without financial support from industrialized countries, the majority of developing countries are unlikely to take significant steps toward CCTs, especially CCS development in the foreseeable future. Along with APEC, ADB, WB, and GEF, other international fora such as the Clean Energy Ministerial, the G8, and the Carbon Sequestration Leadership Forum should be leveraged to muster the political will to determine and act upon the best way to elicit such support. The stimulation of CCTs in developing countries could partly be supported through the Clean Development Mechanism of the Kyoto Protocol since SC coal-fired power plants are now eligible under CDM.

More importantly, the global scientific and engineering communities have turned their creative minds to the safe management of carbon in the last 3–4 decades through R&D and contributed substantially to the growth of CCTs; adequate resources must be provided to continue the efforts for bringing down the costs of the technologies and facilitate countries to adopt CCTs and develop related activities. This would allow these countries to not only have secure energy supplies but to effectively contribute to reduction of GHG emissions in the coming crucial decades.

The developed countries need to formulate and implement carbon policies and regulations in the broader global interests with no detriment to their development; such a situation enables all countries get the advantage of implementing clean coal processes so that inexpensive coal continues as a fuel source in a carbon constrained world, while answering climate change concerns.

REFERENCES

Barnes, I. (2011): Next generation Coal gasification technology, CCC/187, IEA Clean Coal center, September 2011.

Burnard, K., & Bhattacharya, S. (2011): Power generation from Coal: Ongoing developments and Outlook, Information paper, October 2011, IEA, Paris, France.

BP (2013): BP Energy Outlook 2030, London, January 2013.

Clemente, J. (2012): China leads global race to Cleaner coal, Power magazine, December 1, 2012, at www.powermag.com/coal/china-leads-the-Global-Race-to-cleaner-coal_5192.html.

Dalton, S. (2006): Ultra-supercritical technology progress in the US and in coal fleet for tomorrow, 2nd Annual conference of USC Thermal Power Technology Network, 26–28, October 2006.

Gale, J., & Freund, P. (2001): Coal-Bed Methane Enhancement with CO_2 Sequestration Worldwide Potential, *Environmental Geosciences*, 8(3), 210–217.

Gierschner, G. (2008): COMTES700: On Track towards the 50plus Power plant, presentation at New Build Europe 2008, Dusseldorf, 4–5 March.

Godec, Advanced Resources International, Inc. (2012): Knowledge and Status of Research on the Enhanced Recovery and CO_2 Storage Potential in Coals and Shales. March 13, 2012.

Henderson, C. (2008): Future developments in IGCC, CCC/143, London, IEA Clean Coal Center, December 2008.

Herzog, H.J. (2009): A Research program for Providing Retrofit technologies, Paper prepared for MIT Symposium on retrofitting of Coal-fired power plants for carbon capture, MIT, March 23, 2009.

IEA (2008): Clean coal technologies – Accelarating Commercial and Policy drivers for deployment, Paris, OECD/IEA.

IEA (2010): Power Generation from Coal – Measuring & Reporting Efficiency Performance and CO_2 emissions, Coal Industry Advisory Board, IEA.

IEA (2011): IEA – World Energy Outlook 2011, IEA, Paris, France.

IEA (2012): IEA – World Energy Outlook 2012, IEA, Paris, France, released November 2012.

IPCC (2007): Climate Change 2007, Synthesis report, IPCC 4th Assessment Report, IPCC, Geneva, Switzerland.

Jantti, T., Lampenius, H., Ruskannen, M., & Parkonnen, R. (2009): Supercritical OUT CFB projects – Lagisza 460 MW$_e$ and Novocherkasskaya 330 MW$_e$ – presented at *Russia Power 2009*, Moscow, 28–30 April.

Landesman, L.: Alternative Uses for Algae Produced for Photosynthetic CO_2 mitigation.

Minchener, A.J. (2005): Coal gasification for Advanced Power generation, *Fuel*, 84(17), 2222–2235.

National Coal Council (2012): Harnessing Coal's Carbon content to Advance the Economy, Environment and Energy Security, June 22, 2012; Study chair: Richard Bajura, National Coal Council, Washington, DC.

Shaddix, C.R. (2012): Coal combustion, gasification, and beyond: Developing new technologies for a changing world, *Combustion and Flame*, 159, 3003–3006.

USDOE CCPI (2012): Clean Coal Power Initiative Round 1 Demonstration projects, Clean Coal technology, Topical Report No. 27, DOE:OFE, NETL, June 2012; at www.netl.doe.gov/technologied/coalpowercctc/topicalreports/pdfs/CCT-Topica-Report-27.pdf.

USEIA (2011): Energy Information Administration – International Energy Outlook 2011, DOE, Washington DC, September 2011.

USEIA (2012): Annual Energy Outlook 2013 Early Release, EIA, USDOE, Washington DC, 17 November 2012.

VGB (2012/2013): VGB Electricity Generation, Figures & Facts, 2012/2013.

Wald, M. (2012): Turning CO_2 into Fuel, New York Times, March 2012.

Weitzel, P.a.M.P (2004): cited by Wiswqanathan *et al.*, Power, April 2004.

Wikipedia, free encyclopedia (2013): Integrated gasification Combined Cycle, at http://en.wikipedia.org/wiki/integrated_gasification_combined_cycle

Wright, S., *et al.* (2011): Overview of Supercritical CO_2 Power Cycle Development at Sandia National Laboratories, DOE/NETL 2011 UTSR Workshop, October 25–27, 2011, Columbus, OH.

Super heater, reheater, air preheaters, furnaces

Super heater

The purpose of the super heater is to increase the capacity of the plant, to eliminate corrosion of the steam turbine, and to reduce steam consumption of the steam turbine. There are different types of super heaters: Plate Super heaters; Pendant Super heaters; Radiant Super heaters and Final Super heaters. The super heater steam temperature is controlled by spraying water. Control methods used according to the need and design are: (a) Excess Air Control, (b) Flue Gas Recirculation, (c) Gas by-pass Control, and (d) Adjustable Burner Control.

(a) Excess Air Control: The steam outlet temperature of a convection superheater may be increased at partial load by increasing the excess air supply. The reduced gas temperature decreases the furnace heat absorption for the same steam production. The increased gas mass flow with its increased total heat content serves to increase the degree of superheat.

(b) Flue Gas Recirculation: The recirculation of some percentage of the combustion gases serves to control steam temperature in the same manner as does an increase in excess air. By introducing the hot gases below the combustion zone, relatively high efficiency may be maintained.

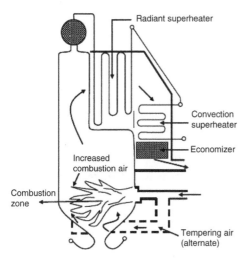

Excess Air control.

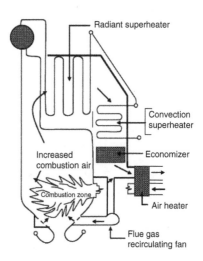

Flue gas recirculation.

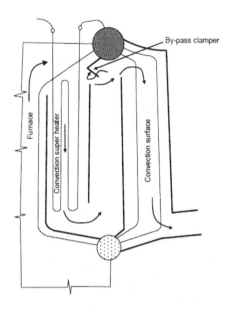

Gas by-pass control.

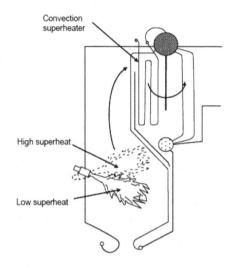

Burner tilt control.

(c) Gas By-pass Control: The boiler convection banks can be arranged in such a manner that portion of the gases can be by-passed around the superheater elements. The superheater is oversized so that it will produce the required degree of superheat at partial load conditions. As the load increases, some of the flue gases are by-passed.

(d) Adjustable Burner Control (Burner tilt): With a multiple burner furnace it is possible to distribute the burners over a considerable burner wall height. This control is obtained by selective firing. Tiltable furnace may be adjusted to shift the position of the combustion zone.

Final superheater temperatures are in the range, 540–570°C for large power plants, and superheated steam pressures are around 175 bar.

Reheater

Steam from the exhaust of the first stage turbine goes back to the boiler for reheating and is returned to the second stage. Reheater coils in the flue gas path does the reheating of the returned steam. The reheat steam is at a much lower pressure than the super heated steam but the final reheater temperature is the same as the superheated steam temperature. Reheating to high temperatures improves the output and efficiency of the Power Plant. Final reheater temperatures are normally in the range, 560 to 600°C. Reheat steam pressures are normally around 45 bar.

Air preheaters

An air preheater (or air heater) is a device designed to heat air before combustion in a boiler in order to increase the thermal efficiency of the process. They may be used alone or to replace a recuperative heat system or to replace a steam coil. The purpose of the air preheater is to recover the heat from the boiler flue gas which increases the thermal efficiency of the boiler by reducing the useful heat lost in the flue gas. As a result, the flue gases are also sent to the flue gas stack at a lower temperature, allowing simplified design of the ducting and the flue gas stack. It also allows control over the temperature of gases leaving the stack.

There are two types of air preheaters for use in steam generators in thermal power stations: (i) a tubular type built into the boiler flue gas ducting, and (ii) a regenerative air preheater. These may be arranged so the gas flows horizontally or vertically across the axis of rotation.

Tubular Type Air Pre-heaters: Tubular preheaters consist of straight tube bundles which pass through the outlet ducting of the boiler and open at each end outside of the ducting. Inside the ducting, the hot furnace gases pass around the preheater tubes, transferring heat from the exhaust gas to the air inside the preheater. Ambient air is forced by a fan through ducting at one end of the preheater tubes and at other end the heated air from inside of the tubes emerges into another set of ducting, which carries it to the boiler furnace for combustion.

The problem with this type is the tubular preheater ductings for cold and hot air require more space and structural supports than a rotating preheater design. Further, due to dust-laden abrasive flue gases, the tubes outside the ducting wear out faster on the side facing the gas current. Many advances have been made to eliminate this problem such as the use of ceramic and hardened steel. Many new circulating fluidized bed (CFB) and bubbling fluidized bed (BFB) steam generators are currently incorporating tubular air heaters offering an advantage with regards to the moving parts of a rotary type.

Dew point corrosion which occurs for a variety of reasons is another issue. The type of fuel used, its sulfur and moisture content are causative factors. However, by far the most significant cause of dew point corrosion is the metal temperature of the tubes. If the metal temperature within the tubes drops below the acid saturation temperature, usually between 88°C and 110°C, but sometimes at temperatures as high as 127°C, then the risk of dew point corrosion damage becomes considerable.

Regenerative Air Pre-heaters: There are two types of regenerative air pre-heaters: the rotating-plate and the stationary-plate regenerative air preheaters. The rotating-plate design consists of a central rotating-plate element installed within a casing that is divided into two (*bi-sector* type), three (*tri-sector* type) or four (*quad-sector* type) sectors containing seals around the element. The seals allow the element to rotate through all the sectors, but keep gas leakage between sectors to a minimum while providing separate gas air and flue gas paths through each sector. Tri-sector types are the most common in modern power generation facilities. In the tri-sector design, the largest sector is connected to the boiler hot gas outlet. The hot exhaust gas flows over the central element, transferring some of its heat to the element, and is then ducted away for further treatment in dust collectors and other equipment before being expelled from the flue gas stack. The second, smaller sector is fed with ambient air by a fan, which passes over the heated element as it rotates into the sector, and is heated before being carried to the boiler furnace for combustion. The third sector is the smallest one and it heats air which is routed into the pulverizers and used to carry the coal-air mixture to coal boiler burners. Thus, the total air heated in the air preheater provides: heating

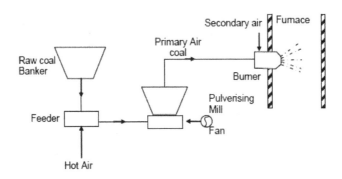

Schematic of Unit system.

air to remove the moisture from the pulverized coal dust, carrier air for transporting the pulverized coal to the boiler burners and the primary air for combustion.

Stationary-plate Regenerative Air Preheater – The heating plate elements in this type of regenerative air preheater are also installed in a casing, but the heating plate elements are stationary rather than rotating. Instead the air ducts in the preheater are rotated so as to alternatively expose sections of the heating plate elements to the upflowing cool air. There are rotating inlet air ducts at the bottom of the stationary plates similar to the rotating outlet air ducts at the top of the stationary plates.

Furnaces

Furnaces are broadly classified into two types, namely, combustion type (using fuels) and electric type, based on the method of generating heat. The furnace should be designed so that in a given time, as much of material as possible can be heated to a uniform temperature as possible with the least possible fuel and labor.

Pulverized coal firing is done by either Unit (Direct) system or Bin (Central) system as shown schematically.

In a *Unit system* the raw coal from the coal bunker drops on to the feeder. Hot air is passed through coal in the feeder to dry the coal. The coal is then transferred to the pulverizing mill where it is pulverized. Primary air is supplied to the mill, by the fan. The mixture of pulverized coal and primary air then flows to burner where secondary air is added. Since each burner or a burner group and pulverizer constitutes a unit, it is called unit system.

The merits of the System: it is simpler than the Central system; there is direct control of combustion from the pulverizing mill, and coal transportation system is simple.

In a *Central System,* crushed coal from the raw coal bunker (bin) is fed by gravity to a dryer where hot air is passed through the coal to dry it. The dryer may use waste flue gases, preheated air or bleeder steam as drying agent. The dry coal is then transferred to the pulverizing mill. The pulverized coal obtained is transferred to the pulverized coal bin. The transporting air is separated from the coal in the cyclone separator. The primary air is mixed with the coal at the feeder and the mixture is supplied to the burner. The merits of this system: the pulverizing mill grinds the coal at a steady rate

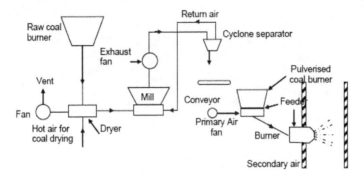

Schematic of Central system.

irrespective of boiler feed; there is always some coal in reserve so that any breakdown in the coal supply will not affect the coal feed to the burner. The system suffers from disadvantages also: (a) the initial cost of the system is high, (b) the coal transportation system is quite complicated, and (c) the system requires more space. However, to a large extent the performance of the pulverized fuel system depends upon the *mill performance*.

Annex 2

Some facts about a coal plant

A typical coal fired power plant of capacity 500 MW is associated with the following:

(1) Around 2 million tons of coal will be required each year to produce the continuous power; and produces 3.5 billion kilowatt-hours per year, enough to power a city of about 140,000 people.

(2) Around 1.6 million cubic meter of air in an hour is delivered by air fans into the furnace for coal combustion.

(3) The boiler for typical 500 MW plant produces around 1600 tons of steam per hour at a temperature of 540–600°C. The steam pressure is in the range of 200 bar. The boiler materials are specially designed to withstand these conditions for operational safety.

(4) Heat transfer from the hot combustion gases to the water in the boiler takes place through radiation and convection.

(5) The electrical generators carry very large electric currents that produce heat and are to be cooled by hydrogen and water.

(6) The steam leaving the turbine is condensed and the water is pumped back for reuse in the boiler. To condense all the steam it will require cooling water, around 50,000 cubic meter per hour, to be circulated from lakes, rivers or the sea. The water is returned to the source with an increase of 3–4°C to prevent any effect to the environment.

(7) Apart from the cooling water, the power plant also requires around 400 m^3 per day of fresh water for making up the losses in the water-steam cycle.

(8) A 500 megawatt coal plant produces the following pollutants:

    (a) 10,000 tons of sulfur dioxide; Sulfur dioxide is the main cause of acid rain, which damages forests, lakes and buildings.

    (b) 10,200 tons of nitrogen oxide; NO_x is a major cause of smog, and also a cause of acid rain.

    (c) 3.7 million tons of carbon dioxide.

    (d) 500 tons of small particles; Small particulates are a health hazard, causing lung damage.

    (e) 220 tons of hydrocarbons; Fossil fuels are made of hydrocarbons; when they don't burn completely, they are released into the air which cause smog.

    (f) 720 tons of carbon monoxide, a poisonous gas.

(g) 125,000 tons of ash and 193,000 tons of sludge from the smokestack scrubber; A scrubber uses powdered limestone and water to remove pollution from the plant's exhaust. This ash and sludge consists of coal ash, limestone, and many pollutants, such as toxic metals like lead and mercury.

(h) 225 pounds of arsenic, 114 pounds of lead, 4 pounds of cadmium, and many other toxic heavy metals; mercury can cause birth defects, brain damage and other ailments. Acid rain also causes mercury poisoning by leaching mercury from rocks and making it available in a form that can be taken up by organisms.

(i) Trace elements of uranium; All but 16 of the 92 naturally occurring elements have been detected in coal, mostly as trace elements below 0.1% (1,000 ppm). A study by DOE's Oak Ridge National Lab found that radioactive emissions from coal combustion are greater than those from nuclear power production.

(j) The 2.2 billion gallons of water it uses for cooling is raised 16°F on average before being discharged into a lake or river. By warming the water year-round it changes the habitat of that body of water.

(Refs: Union of Concerned Scientists & johnzactruba, www.brighthub.com > ... > Energy/Power)

Levelized cost of electricity

Levelized cost represents the per-kilowatt hour cost of building and operating a generating plant over an assumed financial life and duty cycle, and is considered convenient measure of competiveness of different generating technologies. Key inputs to calculating levelized costs include overnight capital costs, fuel costs, fixed and variable operations and maintenance (O&M) costs which vary among technologies. For technologies with significant fuel cost, both fuel cost and overnight cost estimates significantly affect the levelized cost. Further, the inherent uncertainty about future fuel prices and future policies may cause plant owners or investors who finance plants to place a value on portfolio diversification. These are not well represented in calculating levelized cost figures. No incentives extended by the governments are included in the calculations. As with any projection, there is certaint uncertainty about all of these factors and their values can vary regionally and with time as technologies evolve. While levelized costs are a convenient summary measure of the overall competiveness of different generating technologies, actual plant investment decisions are affected by the specific technological and regional characteristics of a project, which involve numerous other considerations such as the projected utilization rate, the existing resource mix and so on. The levelized cost for each utility-scale generation technology is calculated based on a 30-year cost recovery period, using a real after tax weighted average cost of capital (WACC) of 6.8%. However, in the AEO2012 reference case a 3% increase in the cost of capital is added when evaluating investments in greenhouse gas intensive technologies like coal-fired power and coal-to-liquids plants without carbon control and sequestration (CCS). While the 3% adjustment is somewhat arbitrary, in levelized cost terms its impact is similar to that of an emissions fee of $15 per metric ton of carbon dioxide (CO_2) when investing in a new coal plant without CCS, similar to the costs used by utilities and regulators in their resource planning. The adjustment should not be seen as an increase in the actual cost of financing, but rather as representing the implicit problem being added to GHG-intensive projects to account for the possibility they may eventually have to purchase allowances or invest in other GHG emission-reducing projects that offset their emissions. As a result, the levelized capital costs of coal-fired plants without CCS are higher than would otherwise be expected. The levelized cost for each technology is evaluated based on the capacity factor indicated, which generally corresponds to the high end of its likely utilization range. Simple combustion turbines (conventional or advanced technology) that are typically used for peak load duty cycles are evaluated at a 30% capacity factor.

US Averaged Levelized costs (2010 $/Megawatt hour) for plants entering service in 2017.

Plant type	Capacity factor	Levelized capital cost	Fixed O&M	Variable O&M	Transmission investment	System Levelized cost
Conventional Coal	85	64.9	4.0	27.5	1.2	97.7
Advanced Coal	85	74.1	6.6	29.1	1.2	110.9
Advanced Coal with CCS	85	91.8	9.3	36.4	1.2	138.8
Natural gas-fired						
(i) Conventional combined cycle	87	17.2	1.9	45.8	1.2	66.1
(ii) Advanced combined cycle	87	17.5	1.9	42.4	1.2	63.1
(iii) Advanced combinedcycle with CCS	87	34.3	4.0	50.6	1.2	90.1
(iv) Conventional combustion turbine	30	45.3	2.7	76.4	3.6	127.9
(v) Advanced combustion turbine	30	31.0	2.6	64.7	3.6	101.8
Advanced Nuclear	90	87.5	11.3	11.6	1.1	111.4

[1]Costs are expressed in terms of net AC power available to the grid for the installed capacity [Ref: (AEO2012) reference case. [2]. At electricity generation cost, AEO 2012]

The average levelized costs for dispatchable generating technologies that are brought on line in 2017 as represented in the National Energy Modeling System (NEMS) for the *Annual Energy Outlook 2012 (AEO2012)* reference cases are presented here.

Index